Globalization

Globalization

The Transformation of Social Worlds

SECOND EDITION

D. STANLEY EITZEN
Colorado State University

MAXINE BACA ZINN
Michigan State University

WADSWORTH
CENGAGE Learning™

Australia • Brazil • Japan • Korea • Mexico • Singapore • Spain • United Kingdom • United States

WADSWORTH
CENGAGE Learning™

Globalization: The Transformation of Social Worlds, Second Edition
D. Stanley Eitzen and Maxine Baca Zinn

Acquisition Editor: Chris Calderia
Assistant Editor: Tali Beesley
Editorial Assistant: Erin Parkins
Marketing Manager: Nichelle Williams
Marketing Assistant: Ileana Shevlin
Marketing Communications Manager:
    Linda Yip
Project Manager, Editorial Production:
    Samen Iqbal
Creative Director: Rob Hugel
Art Director: Caryl Gorska
Print Buyer: Becky Cross
Permissions Editor: Roberta Broyer
Production Service: Aaron Downey, Matrix
    Productions Inc.
Copy Editor: Tricia Lawrence
Cover Designer: Yuo Riezebos
Cover Image: © Jim Richardson/CORBIS
Compositor: Newgen

For product information and technology assistance, contact us at
Cengage Learning Academic Resource Center, 1-800-423-0563
For permission to use material from this text or product, submit all requests online at **www.cengage.com/permissions**.
Further permissions questions can be e-mailed to **permissionrequest@cengage.com**.

Library of Congress Control Number: 2007939796

Student Edition
ISBN-13: 978-0-495-50432-0
ISBN-10: 0-495-50432-7

Wadsworth Cengage Learning
10 Davis Drive
Belmont, CA 94002-3098
USA

Cengage Learning is a leading provider of customized learning solutions with office locations around the globe, including Singapore, the United Kingdom, Australia, Mexico, Brazil and Japan. Locate your local office at: **international.cengage.com/region**Cengage Learning products are represented in Canada by Nelson Education, Ltd.

For your course and learning solutions, visit **academic.cengage.com**.

Purchase any of our products at your local college store or at our preferred online store **www.ichapters.com**

Printed in the United States of America
1 2 3 4 5 6 7 12 11 10 09 08

Contents

Preface

Sociology as it is practiced, taught, and written about in the United States is, with few exceptions, centered on the United States. Sociological textbooks, for example, give occasional attention to comparative research but rarely examine the global forces that have dramatic and rapid effects everywhere. Of course, the United States is a central player in the world arena with the implementation of its foreign policy, the reach and power of U.S.-based corporations, the spread of its popular culture, and its overuse of natural resources.

Equally important, there are social forces outside its boundaries that have powerful impacts on the United States. The terrorism manifested on September 11, 2001, is an obvious example. So, too, are financial disasters, wars, and extreme weather events that impact U.S. financial markets, agricultural prices, jobs, and the safety of its citizens. This book will consider the reciprocal effects of the United States and the rest of the world. Beyond that, the selections will examine how the world is ever more interconnected, thereby affecting people everywhere around the globe, not just the United States. Some examples:

- Global warming brings climate change, extreme weather conditions worldwide, the migration of people, animals, and diseases as tropical forests are downsized in developing nations combined with excessive carbon use in the developed nations.

- Outbreaks of new diseases and drug-resistant diseases threaten everyone.

- Low-wage economies attract capital, moving jobs from place to place in a "race to the bottom," leaving disarray and unemployment where the jobs are taken from and dislocations and worker exploitation where the jobs are relocated.

- Global market forces are shaping stratification and inequality within societies.

- The ever-larger inequality gap between societies and within them increases the likelihood of social unrest, parochialism, and terrorism worldwide.

- Extreme poverty affects immigration patterns (legal and illegal).

- Transnational criminal networks traffic drugs, sex workers, child labor, and the dumping of hazardous wastes.

As these and other topics are considered in this book, our goal is to weave several themes within the various sections. First, societies and groups within them experience globalization differently. Thus, at the societal level we have selected essays that focus on the consequences of globalization among the developed countries *and* the developing countries. Within these nations, the essays examine how the effects of globalization vary by social class, race, and gender. Second, because the consequences of globalization differ, the selections will highlight the debates on whether the forms that globalization takes are beneficial or not. Third, many of the selections will have a critical edge to them. Implicitly, the authors of the articles (and we) ask: Who benefits and who does not from the changes brought by globalization? Fourth, we want to emphasize that individuals, groups, and societies are not passive actors in the globalization drama. The powerful nations, corporations, and media impose their will from the top, but so too do groups organize and shape globalization from the bottom. A final theme, as elaborated in the introductory essay, highlights the effects of globalization on how sociologists conceptualize the social. In times past, the sociological concepts and processes were framed within the boundaries of the nation-state. That has changed and is changing. Sociologists and other social scientists need to reorient their concepts and theories to reflect the realities of an increasingly interconnected world.

This second edition includes fourteen new selections and one revised piece. The new articles, for the most part, reflect an emphasis on grassroots resistance and social movements aimed at changing the outcomes of globalization toward more humane goals. They also, like the articles retained from the previous edition, combine facts, interpretation, and personal accounts.

We have tried to make this collection of writings reader-friendly. The selection of each article was guided by such questions as: Is it interesting? Is it informative? Is it thought-provoking? Does it communicate without the use of unnecessary jargon and sophisticated methodologies? We hope that the readers of this book approach the subject of globalization motivated to understand the complexities of their changing social world. The essays included in this collection are intended to add to their understanding and to whet their appetites for further exploration into the intricacies and mysteries of transnational social life.

ACKNOWLEDGMENTS

We want to thank the following reviewers for their helpful comments: Robert Graham, Lee University, Cleveland, TN; William J. Haller, Clemson University, Clemson, SC; Annette Schwabe, Florida State University, Tallahassee, FL;

Faye Linda Wachs, Cal Poly Pomona, Pomona, CA; and Jonathan White, Bridgewater State College, Bridgewater, MA.

Thanks also to Zakia Salime and Satoko Motohara of Michigan State University. We are also grateful to Kelly Eitzen Smith, Alan Zinn, and Margaret L. Andersen for their assistance on this project.

Globalization

An Introduction

D. STANLEY EITZEN
MAXINE BACA ZINN

GLOBALIZATION DEFINED

Globalization refers to the greater interconnectedness among the world's people. In its current phase, it is described as "an unprecedented compression
of time and space reflected in the tremendous intensification of social, political,
economic, and cultural interconnections and interdependencies on a global
scale" (Steger, 2002:ix). Globalization is a process whereby goods, information,
people, money, communication, fashion (and other forms of culture) move
across national boundaries. There are several implications of this view of globalization. First, globalization is not a thing or a product but rather a process. It involves such activities as immigration, transnational travel, e-mails and the
Internet, marketing products in one nation that are made elsewhere, the movement of jobs to low-wage economies, transnational investments, satellite broadcasts, the price of oil, coffee, wheat, and other commodities, and finding a
McDonalds and drinking a Coke or Pepsi in virtually every major city in the
world. Second, it follows that globalization is not simply a matter of economics,
but it also has far-reaching political, social, and cultural implications as well.
Third, globalization refers to worldwide changes that are increasingly remolding
the lives of people worldwide. Globalization is not just "something out there"
but it has outcomes for institutions, families, and individuals within societies
(Hytrek and Zentgraf, 2008). And, fourth, not everyone experiences globalization in the same way. For some it expands opportunities and enhances prosperity
while others experience poverty and hopelessness. Periods of rapid social change,

we know, "threaten the familiar, destabilize old boundaries, and upset established traditions. Like the mighty Hindu god Shiva, globalization is not only a great destroyer, but also a powerful creator of new ideas, values, identities, practices, and movements" (Steger, 2002:ix).

GLOBALIZATION THEN

Globalization is not a new phenomenon. For thousands of years people have traveled, traded, and migrated across political boundaries, exchanging food, artifacts, and knowledge. Consider the world around 1000 A.D. (the following is from Sen, 2002; and *U.S. News & World Report*, 1999). The Vikings plundered and traded, establishing settlements in northern France, Britain, Iceland, Greenland, and Russia. Located at the crossroads between East and West, the Byzantine Empire traded with foreigners from the East and the West. India was linked by maritime routes to Africa, the Middle East, and Southeast Asia. China (the Song Empire) used sea routes to trade cotton goods, spices, and horses. The Islamic world was the first civilization to trade with the other major empires in Europe, Asia, and Africa. These cross-boundary interactions involved not only trade but the transfer of inventions, knowledge, and other cultural forms. For example, the ninth-century Arab mathematician Mohammad Ibn Musa-al-Khwarismi, who gave the Western world algorithms and algebra, is "one of many non-Western contributors whose works influenced the European Renaissance and, later, the Enlightenment and the Industrial Revolution" (Sen, 2002:A3).

There have been other periods when globalization processes accelerated. In sixteenth-century Europe, for example, trade and exploration expanded to all parts of the globe with Europeans settling in different regions. The late 1800s and early 1900s was a time characterized by great waves of immigration and high levels of trade and finance across national borders. The period following World War II was the precursor to contemporary globalization. After that war made the current global economy possible, the disintegration of the colonial empires of the British, French, Dutch, Belgians, and Spanish established eighty-eight new nations, and later the Soviet Union collapsed, creating eighteen new countries. These "new" countries were freed to sell their raw materials and products on the world market and to purchase goods. Also they could now establish local industries to compete with those in other countries. Second, following the war, new technological innovations laid the foundation for the transportation and communication advances of the current age. The third change in the post–World War II era has been the emergence of transnational political and financial institutions. In 1945, the World Bank and the International Monetary Fund were created to help rebuild Europe and Japan. These two financial institutions continue to have an important role in the underdeveloped world. The United Nations, with its organizational units such as the World Health Organization and UNESCO, seeks to reduce tensions among nation-states and to find transnational solutions to political and social problems.

The World Court seeks to adjudicate international disputes and to try war criminals.

THE CHARACTERISTICS OF GLOBALIZATION NOW

Global connections have existed for centuries, but they have "figured as a pervasive, major aspect of social life mainly since the 1960s" (Scholte, 2000:19). Since then, the rate of change has been exponential. The speed of movement (travel or via communications technology), and the volume of goods, messages, and symbols has increased dramatically. So, too, have distances gotten shorter as travel and communication time has decreased (Beynon and Dunkerley, 2000:5). The indicators on this increasingly rapid change occur along a number of dimensions (the following is dependent in part on Brecher, Costello, and Smith, 2000:2–4; Scholte, 2000:20–25; and Beynon and Dunkerley, 2000:5–7).

Production

Globalization has transformed the nature of economic activity. From the 1970s forward, transnational corporations built factories in and bought manufactured products from low-wage countries on a vastly expanded scale. What has emerged is a "global assembly line," in which products from athletic shoes to electronics are made by low-wage workers and sold in the developed countries. U.S., Japanese, and European transnational corporations have invested many billions in China and elsewhere to build state-of-the-art factories. Dell, the world's top personal computer maker, for example, has components manufactured by hundreds of suppliers and subsuppliers in Mexico, Taiwan, Malaysia, Korea, and China. Nokia, the Finnish company known primarily for its cellular phones, has components and assembled products produced in ten different countries. The result is the decline of manufacturing in the developed countries and the migration of production jobs to low-wage economies. The United States, where the average manufacturing wage is $16 an hour, lost 3.2 million manufacturing jobs from 2001 to 2006 (Crutsinger, 2007). These jobs migrated to places like China, where the average manufacturing wage is 61 cents an hour. The job losses occur in a domino effect as jobs that once moved to Mexico because of their low wages, for instance, have moved to even lower wage economies. This migration of jobs has been called "the race to the bottom."

Markets

In the past corporations limited their sales to domestic or perhaps regional markets. Now goods and services are marketed to the entire world. Nokia, for example, sells its products in 130 countries. Sometimes a transnational corporation will locate a factory in a country where it markets products. This is the strategy of Japanese automobile manufacturers Honda and Toyota, which have located major plants in the United States, where their products are so popular.

Technology: The Tools of Globalization

New technologies—robotics, fiber optics, container ships, computers, communications satellites, and the Internet—have transformed information storage and retrieval, communication, production, and transportation. Microelectronic-based systems of information, for example, allow for the storage, manipulation, and retrieval of data in huge quantities (the amount of unique information generated worldwide each year is measured in exabytes—one exabyte is 1 followed by 18 zeroes). Information can be sent in microseconds via communications satellites throughout the world. As *BusinessWeek* put it: "Anyone with a computer is a citizen of the world . . ." (1999:71). In short, "Technological advancement in transportation and communications has not merely made the world smaller, *for many purposes it has made geography irrelevant*" (Peoples and Bailey, 2003:36, emphasis added).

Corporate Restructuring

Major corporations have always operated internationally. Beginning in the 1980s, they reorganized internally to take advantage of the global economy. They merged with other corporations and developed strategic alliances with others. They arranged for others in low-wage economies to do various tasks ("corporate outsourcing"). Take General Motors, for example. It has a financial stake in Daewoo Motors (Korea), Shanghai General Motors (China), Jinbei General Motors (China), Holden (Australia), Fiat (Italy), Fuji (Japan), Suzuki (Japan), and Saab (Sweden). "GM is putting the pieces together, tapping the engineering, manufacturing and distribution resources of its global partners. . . . The strategy has the potential to save billions of dollars, dramatically speed up product development and give GM an edge in emerging markets" (Muller, 2004:64). The result is a decentralization of production but a concentration of economic power (Harrison, 1994). The 2,000 largest global corporations in 2006 represented 55 countries, employed 68 million, and had total revenues of $24 trillion, assets of $88 trillion, and profits of $1.7 trillion (*Forbes*, 2006).

Neoimperialism

Following World War II the imperialist powers gave political independence back to their third-world colonies. Globalization, however, has kept these countries dependent economically on Western Europe, Japan, and the United States. "Globalization has taken from poor countries control of their own economic policies and concentrated their assets in the hands of first world investors. Although it has enriched some third world elites, it has subordinated them to foreign corporations, international institutions, and dominant states" (Brecher, Costello, and Smith, 2000:3–4).

Changing Structure of Work

With globalization, worker security everywhere has declined. "All over the world, employers have downsized, outsourced, and made permanent jobs into contingent ones. Employers have attacked job security requirements, work rules, worker representation, healthcare, pensions, and other social benefits, and anything else that defined workers as human beings and employers as partners in a social relationship, rather than simply as buyers and sellers of labor power" (Brecher, Costello, and Smith, 2000:3; Hacker, 2006). In this pro-employer environment, labor unions have lost their power. Employers, when faced with employee demands for higher wages or better benefits, can simply threaten to move the operation to a setting where wages and benefits are lower.

Movement of People

Although people have always crossed national borders, immigration has accelerated under the current conditions of globalization. Worldwide, more than 200 million people are living outside their country of birth or citizenship. Wars, droughts, floods, and changing climates are pushing people out of their homelands. So, too, is the hope of jobs luring them elsewhere.

This migration is typically from poor countries to rich ones. Over half of the world's legal and illegal immigrants are women. This "feminization of migration" reflects a worldwide gender revolution in which millions of women migrate across the globe to serve as nannies, maids, and sex workers. Work that native-born women once performed for free is now purchased in the global marketplace (Ehrenreich and Hochschild, 2002; Hondagneu-Sotelo, 2003).

One consequence of this flow of people across borders is the reverse flow of money, as many immigrants send money back to relatives in their native land. An estimated $20 billion annually, for example, is sent back to Mexico by Mexicans living legally or illegally in the United States. In the case of Mexico, this is more money than the revenues received from its petroleum industry.

Global Institutions

Organizations such as the World Trade Organization (WTO), the World Bank, and the International Monetary Fund (IMF) are involved in fostering transnational trade and providing economic development in underdeveloped countries. These new organization forms are powerful forces accelerating the globalization process. Whether their contributions have positive results is open to debate, as noted throughout this book.

Neoliberal Ideology and Policies

Contemporary globalization is fueled by the prevailing ideology known as neoliberalism, or the Washington consensus. This ideology dates back to John Locke and Adam Smith, arguing that market forces will bring prosperity, liberty, and democracy if left unhindered by government intervention. In terms of policy, neoliberals promote privatization, deregulation, and dismantling of the welfare state. Most significant, this ideology promotes free trade, that is, state borders should be open to trade without tariffs and other restrictions. This ideology is behind such agreements as NAFTA (the North American Free Trade Act), and informs the policies of the World Trade Organization and the International Monetary Fund.

Governance

The sovereignty of the nation-state has for the most part been diminished by globalization (an exception is the United States, which resists efforts by international organizations to control it). There are suprastate organizations that regulate transnational trade and international law. As a consequence of accepting neoliberal ideology, national governments do not hinder corporate decisions regarding outsourcing and the movement of capital, even though these decisions go against the welfare of their citizens. In effect, economic, political, and cultural change is now beyond the control of any national government (Beynon and Dunkerley, 2000:6).

Permeable Borders

Political and geographical boundaries have been crossed by trade, tourism, immigration, and electronic communications. Insularity is no longer possible as environmental pollution through the air, water, or food supply anywhere affects people elsewhere. So, too, diseases are difficult to contain as evidenced by the AIDS pandemic or the spread of mad cow disease. Criminal networks easily function and flourish when borders are permeable. They engage in the distribution of illegal drugs, prostitution, human traffic in slaves, and sweatshops. Terrorism also becomes transnational when borders are porous.

Global Culture

National culture, traditionally, has been tied to place and time. The knowledge, symbols, and stories people share within a national consciousness give identity to a nation and its people. Global culture, on the other hand, is de-ethnicized and de-territorialized, existing outside the usual reference to geographical territory. It is created and sustained by the media, corporate advertising, and the entertainment industry. The result is a single world culture "centered on consumerism, mass media, Americana, and the English language. Depending on one's perspective, this homogenization entails either progressive cosmopolitanism or oppressive imperialism" (Scholte, 2000:23). The westernized consumer lifestyle is symbolized by similar products (Nike shoes, fashion, pop music, Disney products, movies, Coca-Cola, McDonald's, ESPN, and CNN) found everywhere (Beynon and Dunkerly, 2000:13–21).

The global culture is not as uniform and universal as it would seem. There are often clashes between local and global cultures. Cultural diversity abounds. Religious fundamentalists in many parts of the world, most notably the Middle East, passionately resist modernity in general and the intrusion of the West in particular. Many people embrace their national identity and culture. Moreover, global communications and markets are often adapted to fit diverse local contexts. "Through so-called 'glocalization,' global news reports, global products, global social movements and the like take different forms and make different impacts depending on local particularities" (Scholte, 2000:23).

GLOBALIZATION: RECONFIGURING THE SOCIAL

The discipline of sociology emerged in the eighteenth and nineteenth centuries as the scholarly study of society. Understandably, since the world was divided into nations during this period, the focus of sociologists was on society as the nation-state, with geographical boundaries and social institutions unique to that society. Discussions of place included the local (community), urban and rural, and society. Social problems were examined and solutions offered for problems at the local and societal levels. Interaction was face-to-face, leading to primary groups. Secondary groups and bureaucracies were based locally, regionally, and nationally. Hierarchies of social inequality (class, gender, and race), too, were described and explained at the local, regional, and societal levels.

Globalization—the transformation of world society in terms of flows of people, goods, capital, and ideas across national boundaries, linkages, institutions, culture, and consciousness (Lechner and Boli, 2000:2)—accelerated in the last decades of the twentieth century, resulting in sociologists and other social scientists beginning to think globally. Just as the sources of social problems are located in the ordinary, everyday, normal workings of social institutions (e.g., institutional racism and sexism) and class relations, sociologists are beginning to explore the ways that globalization, most particularly the world capitalist system, contributes to problems of people within and across national boundaries. For example, the various manifestations of inequality across and within societies and the degradation of the environment are consequences of the actions by transnational corporations. The pace of globalization has quickened at the beginning of the twenty-first century, and many sociologists are rethinking their eighteenth- and nineteenth-century roots to confront and understand the globalized and globalizing world. This reconceptualizing of the social is a shift in worldview (literally). The selections in this reader will implicitly provide an overview of the new sociology as it seeks to understand the new social worlds resulting from globalization. This requires that we look at globalization at many different levels—from the largely invisible but powerful processes operating at the macro level to global practices at institutional levels, to their impact on the daily experiences of women, men, and children of different classes, races/ethnicities, and international contexts.

STRUCTURE OF THE BOOK

Chapter 1 defines globalization. Chapter 2 describes the various dimensions of globalization. While the process of globalization is clear, its consequences are not. Chapter 3 examines the debates surrounding globalization such as: Does globalization bring diverse people together or divide them further? Does globalization increase or decrease global inequality? Should trade across borders be free? Are profit-seeking global corporations the source of good or ill? (Put another way, is world capitalism the answer?) Chapter 4 looks at the economic side of globalization, engaging in such topics as the neoliberal ideology, transnational commerce, and the flow of work and jobs across borders. Chapter 5 examines global power and politics, focusing on the declining sovereignty and significance of nation-states, institutions of transnational governance, and the new world order. Chapter 6 centers on cultural globalization as manifested through the media, consumerism, and tourism. Chapter 7 considers the restructuring of social institutions and social arrangements resulting from globalization, focusing on gender, families, and relationships. Chapter 8 looks at the globalization of terror—the threats, the tools of global terrorism, and why the United States is so vulnerable. Chapter 9 examines globalization and other social problems—environmental degradation, worker exploitation, and transnational criminal networks. The concluding chapter considers efforts by groups and movements organized to change the negative consequences of globalization.

REFERENCES

Beynon, John, and David Dunkerley (eds.). 2000. *Globalization: The Reader* (New York: Routledge).

Brecher, Jeremy, Tim Costello, and Brendan Smith. 2000. *Globalization from Below: The Power of Solidarity* (Cambridge, MA: South End Press).

BusinessWeek. 1999. "The Internet Age," (October 4):71.

Clifford, Mark L. 2002. "How Low Can Prices Go?" *BusinessWeek* (December 2):60–61.

Crutsinger, Martin. 2007. "Loss of U.S. Manufacturing Jobs Accelerating."Associated Press (April 30).

Ehrenreich, Barbara and Arlie Russell Hochschild. 2002. "Introduction." In *Global Women*, Barbara Ehrenreich and Arlie Russell Hochschild (eds.). (New York: Metropolitan Books).

Forbes 2006. "The Global 2000," (April 17):158–161.

Hacker, Jacob S. 2006. *The Great Risk Shift* (New York: Oxford University Press).

Harrison, Bennett. 1994. *Lean and Mean: The Changing Landscape of Corporate Power in the Age of Flexibility* (New York: Basic Books).

Hondagneu-Sotelo, Pierrette. 2003. "Gender and Immigration: A Retrospective and Introduction." In *Gender and U.S. Immigration: Contemporary Trends*, Pierrette Hondagneu-Sotelo (ed.). (Berkeley: University of California Press), 3–19.

Hytrek, Gary, and Kristine M. Zentgraf. 2008. *America Transformed: Globalization, Inequality, and Power* (New York: Oxford University Press).

Lechner, Frank J., and John Boli (eds.). 2000. *The Globalization Reader* (Oxford, UK: Blackwell Publishers).

Martin, Philip, and Jonas Widgren. 2002. "International Migration: Facing the Challenge," *Population Bulletin* 57 (March): entire issue.

Muller, Joann. 2004. "Global Motors," *Forbes* (January 12):62–68.

Peoples, James, and Garrick Bailey. 2003. *Humanity: An Introduction to Cultural Anthropology* (Belmont, CA: Wadsworth). The section on globalization was accessed at http://www.wadsworth.com/anthropology_d/resources/terrorism/booklet.html

Sen, Amartya. 2002. "How to Judge Globalism," *The American Prospect* (Winter):A2–A6.

Scholte, Jan Aart. 2000. *Globalization: A Critical Introduction* (New York: Palgrave).

Steger, Manfred B. 2002. *Globalism: The New Market Ideology* (Lanham, MD: Rowman & Littlefield).

U.S. News & World Report 1999. "The Year 1000: What Life was Like in the Last Millennium," a special double issue (August 16 and 23):38–94.

2

Dimensions of Globalization

Typically, we think of globalization in economic terms—the movement of capital and jobs across national boundaries, or the production of consumer goods in one country to be sold in other countries, or the price of gasoline, bread, coffee, and oranges affected by international events. But while the economic dimension of globalization is central, globalization also involves, for example, the transnational flow of information, people, communication, technology, fashion, environmental hazards, terrorism, disease, and criminal activities.

The first essay, by noted British sociologist Anthony Giddens, illustrates the broad sweep of globalization and its consequences. He views globalization as a transformation force, not only "out there" away from the individual but also an "in here" phenomenon, influencing us personally.

Next, *New York Times* columnist and twice Pulitzer Prize recipient Thomas L. Friedman introduces his book *The World Is Flat*, in which he argues that technology is leveling the playing field of global competition. The "flattening" is positive as people everywhere have the ability to connect, cooperate, and compete in a new era of prosperity and innovation.

Jeremy Brecher, Tim Costello, and Brendan Smith, in the introductory chapter to their book *Globalization from Below*, give an overview of globalization by providing the various dimensions of the globalization process.

An important dimension of globalization is international migration. Susan F. Martin, director of the Institute for the Study of International Migration at Georgetown University, illustrates the magnitude of immigration flows and their consequences.

The final contribution to this chapter, by D. Stanley Eitzen, lists a number of current facts regarding immigration, transnational trade and finance, cultural globalization, and the globalization of disease and environmental hazards.

1

Globalisation

ANTHONY GIDDENS

A friend of mine studies village life in central Africa. A few years ago, she paid her first visit to a remote area where she was to carry out her fieldwork. The day she arrived, she was invited to a local home for an evening's entertainment. She expected to find out about the traditional pastimes of this isolated community. Instead, the occasion turned out to be a viewing of *Basic Instinct* on video. The film at that point hadn't even reached the cinemas in London.

Such vignettes reveal something about our world. And what they reveal isn't trivial. It isn't just a matter of people adding modern paraphernalia—videos, television sets, personal computers and so forth—to their existing ways of life. We live in a world of transformations, affecting almost every aspect of what we do. For better or worse, we are being propelled into a global order that no one fully understands, but which is making its effects felt upon all of us.

Globalisation may not be a particularly attractive or elegant word. But absolutely no one who wants to understand our prospects at century's end can ignore it. I travel a lot to speak abroad. I haven't been to a single country recently where globalisation isn't being intensively discussed. In France, the word is *mondialisation*. In Spain and Latin America, it is *globalización*. The Germans say *Globalisierung*.

The global spread of the term is evidence of the very developments to which it refers. Every business guru talks about it. No political speech is complete without reference to it. Yet even in the late 1980s the term was hardly used, either in the academic literature or in everyday language. It has come from nowhere to be almost everywhere.

Given its sudden popularity, we shouldn't be surprised that the meaning of the notion isn't always clear, or that an intellectual reaction has set in against it. Globalisation has something to do with the thesis that we now all live in one world—but in what ways exactly, and is the idea really valid? Different thinkers have taken almost completely opposite views about globalisation in debates that have sprung up over the past few years. Some dispute the whole thing. I'll call them the sceptics.

According to the sceptics, all the talk about globalisation is only that—just talk. Whatever its benefits, its trials and tribulations, the global economy isn't

SOURCE: Copyright 2000 "Globalisation" from *Runaway World: How Globalization Is Reshaping Our Lives* by Lord Anthony Giddens. Reproduced by permission of Routledge Taylor & Francis Books, Inc. and Lord Anthony Giddens.

especially different from that which existed at previous periods. The world carries on much the same as it has done for many years.

Most countries, the sceptics argue, gain only a small amount of their income from external trade. Moreover, a good deal of economic exchange is between regions, rather than being truly world-wide. The countries of the European Union, for example, mostly trade among themselves. The same is true of the other main trading blocs, such as those of Asia–Pacific or North America.

Others take a very different position. I'll label them the radicals. The radicals argue that not only is globalisation very real, but that its consequences can be felt everywhere. The global market-place, they say, is much more developed than even in the 1960s and 1970s and is indifferent to national borders. Nations have lost most of the sovereignty they once had, and politicians have lost most of their capability to influence events. It isn't surprising that no one respects political leaders any more, or has much interest in what they have to say. The era of the nation-state is over. Nations, as the Japanese business writer Kenichi Ohmae puts it, have become mere "fictions." Authors such as Ohmae see the economic difficulties of the 1998 Asian crisis as demonstrating the reality of globalisation, albeit seen from its disruptive side.

The sceptics tend to be on the political left, especially the old left. For if all of this is essentially a myth, governments can still control economic life and the welfare state remain intact. The notion of globalisation, according to the sceptics, is an ideology put about by free-marketeers who wish to dismantle welfare systems and cut back on state expenditures. What has happened is at most a reversion to how the world was a century ago. In the late nineteenth century there was already an open global economy, with a great deal of trade, including trade in currencies.

Well, who is right in this debate? I think it is the radicals. The level of world trade today is much higher than it ever was before, and involves a much wider range of goods and services. But the biggest difference is in the level of finance and capital flows. Geared as it is to electronic money—money that exists only as digits in computers—the current world economy has no parallels in earlier times.

In the new global electronic economy, fund managers, banks, corporations, as well as millions of individual investors, can transfer vast amounts of capital from one side of the world to another at the click of a mouse. As they do so, they can destabilize what might have seemed rock-solid economies—as happened in the events in Asia.

The volume of world financial transactions is usually measured in U.S. dollars. A million dollars is a lot of money for most people. Measured as a stack of hundred-dollar notes, it would be eight inches high. A billion dollars—in other words, a thousand million—would stand higher than St. Paul's Cathedral. A trillion dollars—a million million—would be over 120 miles high, 20 times higher than Mount Everest.

Yet far more than a trillion dollars is now turned over *each day* on global currency markets. This is a massive increase from only the late 1980s, let alone the more distant past. The value of whatever money we may have in our pockets, or our bank accounts, shifts from moment to moment according to fluctuations in such markets.

I would have no hesitation, therefore, in saying that globalisation, as we are experiencing it, is in many respects not only new, but also revolutionary. Yet I don't believe that either the sceptics or the radicals have properly understood either what it is or its implications for us. Both groups see the phenomenon almost solely in economic terms. This is a mistake. Globalisation is political, technological and cultural, as well as economic. It has been influenced above all by developments in systems of communication, dating back only to the late 1960s.

In the mid-nineteenth century, a Massachusetts portrait painter, Samuel Morse, transmitted the first message, "What hath God wrought?" by electric telegraph. In so doing, he initiated a new phase in world history. Never before could a message be sent without someone going somewhere to carry it. Yet the advent of satellite communications marks every bit as dramatic a break with the past. The first commercial satellite was launched only in 1969. Now there are more than 200 such satellites above the earth, each carrying a vast range of information. For the first time ever, instantaneous communication is possible from one side of the world to the other. Other types of electronic communication, more and more integrated with satellite transmission, have also accelerated over the past few years. No dedicated transatlantic or transpacific cables existed at all until the late 1950s. The first held fewer than 100 voice paths. Those of today carry more than a million.

On 1 February 1999, about 150 years after Morse invented his system of dots and dashes, Morse Code finally disappeared from the world stage. It was discontinued as a means of communication for the sea. In its place has come a system using satellite technology, whereby any ship in distress can be pinpointed immediately. Most countries prepared for the transition some while before. The French, for example, stopped using Morse Code in their local waters in 1997, signing off with a Gallic flourish: "Calling all. This our last cry before our eternal silence."

Instantaneous electronic communication isn't just a way in which news or information is conveyed more quickly. Its existence alters the very texture of our lives, rich and poor alike. When the image of Nelson Mandela may be more familiar to us than the face of our next-door neighbour, something has changed in the nature of our everyday experience.

Nelson Mandela is a global celebrity, and celebrity itself is largely a product of new communications technology. The reach of media technologies is growing with each wave of innovation. It took 40 years for radio in the United States to gain an audience of 50 million. The same number was using personal computers only 15 years after the personal computer was introduced. It needed a mere 4 years, after it was made available, for 50 million Americans to be regularly using the Internet.

It is wrong to think of globalisation as just concerning the big systems, like the world financial order. Globalisation isn't only about what is "out there," remote and far away from the individual. It is an "in here" phenomenon too, influencing intimate and personal aspects of our lives. The debate about family values, for example, that is going on in many countries might seem far removed from globalising influences. It isn't. Traditional family systems are becoming transformed, or are under strain, in many parts of the world, particularly as women stake claim to greater equality. There has never before been a society, so far as we know from the historical record, in which women have been even approximately equal to men. This is a truly global revolution in everyday life, whose consequences are being felt around the world in spheres from work to politics.

Globalisation thus is a complex set of processes, not a single one. And these operate in a contradictory or oppositional fashion. Most people think of globalisation as simply "pulling away" power or influence from local communities and nations into the global arena. And indeed this is one of its consequences. Nations do lose some of the economic power they once had. Yet it also has an opposite effect. Globalisation not only pulls upwards, but also pushes downwards, creating new pressures for local autonomy. The American sociologist Daniel Bell describes this very well when he says that the nation becomes not only too small to solve the big problems, but also too large to solve the small ones.

Globalisation is the reason for the revival of local cultural identities in different parts of the world. If one asks, for example, why the Scots want more independence in the U.K. or why there is a strong separatist movement in Quebec, the answer is not to be found only in their cultural history. Local nationalisms spring up as a response to globalising tendencies, as the hold of older nation-states weakens.

Globalisation also squeezes sideways. It creates new economic and cultural zones within and across nations. Examples are the Hong Kong region, northern Italy, and Silicon Valley in California. Or consider the Barcelona region. The area around Barcelona in northern Spain extends into France. Catalonia, where Barcelona is located, is closely integrated into the European Union. It is part of Spain, yet also looks outwards.

These changes are being propelled by a range of factors, some structural, others more specific and historical. Economic influences are certainly among the driving forces—especially the global financial system. Yet they aren't like forces of nature. They have been shaped by technology, and cultural diffusion, as well as by the decisions of governments to liberalise and deregulate their national economies.

The collapse of Soviet communism has added further weight to such developments, since no significant group of countries any longer stands outside. That collapse wasn't just something that just happened to occur. Globalisation explains both why and how Soviet communism met its end. The former Soviet Union and the East European countries were comparable to the West in terms of growth rates until somewhere around the early 1970s. After that point, they fell rapidly behind. Soviet communism, with its emphasis upon state-run enterprise and heavy industry, could not compete in the global electronic economy. The ideological and cultural control upon which communist political authority was based similarly could not survive in an era of global media.

The Soviet and the East European regimes were unable to prevent the reception of Western radio and television broadcasts. Television played a direct role in the 1989 revolutions, which have rightly been called the first "television revolution." Street protests taking place in one country were watched by television audiences in others, large numbers of whom then took to the streets themselves.

Globalisation, of course, isn't developing in an even-handed way, and is by no means wholly benign in its consequences. To many living outside Europe and North America, it looks uncomfortably like Westernisation—or, perhaps, Americanisation, since the U.S. is now the sole superpower, with a dominant economic, cultural and military position in the global order. Many of the most visible cultural expressions of globalisation are American—Coca-Cola, McDonald's, CNN.

Most of the giant multinational companies are based in the U.S. too. Those that aren't all come from the rich countries, not the poorer areas of the world. A pessimistic view of globalisation would consider it largely an affair of the industrial North, in which the developing societies of the South play little or no active part. It would see it as destroying local cultures, widening world inequalities and worsening the lot of the impoverished. Globalisation, some argue, creates a world of winners and losers, a few on the fast track to prosperity, the majority condemned to a life of misery and despair.

Indeed, the statistics are daunting. The share of the poorest fifth of the world's population in global income has dropped, from 2.3 per cent to 1.4 per cent between 1989 and 1998. The proportion taken by the richest fifth, on the other hand, has risen. In sub-Saharan Africa, 20 countries have lower incomes per head in real terms than they had in the late 1970s. In many less developed countries, safety and environmental regulations are low or virtually non-existent. Some transnational companies sell goods there that are controlled or banned in the industrial countries—poor-quality medical drugs, destructive pesticides or high tar and nicotine content cigarettes. Rather than a global village, one might say, this is more like global pillage.

Along with ecological risk, to which it is related, expanding inequality is the most serious problem facing world society. It will not do, however, merely to blame it on the wealthy. It is fundamental to my argument that globalisation today is only partly Westernisation. Of course the Western nations, and more generally the industrial countries, still have far more influence over world affairs than do the poorer states. But globalisation is becoming increasingly decentred—not under the control of any group of nations, and still less of the large corporations. Its effects are felt as much in Western countries as elsewhere.

This is true of the global financial system, and of changes affecting the nature of government itself. What one could call "reverse colonization" is becoming more and more common. Reverse colonization means that non-Western countries influence developments in the West. Examples abound—such as the latinising of Los Angeles, the emergence of a globally oriented high-tech sector in India, or the selling of Brazilian television programmes to Portugal.

Is globalisation a force promoting the general good? The question can't be answered in a simple way, given the complexity of the phenomenon. People who ask it, and who blame globalisation for deepening world inequalities, usually have in mind economic globalisation and, within that, free trade. Now, it is surely obvious that free trade is not an unalloyed benefit. This is especially so as concerns the less developed countries. Opening up a country, or regions within it, to free trade can undermine a local subsistence economy. An area that becomes dependent upon a few products sold on world markets is very vulnerable to shifts in prices as well as to technological change.

Trade always needs a framework of institutions, as do other forms of economic development. Markets cannot be created by purely economic means, and how far a given economy should be exposed to the world market-place must depend upon a range of criteria. Yet to oppose economic globalisation, and to opt for economic protectionism, would be a misplaced tactic for rich and poor nations alike. Protectionism may be a necessary strategy at some times and in some countries. In my view, for example, Malaysia was correct to introduce controls in 1998, to stem the flood of capital from the country. But more permanent forms of protectionism will not help the development of the poor countries, and among the rich would lead to warring trade blocs.

The debates about globalisation I mentioned at the beginning have concentrated mainly upon its implications for the nation-state. Are nation-states, and hence national political leaders, still powerful, or are they becoming largely irrelevant to the forces shaping the world? Nation-states are indeed still powerful and political leaders have a large role to play in the world. Yet at the same time the nation-state is being reshaped before our eyes. National economic policy can't be as effective as it once was. More importantly, nations have to rethink their identities now the older forms of geopolitics are becoming obsolete. Although this is a contentious point, I would say that, following the dissolving of the Cold War, most nations no longer have enemies. Who are the enemies of Britain, or France, or Brazil? The war in Kosovo didn't pit nation against nation. It was a conflict between old-style territorial nationalism and a new, ethically driven interventionism.

Nations today face risks and dangers rather than enemies, a massive shift in their very nature. It isn't only of the nation that such comments could be made. Everywhere we look, we see institutions that appear the same as they used to be from the outside, and carry the same names, but inside have become quite different. We continue to talk of the nation, the family, work, tradition, nature, as if they were all the same as in the past. They are not. The outer shell remains, but inside they have changed—and this is happening not only in the U.S., Britain, or France, but almost everywhere. They are what I call "shell institutions." They are institutions that have become inadequate to the tasks they are called upon to perform.

As the changes I have described in this chapter gather weight, they are creating something that has never existed before, a global cosmopolitan society. We are the first generation to live in this society, whose contours we can as yet only dimly see. It is shaking up our existing ways of life, no matter where we happen to be. This is not—at least at the moment—a global order driven by collective human will. Instead, it is emerging in an anarchic, haphazard fashion, carried along by a mixture of influences.

It is not settled or secure, but fraught with anxieties, as well as scarred by deep divisions. Many of us feel in the grip of forces over which we have no power. Can we reimpose our will upon them? I believe we can. The powerlessness we experience is not a sign of personal failings, but reflects the incapacities of our institutions. We need to reconstruct those we have, or create new ones. For globalisation is not incidental to our lives today. It is a shift in our very life circumstances. It is the way we now live.

2

The World Is Flat

Columbus reported to his king and queen that the world was round, and he went down in history as the man who first made this discovery. I returned home and shared my discovery only with my wife, and only in a whisper.

"Honey," I confided, "I think the world is flat."

How did I come to this conclusion? I guess you could say it all started in Nandan Nilekani's conference room at Infosys Technologies Limited. Infosys is one of the jewels of the Indian information technology world, and Nilekani, the company's CEO, is one of the most thoughtful and respected captains of Indian industry. I drove with the Discovery Times crew out to the Infosys campus, about forty minutes from the heart of Bangalore, to tour the facility and interview Nilekani. The Infosys campus is reached by a pockmarked road, with sacred cows, horse-drawn carts, and motorized rickshaws all jostling alongside our vans. Once you enter the gates of Infosys, though, you are in a different world. A massive resort-size swimming pool nestles amid boulders and manicured lawns, adjacent to a huge putting green. There are multiple restaurants and a fabulous health club. Glass-and-steel buildings seem to sprout up like weeds each week. In some of those buildings, Infosys employees are writing specific software programs for American or European companies; in others, they are running the back rooms of major American and European-based multinationals—everything from computer maintenance to specific research projects to answering customer calls routed there from all over the world. Security is tight, cameras monitor the doors, and if you are working for American Express, you cannot get into the building that is managing services and research for General Electric. Young Indian engineers, men and women, walk briskly from building to building, dangling ID badges. One looked like he could do my taxes. Another looked like she could take my computer apart. And a third looked like she designed it!

After sitting for an interview, Nilekani gave our TV crew a tour of Infosys's global conferencing center—ground zero of the Indian outsourcing industry. It was a cavernous wood-paneled room that looked like a tiered classroom from an Ivy League law school. On one end was a massive wall-size screen and overhead there were cameras in the ceiling for teleconferencing. "So this is our conference room, probably the largest screen in Asia—this is forty digital screens [put to-

SOURCE: Excerpt from Thomas L. Friedman, *The World Is Flat*, New York: Farrar, Straus and Giroux, 2005, pp. 5–11.

gether]," Nilekani explained proudly, pointing to the biggest flat-screen TV I had ever seen. Infosys, he said, can hold a virtual meeting of the key players from its entire global supply chain for any project at any time on that supersize screen. So their American designers could be on the screen speaking with their Indian software writers and their Asian manufacturers all at once. "We could be sitting here, somebody from New York, London, Boston, San Francisco, all live. And maybe the implementation is in Singapore, so the Singapore person could also be live here . . . That's globalization," said Nilekani. Above the screen there were eight clocks that pretty well summed up the Infosys workday: 24/7/365. The clocks were labeled US West, US East, GMT, India, Singapore, Hong Kong, Japan, Australia.

"Outsourcing is just one dimension of a much more fundamental thing happening today in the world," Nilekani explained. "What happened over the last [few] years is that there was a massive investment in technology, especially in the bubble era, when hundreds of millions of dollars were invested in putting broadband connectivity around the world, undersea cables, all those things." At the same time, he added, computers became cheaper and dispersed all over the world, and there was an explosion of software—e-mail, search engines like Google, and proprietary software that can chop up any piece of work and send one part to Boston, one part to Bangalore, and one part to Beijing, making it easy for anyone to do remote development. When all of these things suddenly came together around 2000, added Nilekani, they "created a platform where intellectual work, intellectual capital, could be delivered from anywhere. It could be disaggregated, delivered, distributed, produced, and put back together again— and this gave a whole new degree of freedom to the way we do work, especially work of an intellectual nature . . . And what you are seeing in Bangalore today is really the culmination of all these things coming together."

We were sitting on the couch outside of Nilekani's office, waiting for the TV crew to set up its cameras. At one point, summing up the implications of all this, Nilekani uttered a phrase that rang in my ear. He said to me, "Tom, the playing field is being leveled." He meant that countries like India are now able to compete for global knowledge work as never before–and that America had better get ready for this. America was going to be challenged, but, he insisted, the challenge would be good for America because we are always at our best when we are being challenged. As I left the Infosys campus that evening and bounced along the road back to Bangalore, I kept chewing on that phrase: "The playing field is being leveled."

What Nandan is saying, I thought, is that the playing field is being flattened . . . Flattened? Flattened? My God, he's telling me the world is flat!

Here I was in Bangalore—more than five hundred years after Columbus sailed over the horizon, using the rudimentary navigational technologies of his day, and returned safely to prove definitively that the world was round—and one of India's smartest engineers, trained at his country's top technical institute and backed by the most modern technologies of his day, was essentially telling me that the world was *flat*—as flat as that screen on which he can host a meeting of his whole global supply chain. Even more interesting, he was citing this development as a good thing, as a new milestone in human progress and a great opportunity for India and the world—the fact that we had made our world flat!

In the back of that van, I scribbled down four words in my notebook: "The world is flat." As soon as I wrote them, I realized that this was the underlying message of everything that I had seen and heard in Bangalore in two weeks of filming. The global competitive playing field was being leveled. The world was being flattened.

As I came to this realization, I was filled with both excitement and dread. The journalist in me was excited at having found a framework to better understand the morning headlines and to explain what was happening in the world today. Clearly, it is now possible for more people than ever to collaborate and compete in real time with more other people on more different kinds of work from more different corners of the planet and on a more equal footing than at any previous time in the history of the world—using computers, e-mail, networks, teleconferencing, and dynamic new software. That is what Nandan was telling me. That was what I discovered on my journey to India and beyond. And that is what this book is about. When you start to think of the world as flat, a lot of things make sense in ways they did not before. But I was also excited personally, because what the flattening of the world means is that we are now connecting all the knowledge centers on the planet together into a single global network, which—if politics and terrorism do not get in the way—could usher in an amazing era of prosperity and innovation.

But contemplating the flat world also left me filled with dread, professional and personal. My personal dread derived from the obvious fact that it's not only the software writers and computer geeks who get empowered to collaborate on work in a flat world. It's also al-Qaeda and other terrorist networks. The playing field is not being leveled only in ways that draw in and superempower a whole new group of innovators. It's being leveled in a way that draws in and superempowers a whole new group of angry, frustrated, and humiliated men and women.

Professionally, the recognition that the world was flat was unnerving because I realized that this flattening had been taking place while I was sleeping, and I had missed it. I wasn't really sleeping, but I was otherwise engaged. Before 9/11, I was focused on tracking globalization and exploring the tension between the "Lexus" forces of economic integration and the "Olive Tree" forces of identity and nationalism—hence my 1999 book, *The Lexus and the Olive Tree*. But after 9/11, the olive tree wars became all-consuming for me. I spent almost all my time traveling in the Arab and Muslim worlds. During those years I lost the trail of globalization.

I found that trail again on my journey to Bangalore in February 2004. Once I did, I realized that something really important had happened while I was fixated on the olive groves of Kabul and Baghdad. Globalization had gone to a whole new level. If you put *The Lexus and the Olive Tree* and this book together, the broad historical argument you end up with is that there have been three great eras of globalization. The first lasted from 1492—when Columbus set sail, opening trade between the Old World and the New World—until around 1800. I would call this era Globalization 1.0. It shrank the world from a size large to a size medium. Globalization 1.0 was about countries and muscles. That is, in

Globalizational 1.0 the key agent of change, the dynamic force driving the process of global integration was how much brawn—how much muscle, how much horsepower, wind power, or, later, steam power—your country had and how creatively you could deploy it. In this era, countries and governments (often inspired by religion or imperialism or a combination of both) led the way in breaking down walls and knitting the world together, driving global integration. In Globalization 1.0, the primary questions were: Where does my country fit into global competition and opportunities? How can I go global and collaborate with others through my country?

The second great era, Globalization 2.0, lasted roughly from 1800 to 2000, interrupted by the Great Depression and World Wars I and II. This era shrank the world from a size medium to a size small. In Globalization 2.0, the key agent of change, the dynamic force driving global integration, was multinational companies. These multinationals went global for markets and labor, spearheaded first by the expansion of the Dutch and English joint-stock companies and the Industrial Revolution. In the first half of this era, global integration was powered by falling transportation costs, thanks to the steam engine and the railroad, and in the second half by falling telecommunication costs—thanks to the diffusion of the telegraph, telephones, the PC, satellites, fiber-optic cable, and the early version of the World Wide Web. It was during this era that we really saw the birth and maturation of a global economy, in the sense that there was enough movement of goods and information from continent to continent for there to be a global market, with global arbitrage in products and labor. The dynamic forces behind this era of globalization were breakthroughs in hardware—from steamships and railroads in the beginning to telephones and mainframe computers toward the end. And the big questions in this era were: Where does my company fit into the global economy? How does it take advantage of the opportunities? How can I go global and collaborate with others through my company? *The Lexus and the Olive Tree* was primarily about the climax of this era, an era when the walls started falling all around the world, and integration, and the backlash to it, went to a whole new level. But even as the walls fell, there were still a lot of barriers to seamless global integration. Remember, when Bill Clinton was elected president in 1992, virtually no one outside of government and the academy had e-mail, and when I was writing *The Lexus and the Olive Tree* in 1998, the Internet and e-commerce were just taking off.

Well, they took off—along with a lot of other things that came together while I was sleeping. And that is why I argue in this book that around the year 2000 we entered a whole new era: Globalization 3.0. Globalization 3.0 is shrinking the world from a size small to a size tiny and flattening the playing field at the same time. And while the dynamic force in Globalization 1.0 was countries globalizing and the dynamic force in Globalization 2.0 was companies globalizing, the dynamic force in Globalization 3.0—the thing that gives it its unique character—is the newfound power for *individuals* to collaborate and compete globally. And the lever that is enabling individuals and groups to go global so easily and so seamlessly is not horsepower, and not hardware, but software—all sorts of new applications—in conjunction with the creation of a global fiber-optic network that has made us all next-door neighbors. Individuals must, and can, now ask, Where do *I* fit into the global competition and opportunities of the day, and how can *I*, on my own, collaborate with others globally?

But Globalization 3.0 not only differs from the previous eras in how it is shrinking and flattening the world and in how it is empowering individuals. It is different in that Globalization 1.0 and 2.0 were driven primarily by European and American individuals and businesses. Even though China actually had the biggest economy in the world in the eighteenth century, it was Western countries, companies, and explorers who were doing most of the globalizing and shaping of the system. But going forward, this will be less and less true. Because it is flattening and shrinking the world, Globalization 3.0 is going to be more and more driven not only by individuals but also by a much more diverse—non-Western, non-white—group of individuals. Individuals from every corner of the flat world are being empowered. Globalization 3.0 makes it possible for so many more people to plug and play, and you are going to see every color of the human rainbow take part.

(While this empowerment of individuals to act globally is the most important new feature of Globalization 3.0, companies—large and small—have been newly empowered in this era as well. I discuss both in detail later in the book.)

Needless to say, I had only the vaguest appreciation of all this as I left Nandan's office that day in Bangalore. But as I sat contemplating these changes on the balcony of my hotel room that evening, I did know one thing: I wanted to drop everything and write a book that would enable me to understand how this flattening process happened and what its implications might be for countries, companies, and individuals. So I picked up the phone and called my wife, Ann, and told her, "I am going to write a book called *The World Is Flat*." She was both amused and curious—well, maybe *more* amused than curious! Eventually, I was able to bring her around, and I hope I will be able to do the same with you, dear reader. Let me start by taking you back to the beginning of my journey to India, and other points east, and share with you some of the encounters that led me to conclude the world was no longer round—but flat.

3

Globalization and Its Specter

JEREMY BRECHER
TIM COSTELLO
BRENDAN SMITH

GLOBALIZATION FROM ABOVE

Epochal changes can be difficult to grasp—especially when you are in their midst. Those who lived through the rise of capitalism or the industrial revolution knew something momentous was happening, but just what was new and what it meant were subjects of confusion and debate.

In a sense, there has been a global economy for 500 years. But the last quarter of the 20th century saw global economic integration take new forms. At first, globalization manisfested itself as apparently separate and rather marginal phenomena: the emergence of the "Eurodollar market," "off-shore export platforms," and "supply-side economics," for example. It was easy to separate out one or another aspect of globalization—such as the growth of trade or of international economic institutions—and see it as an isolated phenomenon. These seemingly peripheral developments, however, gradually interacted in ways that changed virtually every aspect of life and defined globalization as a new global configuration.[1]

Globalization was not the result of a plot or even a plan. It was caused by people acting with intent—seeking new economic opportunities, creating new institutions, trying to outflank political and economic opponents. But it resulted not just from their intent, but also from unintended side effects of their actions and the consequences of unintended interactions.[2] Future historians will note at least the following aspects of the globalization process:

Production: In the 1970s, corporations began building factories and buying manufactured products in low-wage countries in the third world on a vastly expanded scale. Such off-shore production grew into today's "global assembly line," in which the components of a shirt or car may be made and assembled in a dozen or more different countries. Direct investment abroad by "American" companies has grown so rapidly that the value of the goods and services they produce and sell outside the United States is now three times the total value of all American exports.[3]

Markets: Corporations came increasingly to view the entire world as a single market in which they buy and sell goods, services, and labor. . . .

SOURCE: From *Globalization from Below: The Power of Soldarity* (Cabridge, MA: South End Press, 2000), pp. 1–9. Note: Endnotes for this article have been renumbered—ED.

Finance: Starting with the rise of the Eurodollar market in the 1970s, international capital markets have globalized at an accelerating rate. . . .

Technology: New information, communication, and transportation technologies—computers, satellite communications, containerized shipping, and, increasingly, the Internet—have reduced distance as a barrier to economic integration. Furthermore, the process of creating new technologies has itself become globalized.[4]

Global institutions: The World Trade Organization (WTO), the International Monetary Fund (IMF), the World Bank, and similar institutions at a regional level have developed far greater powers and have used them to accelerate the globalization process.

Corporate restructuring: While corporations have always operated internationally, starting in the 1980s they began to restructure in order to operate in a global economy. New corporate forms—strategic alliances, global outsourcing, captive suppliers, supplier chains, and, increasingly, transnational mergers—allowed for what the economist Bennett Harrison has called the "concentration of control [with] the decentralization of production."[5]

Changing structure of work: Globalization has been characterized by a "re-commodification of labor" in which workers have increasingly lost all rights except the right to sell their labor power. All over the world, employers have downsized, outsourced, and made permanent jobs into contingent ones. Employers have attacked job security requirements, work rules, worker representation, healthcare, pensions, and other social benefits, and anything else that defined workers as human beings and employers as partners in a social relationship, rather than simply as buyers and sellers of labor power.

Neoliberal ideology and policies: Starting with monetarism and supply-side economics, globalization has been accompanied—and accelerated—by an emerging ideology now generally known as neoliberalism or the Washington consensus. It argues that markets are efficient and that government intervention in them is almost always bad. The policy implications—privatization, deregulation, open markets, balanced budgets, deflationary austerity, and dismantling of the welfare state—were accepted by or imposed on governments all over the world.

Changing role of the state: While some governments actively encouraged globalization and most acquiesced, globalization considerably reduced the power of nation states, particularly their power to serve the interests of their own people. Capital mobility undermined the power of national governments to pursue full employment policies or regulate corporations. International organizations and agreements increasingly restricted environmental and social protections. Neoliberal ideology reshaped beliefs about what government should do and what it is able to accomplish.

Neo-imperialism: Globalization reversed the post–World War II movement of third world countries out of colonialism toward economic independence. Globalization has restored much of the global dominance of the former imperialist powers, such as Western Europe, Japan, and, above all, the United States. With the collapse of Communism, that dominance has also spread to much of the formerly communist world. Globalization has taken from poor countries control of their own economic policies and concentrated their assets in the hands of first world investors. While it has enriched some third world elites, it has subordinated them to foreign corporations, international institutions, and dominant states. It has intensified economic rivalry among the rich powers.[6]

Movement of people: While people have always crossed national borders, the economic disruptions and reduction of national barriers caused by globalization are accelerating migration. International travel and tourism have become huge industries in their own right.

Cultural homogenization: Globalization has undermined the economic base of diverse local and indigenous communities all over the world. Growing domination of global media by a few countries and companies has led not to greater diversity, but to an increasingly uniform culture of corporate globalism.

As *New York Times* columnist and globalization advocate Thomas Friedman summed up, we are in a new international system: "Globalization is not just a trend, not just a phenomenon, not just an economic fad. It is the international system that has replaced the cold-war system." The driving force behind globalization is free-market capitalism: "Globalization means the spread of free-market capitalism to virtually every country in the world."[7]

THE CONTRADICTIONS OF GLOBALIZATION
FROM ABOVE

The proponents of globalization promised that it would benefit all: that it would "raise all the boats." Workers and communities around the globe were told that if they downsized, deregulated, eliminated social services, and generally became more competitive, the benefits of globalization would bless them. The poorest and most desperate were promised that they would see their standard of living increase if they accepted neoliberal austerity measures. They kept their end of the bargain, but globalization from above did not reciprocate. Instead, it is aggravating old and creating new problems for people and the environment.

Even conventional economic theory recognizes that the "hidden hand" of the market doesn't always work. Unregulated markets regularly produce unintended side effects or "externalities"—such as ecological pollution for which the producer doesn't have to pay, or the devastation of communities when corporations move away. Unregulated markets also produce unintended interaction effects, such as the downward spirals of depressions and trade wars. Unregulated markets do nothing to correct inequalities of wealth; indeed, they often intensify the concentration of wealth, leading to expanding gaps between rich and poor.[8] Globalization from above has globalized these problems, while dismantling at every level the non-market institutions that once addressed them.

Globalization promotes a destructive competition in which workers, communities, and entire countries are forced to cut labor, social, and environmental costs to attract mobile capital. When many countries each do so, the result is a disastrous "race to the bottom."

The race to the bottom occurs not just between developing and developed worlds, but increasingly among the countries of the third world. Consider the case of Argentina and Brazil. Early in 1999, Brazil devalued its currency by 40 percent. A *New York Times* reporter in Argentina found that "[a]bout 60 manufacturing companies have moved to Brazil in recent months, seeking lower labor costs and offers of tax breaks and other government subsidies." Companies closing Argentine factories to supply the Argentine market from Brazil included Tupperware, Goodyear, and Royal Philips Electronics. The Argentine auto and auto-parts industries suffered a 33 percent loss of production and a 59 percent fall in exports in 1999. "General Motors, Ford Motor and Fiat are all transferring production to Brazil."[9]

Argentine President Fernando De la Rua commented, "If you ask me what is my chief concern in a word, that word is 'competitiveness.'" The measures he has taken to become more "competitive" exemplify the race to the bottom. "The crown jewel of the De la Rua economic policy is his labor reform" intended to "reduce the bargaining power of labor unions and help businesses more easily hire and fire new workers."

But this gutting of labor rights was not enough to protect Argentine manufacturers against products from lower-wage countries. A shoe manufacturer who expected the new labor law to cut his labor costs by 10 percent "felt constrained because of the competitive disadvantage he continued to suffer in relation to Brazilian shoe producers who pay their workers one-third the wage an Argentine shoemaker earns." The director of a medical supply company who was considering closing his plant observed that it was impossible to compete with the flood of cheap Korean and Chinese syringes in recent years and that Brazilian officials were offering a package of tax breaks and subsidized loans to relocate to Brazil.

The role of international institutions in promoting the race to the bottom is illustrated by the fact that both Brazil and Argentina were shaping their economic policies in accord with loan agreements they had made with the IMF.

First world countries are also engaged in the race to the bottom. Over the past two decades, for example, the United States made huge cuts in corporate taxes while slashing federal funding for health, education, and community development. Canada, which did not make equivalent cuts, found that its tax structure was "making it difficult for companies to compete internationally. Many businesses have simply moved across the border to the U.S." In response, Canada decided in early 2000 to lower its corporate tax rate from 28 percent to 21 percent. In a fit of ingratitude, the Business Council on National Issues, representing Canada's 150 largest companies, condemned the cuts as "timid." The Business Council's president opined that "[t]he strategy should be to provide an environment more attractive than the U.S. now." The disappointed chief executive of an e-commerce services company said he had been planning to open offices in Calgary, Alberta, and Vancouver, British Columbia, but that after the inadequate tax cuts he was leaning toward Chicago or Minneapolis instead. The director of the Canadian Taxpayers Federation observed, "There's competition for tax cuts, just like everything else."[10]

The race to the bottom brings with it the dubious blessings of impoverishment, growing inequality, economic volatility, the degradation of democracy, and destruction of the environment.

Impoverishment: The past quarter-century of globalization has seen not a reduction but a vast increase in poverty. According to the 1999 U.N. *Human Development Report*, more than 80 countries have per capita incomes lower than they were a decade or more ago.[11] James Wolfensohn, president of the World Bank, says that, rather than improving, "global poverty is getting worse. Some 1.2 billion people now live in extreme poverty."[12] Global unemployment is approaching 1 billion.[13]

In the United States, the downward pressures of globalization are manifested in the stagnation of wages despite the longest period of economic growth in American history. Real average wages were $9 per hour in 1973; 25 years later, they were $8 per hour. The typical married-couple family worked 247 more hours per year in 1996 than in 1989—more than six weeks' worth of additional work each year.[14]

Inequality: Globalization has contributed to an enormous increase in the concentration of wealth and the growth of poverty both within countries and worldwide. Four hundred and forty-seven billionaires have wealth greater than the income of the poorest half of humanity. In the United States, the richest man has wealth equal to that of the poorest 40 percent of the American people.[15] The net worth of the world's 200 richest people increased from $440 billion to more than $1 trillion in just the four years from 1994 to 1998. The assets of the three richest people were more than the combined GNP of the 48 least developed countries.[16]

The downward pressures of globalization have been focused most intensively on discriminated-against groups that have the least power to resist, including women, racial and ethnic minorities, and indigenous peoples. Women have been the prime victims of exploitation in export industries and have suffered the brunt of cutbacks in public services and support for basic needs. Immigrants and racial and ethnic minorities in many parts of the world have not only been subject to exploitation, but have been abused as scapegoats for the economic troubles caused by globalization from above. Indigenous people have had their traditional ways of life disrupted and their economic resources plundered by global corporations and governments doing their bidding.

Volatility: Global financial deregulation has reduced barriers to the international flow of capital. More than $1.5 trillion now flows across international borders daily in the foreign currency market alone. These huge flows easily swamp national economies. The result is a world economy marked by dangerous and disruptive financial volatility.

In 1998, for example, an apparently local crisis in Thailand rapidly spread around the globe. In two years, Malaysia's economy shrunk by 25 percent, South Korea's by 45 percent, and Thailand's by 50 percent. Indonesia's economy shrunk by 80 percent; its per capita gross domestic product dropped from $3,500 to less than $750; and 100 million people—nearly half of the population—sank below the poverty line.[17] According to former World Bank chief economist Joseph Stiglitz,

> Capital market liberalization has not only not brought people the prosperity they were promised, but it has also brought these crises, with wages falling 20 or 30 percent, and unemployment going up by a factor of two, three, four or ten.[18]

Degradation of democracy: Globalization has reduced the power of individuals and peoples to shape their destinies through participation in democratic processes.

Of the 100 largest economies in the world, 51 today are corporations, not countries.[19] Globalization has greatly increased the power of global corporations relative to local, state, and national governments. The ability of governments to pursue development, full employment, or other national economic goals has been undermined by the growing ability of capital to simply pick up and leave.

There are few international equivalents to the anti-trust, consumer protection, and other laws that provide a degree of corporate accountability at the national level. As a result, corporations are able to dictate policy to governments, backed by the threat that they will relocate.

Governmental authority has been undermined by trade agreements such as NAFTA and WTO and by international financial institutions such as the IMF and World Bank, which restrict the power of national, state, and local governments to govern their own economies. These institutions are all too often themselves complicit in the denial of human rights. (At a time when 100 to 200 Algerians were having their throats cut every week, the IMF stated, "Directors agree that Algeria's exemplary adjustment and reform efforts deserve continued support of the international financial community.")[20] They make decisions affecting billions of people, but they are largely free of democratic control and accountability. As one unnamed WTO official was quoted by the *Financial Times*, "The WTO is the place where governments collude in private against their domestic pressure groups."[21]

Environmental destruction: Globalization is accelerating ecological catastrophe both globally and locally. Countries are forced to compete for investment by lowering environmental protections in an ecological race to the bottom. (Seventy countries have rewritten their mining codes in recent years to encourage investment.[22]) Neoliberal policies imposed by international institutions or voluntarily accepted by national governments restrict environmental regulation. Worldwide, corporations promote untested technologies, such as pesticides and genetic engineering, turning the planet into a testing lab and its people into guinea pigs. Growing poverty leads to desperate overharvesting of natural resources.

Global corporations' oil refineries, chemical plants, steel mills, and other factories are the main source of greenhouse gases, ozone-depleting chemicals, and toxic pollutants. Overfishing of the world's waters, overcutting of forests, and abuse of agricultural land result from the search for higher corporate profits, the drive to increase exports, and the increase in poverty.

Globally, environmental destruction is changing the basic balances on which life depends. Carbon dioxide has reached record levels in the atmosphere.[23] Global warming is already resulting in the melting of glaciers, the dying of coral reefs, climate instability, and "a disturbing change in disease patterns."[24] An estimated one-quarter of the world's mammal species and 13 percent of plant species are threatened with extinction in the worst period of mass extinction of species in 65 million years.[25]

ENDNOTES

1. Some analysts still debate to what extent globalization is genuine or significant. See, for example, Doug Henwood, "What Is Globalization Anyway?" ZNet Commentary, November 26, 1999, maintaining that globalization is exaggerated (http://zmag.org/ZSustainers/ZDaily/1999-11/26henwood.htm), and Richard Du Boff and Edward Herman, "Questioning Henwood on Globalization," ZNet Commentary, December 1, 1999, finding Henwood's arguments "incomplete and unconvincing" (http://zmag.org/ZSustainers/ZDaily/1999-12/01herman.htm).

2. For a more detailed analysis of the first phase of globalization, see Brecher and Costello, *Global Village or Global Pillage*, and works cited there. For a review of developments up to 1998, see the Introduction to the second edition.

3. John Tagliaubue, "For Americans, an Indirect Route to the Party," *New York Times*, June 14, 1998, p. 3, 4, citing James E. Carlson, economist at Merrill Lynch in New York.

4. For the globalization of the process of technological change, see Peter Dorman, "Actually Existing Globalization," prepared for "Globalization and Its Dis-Contents," Michigan State University, April 3, 1998, pp. 4–7.

5. Bennett Harrison, *Lean and Mean: The Changing Landscape of Corporate Power in the Age of Flexibility* (New York: Basic Books, 1994), pp. 9, 171. For further discussion of changing corporate and work structures and their implications for labor organization, see Jeremy Brecher and Tim Costello, "Labor and the Dis-Integrated Corporation," *New Labor Forum* 2 (Spring/Summer 1998): 5ff.

6. For a view emphasizing the similarity between contemporary globalization and previous periods of shift in global hegemony, see Giovanni Arrighi and Beverly J. Silver, *Chaos and Governance in the Modern World System* (Minneapolis: University of Minnesota Press, 1999). For a view that globalization is replacing traditional national imperialism with a universal, non-national system of empire, see Michael Hardt and Antonio Negri, *Empire* (Cambridge: Harvard UP, 2000).

7. *New York Times Magazine*, March 28, 1999.

8. Economists often discuss such failings under the heading "market failures." For further discussion of market failures and "political failures," see Jeremy Brecher, "*Can'st Thou Draw Out Leviathan with a Fishhook?*" (Washington: Grassroots Policy Project, 1995), and Charles E. Lindblom, *Politics and Markets* (New York: Basic Books, 1977), Chapters 5 and 6.

9. Clifford Krauss, "Injecting Change Into Argentina: New President Tries to Keep Industry from Leaving the Country," *New York Times*, March 8, 2000, p. Cl. See

also Craig Torres and Matt Moffett, "Neighbor-Bashing: Argentina Cries Foul as Choice Employers Beat a Path Next Door," *Wall Street Journal*, May 2, 2000, p. 1.

10. "Canada's Tax Cut Underwhelms Businesses: CEOs Fault Slowness of Phase-In and Look Abroad for Expansion," *Wall Street Journal*, May 2, 2000, p. A23.

11. United Nations Development Program (UNDP), *Human Development Report 1999* (New York: Oxford UP, 1999).

12. James D. Wolfensohn, "Let's Hear Everyone and Get on with Imaginative Solutions." *International Herald Tribune*, January 28, 2000 (http://www.iht.com/).

13. *World Employment Report, 1996/97* (Geneva: ILO, 1996).

14. Lawrence Mishel, Jared Bernstein, and John Schmitt, *The State of Working America 1998–99* (Ithaca: Economic Policy Institute/Cornell UP, 1999).

15. Anderson et al., *Field Guide to the Global Economy*, p. 53. *BusinessWeek*, April 20, 1998.

16. UNDP, *Human Development Report 1999*, p. 37.

17. Fareed Zakaria, "Will Asia Turn Against the West?" *New York Times*, July 10, 1998, p. A15.

18. Reuters, "Worker Rights Key to Development," January 8, 2000.

19. Anderson et al., *Field Guide to the Global Economy*, p. 69.

20. Jan Pronk, "Globalization: A Developmental Approach," in Jan Nederveen Pieterse ed., *Global Futures: Shaping Globalization* (London: Zed Books, 2000), p. 48.

21. "Network Guerrillas," *Financial Times*, April 30, 1998, p. 20.

22. Hilary French, *Vanishing Borders: Protecting the Planet in the Age of Globalization* (New York: Norton, 2000), p. 27.

23. Hilary French, *Vanishing Borders*, p. 8.

24. "There are strong indications that a disturbing change in disease patterns has begun and that global warming is contributing to them," according to Paul Epstein, associate director of Harvard Medical School's Center for Health and the Global Environment. Quoted in Hilary French, *Vanishing Borders*, p. 46.

25. Hilary French, *Vanishing Borders*, pp. 8–9.

4

Heavy Traffic

International Migration in an Era of Globalization

SUSAN F. MARTIN

A t the start of the new millennium, some 150 million people, or 2.5 percent of the world's population, live outside their country of birth. That number has doubled since 1965. With poverty, political repression, human rights abuses, and conflict pushing more and more people out of their home countries while economic opportunies, political freedom, physical safety, and security pull both highly skilled and unskilled workers into new lands, the pace of international migration is unlikely to slow any time soon.

Few countries remain untouched by migration. Nations as varied as Haiti, India, and the former Yugoslavia feed international flows. The United States receives by far the most international migrants, but migrants also pour into Germany, France, Canada, Saudi Arabia, and Iran. Some countries, such as Mexico, send emigrants to other lands, but also receive immigrants—both those planning to settle and those on their way elsewhere.

Institutions and laws for achieving cooperation among receiving, source, and transit countries are in their infancy. The World Trade Organization oversees the movement of goods worldwide and the International Monetary Fund monitors the global movement of capital, but no comparable institution regulates the movements of people. Nor does a common understanding exist among states, or experts for that matter, as to the costs and benefits of freer or more restrictive immigration policies.

The surge in international migration, though, is prompting states everywhere to recognize the need for greater harmonization of policies and approaches. Bilateral discussions of migration issues—between the United States and Mexico, for example, over an expanded guestworker program and amnesty for unauthorized Mexican workers—have become more commonplace. During the past decade, regional groups have been set up in the Americas, Europe, East Asia, Africa, and elsewhere to allow receiving, source, and transit countries to address issues of mutual concern.

SOURCE: Susan F. Martin, "Heavy Traffic: International Migration in an Era of Globalization," *Brookings Review* 19 (Fall 2001), pp. 41–44. Reprinted with permission from Brookings Institution Press.

NEW TRENDS

Economic, geopolitical, and demographic trends reinforce the need to consolidate these regional institutions and begin to develop a global regime for managing migration.

Economic trends influence migration patterns in many ways. Multinational corporations, for example, press governments to ease movements of executives, managers, and other key personnel from one country to another. When labor shortages appear, whether in information technology or seasonal agriculture, companies also seek to import foreign workers to fill jobs. Although the rules for admitting foreign workers are largely governed by national legislation, regional and international trade regimes such as the North American Free Trade Agreement and the General Agreement on Trade in Services include provisions for admitting foreign executives, managers, and professionals. Under NAFTA, for example, U.S., Canadian, and Mexican (as of 2004) professionals in designated occupations may work in the other NAFTA countries without regard to numerical limits imposed on other foreign nationals.

The growth in global trade and investment also affects source countries. Economic development has long been regarded as the best long-term solution to emigration pressures arising from the lack of economic opportunities in developing countries. Almost uniformly, however, experts caution that emigration pressures are likely to remain and, possibly, increase before the long-term benefits accrue. Wayne Cornelius and Philip Martin postulate that as developing countries' incomes begin to rise and opportunities to leave home increase, emigration first increases and declines only later as wage differentials between emigration and immigration countries fall. Italy and Korea, in moving from emigration to immigration countries, give credence to that theory.

Geopolitical changes since the Cold War era offer both opportunities and challenges for managing international migration, particularly refugee movements. During the Cold War, the United States and other Western countries saw refugee policy as an instrument of foreign policy. The Cold War made it all but impossible to address the roots of refugee movements, which often resulted from surrogate conflicts in Southeast Asia, Central America, Afghanistan, and the Horn of Africa. Few refugees were able or willing to return to lands still dominated by conflict or Communism. With the end of the Cold War, new opportunities to return emerged as decades-old conflicts came to an end. Democratization and increased respect for human rights took hold in many countries, as witnessed in the formerly Communist countries of East Europe, making repatriation a reality for millions of refugees who had been displaced for years.

At the same time, rabid nationalism fueled new conflicts that have led to massive displacement in places such as the former Yugoslavia and the Great Lakes region of Africa. When the displacements spilled over to other countries, or became humanitarian crises that threatened the lives of millions, governments proved willing to intervene—even with military force—on behalf of victims. Faced with crises in northern Iraq, Bosnia, Kosovo, and east Timor, classic notions of sovereignty that would once have precluded such intervention came under considerable pressure. On the positive side, people who once would have had to cross international borders to find aid could now find it at home. On the negative side, however, the so-called safe zones established in Bosnia, Iraq, and elsewhere often proved far from secure, leaving internal refugees more vulnerable than those able to cross into neighboring countries.

Demographic trends also reinforce arguments for a global regime to manage migration. Worldwide, fertility rates are falling, although developing countries continue to see rapid population growth. In most industrialized countries, fertility levels are well below replacement rates. In Europe, the average number of children born per woman is 1.4; Italy's fertility rate is 1.2. Countries with declining fertility face the likelihood of a fall in total population, leading some demographers to see a looming population implosion. Such nations can also expect an aging population, with fewer working-age people for each older person. Although immigration will not solve the problem, it will help ease labor shortages and redress somewhat the aging of the society.

Demographic trends also help explain emigration pressures in Africa, Latin America, and some parts of Asia, where fertility rates are high. Rapidly growing societies often cannot generate enough jobs to keep pace with new entries into the labor force. Growth may also cause environmental degradation, particularly when land use policies do not protect fragile ecosystems. Natural disasters also wreak havoc on densely populated areas in poor countries. Recent hurricanes in Honduras and Nicaragua and earthquakes in El Salvador and India displaced huge numbers of people from ravaged homes and communities.

NEW RESPONSES

In 1997, United Nations Secretary General Kofi Annan addressed the possibility of convening a conference on international migration and development. Upon consulting with U.N. member governments, he found insufficient consensus about what such a conference could accomplish and reported: "The disparate experiences of countries or subregions with regard to international migration suggest that, if practical solutions are to be found, they are likely to arise from the consideration of the particular situation of groups of countries sharing similar positions or concerns with the global international migration system. . . . In the light of this, it may be expedient to pursue regional or subregional approaches whenever possible."

Since 1997, regional processes have matured. Perhaps most developed is the Regional Migration Conference, the so-called Puebla Group, which brings together all the countries of Central and North America for regular dialogue on migration issues, including an annual session at the vice-ministerial level. The Puebla Group's Plan of Action calls for cooperation in exchanging information on migration policy, exploring links between development and migration, combating migrant trafficking, returning extra-regional migrants, and ensuring full respect for the human rights of migrants, as well as reintegrating repatriated migrants, equipping and modernizing immigration control systems, and training officials in migration policy and procedures. Discussions have led law enforcement officials of the United States, Mexico, and several Central American countries to cooperate in arresting and prosecuting members of large-scale smuggling and trafficking operations that move migrants illegally across borders and then force them to work in prostitution, sweatshops, and other exploitive activities.

Similar regional groups are working in East and Southeast Asia. The "Manila Process" focuses on unauthorized migration and trafficking in East and Southeast Asia. Since 1996, it has brought together each year 17 countries for regular exchange of information. The Asia-Pacific Consultations include governments in Asia and Oceania and focus on a broad range of population movements in the region. Both ongoing dialogues were strengthened by a 1999 International Symposium on Migration hosted by the Royal Thai government. In the resulting Bangkok Declaration on Irregular Migration, 19 Asian countries agreed to cooperate to combat smuggling and trafficking.

Other such groups are in the making in the Southern Cone of South America, in western and southern Africa, and in the Mediterranean. The intent is to bring together the governments of all countries involved in migration, whether origin, transit, or receiving. At present, the groups are forums for exchanging information and perspectives, although the more developed ones, such as the Puebla Group, are leading to joint action as well. Given the lack of shared information or consensus about migration policies and practices, the discussion stage is a necessary first step in developing the capacity for joint efforts.

STEPS TOWARD A GLOBAL REGIME

Will these regional processes lead to a global migration regime? It is too early to know. Three issues must first be addressed. First, states must reach a consensus that harmonizing policies will make migration more orderly, safe, and manageable. And in fact signs exist of growing convergence among regional groups in setting out an agenda for harmonization. Many items on regional agendas relate to unauthorized migration—and how best to deter it, consistent with respect for the rule of law and the human rights of migrants. Although source and destination countries may differ still about what causes unauthorized migration, agreement is growing on some approaches to it—for example, that curbing alien smuggling and trafficking (a global enterprise that nets an estimated $7–10 billion a year) requires international cooperation.

Other issues arise in addressing forced migration. How, for example, should states protect people fleeing repression and conflict? When conflicts end and migrants no longer need protection, when and how should they be required to return? The growing use of temporary protection, as in the crises in Bosnia and Kosovo, has led European Union member states to place a high priority on harmonizing temporary protection policies and mechanisms for burden sharing. At the same time, the Puebla Group has focused on temporary protection of victims of natural disasters in the aftermath of Hurricane Mitch.

More recently, issues involving legal admissions have appeared on regional agendas. When and to whom should visa restrictions apply? Under what circumstances should family reunification be guaranteed? Who should be eligible for work and residence permits? What rights should accrue to those legally admitted for work or family purpose? Answers to these questions will take time because states still differ widely in their attitudes and policies toward legal immigration. But signs of change can be seen even here. The European Union has led the way, with free movement of labor a long-held principle for its own nationals. The 1997 Amsterdam Treaty takes the EU the next step, mandating a common immigration policy for participating states.

Second, a global migration regime will require standards, policies, and new legal frameworks. A legal framework already exists for refugee movements, with most countries now signatories to the 1951 U.N. Convention on the Status of Refugees or its 1967 Protocol. Most important, signatories agree that they will not return (*refoule*) persons to countries where they have a well-founded fear of persecution. International agreements also apply to the rights of migrant workers, but few states ratified the 1990 U.N. Convention on the Protection of the Rights of All Migrant Workers and Members of Their Families, showing that they were unwilling to take on a comprehensive set of obligations toward migrant workers and their families. No body of international law or policy governs responses to other forms of international migration. But with growing economic integration, global trade agreements may become vehicles for formulating such policies. Ongoing negotiations on the General Agreement on Trade in Services involve new agreements—under the rubric of the "movement of natural persons"—to ease the admission of executives, managers, and professionals providing services.

The third issue to be settled before a global migration regime can come into being is organizational responsibilities. At the heart of the refugee regime is the U.N. High Commissioner for Refugees, whose mandate dates back to 1950. No comparable institution exists for other migration matters. The International Organization for Migration (IOM), a Geneva-based intergovernmental body, with 86 member states and 41 observer states, comes closest. Since its founding in 1951 to help resettle refugees and displaced persons from World War II and the Cold War, IOM has taken on a broad set of responsibilities, including counter trafficking programs, migrant health and medical services, technical assistance and capacity building, and assisted return programs. In particular, it serves as the secretariat for some of the regional processes discussing international migration. IOM is not a part of the U.N. system (although it cooperates with U.N. agencies), and it represents far fewer states. To become the focal point of a new migration regime, it would need substantial new resources and government support, particularly in helping governments formulate new global migration policies. Its governing board, composed of government representatives, has already requested and provided funds to IOM to strengthen its capacity to advise governments on best practices in migration management. Both source and destination countries recognize the need for improved responses, and IOM's broad membership helps it devise policies that balance varied governmental interests.

A MULTILATERAL APPROACH

In an increasingly interconnected world, governments are unlikely to be able to solve the many problems posed by international migration through unilateral approaches only. Source, receiving, and transit countries must all cooperate to manage international migration.

Reaching agreement will be relatively straightforward when countries share similar interests and problems—for example, in combating the most exploitive type of human trafficking. Often, however, interests will diverge. Source countries such as Mexico will press for easier access to the labor markets of wealthier countries. Receiving countries such as the United States will face public concerns about seemingly uncontrolled movements into their territories. Agreement will be difficult. Still, these issues will not go away just because different countries see them differently. Sheer necessity is likely to move governments toward a global migration regime.

5

Dimensions of Globalization

D. STANLEY EITZEN

Below is a list of facts that highlight the scope and variety of the interconnectedness among the world's people.

TRANSNATIONAL MIGRATION

- More than 200 million people are living outside their country of birth or citizenship (about 3 percent of the world's population).

- In 2005 there were 37 million foreign-born U.S. residents (about 12 percent of the population).

- Worldwide, 32.9 million people have been driven from home by war, persecution, and poverty. The countries where this is most serious are Iraq, Afghanistan, Sudan, Somalia, and the Democratic Republic of Congo. Most notably, because of the war in Iraq about 2.2 million Iraqis have fled to neighboring countries.

- By 2010 an estimated 50 million people will have migrated for environmental reasons (floods, droughts, climate change, and rising sea levels). By 2050 global warming could displace as many as 1 billion people.

- An estimated 12 million unauthorized foreigners reside in the United States.

- The United Nations estimates that 4 million persons are moved illegally from one country to another and within countries for prostitution. The Central Intelligence Agency estimates that between 18,000 and 20,000 people are trafficked annually into the United States as sex slaves. Another 30,000 women are trafficked to the United States to work in sweatshops or as domestic servants.

- Of the 849 baseball players who opened the 2007 major league season, 29 percent were born outside the United States; almost half of the players signed to minor league baseball contracts were born outside the United States.

SOURCE: D. Stanley Eitzen. Compiled expressly for this book.

- In 1998 there was only one South Korean playing on the U.S. women's professional golf tour (LPGA). In 2003 there were 18. In 2007 there were 45.

- In 2003 Americans adopted 21,616 children from foreign countries (China, Russia, and Guatemala were the three most common countries of origin).

- More than two-thirds of IBM workers, both foreign nationals and U.S. citizens, work outside the United States.

- In 2005, 5 million trucks and 92 million personal vehicles crossed the border from Mexico to the United States.

TRANSNATIONAL TRADE

- The United States imported goods worth $278.8 billion from China in 2006 (up from $3.9 billion in 1985). In that year China exported $91.7 billion to Japan, $44.5 billion to South Korea, $40.3 billion to Germany, $30.8 billion to the Netherlands, and $24.2 billion to Britain.

- In 2006, the United States imported $765.3 billion more in goods and services than it exported. The imbalance with China was $232.5 billion.

- In mid-2007 foreign investors owned a record 80 percent of U.S. Treasury notes due from 3 to 10 years.

- Chinese food products are exported to more than 200 countries.

- Shrimp is the most popular seafood in the United States. In 2006 almost all of the shrimp consumed in the United States came from Thailand, China, Indonesia, Ecuador, and Vietnam.

- The ingredients of Twinkies include thiamine mononitrate from China, palm oil from Malaysia, colorants from Peru, niacin from Switzerland, and wheat gluten from Europe.

- Tricon Corporation, owner of Kentucky Fried Chicken, Pizza Hut, and Taco Bell, was formed in 1997 (a spinoff from Pepsi). Since then it has opened more than 3,200 restaurants in over 100 countries and plans to open more than 1,000 more overseas each year for the foreseeable future.

- The United States depends on imports for 55 percent of its petroleum needs.

- More than 8,000 foreign companies outsource information technology work to the Philippines.

- American Express outsourced more than 50 percent of its software development to India, where the average annual salary for a software engineer is $5,880.

- Baseballs for the major leagues in the United States are handcrafted by workers in Turrialba, Costa Rica. The workers are paid 30 cents per ball, and Rawlings Sporting Goods sells the balls for $14.99 at retail in the United States.

- More than half of all legal Colombian exports go to the United States—if you add the value of cocaine and heroin, the percentage rises to 80.

- The United States, the world's largest user of heroin, receives its illicit supply primarily from Colombia, Mexico, Peru, Burma, and Afghanistan.

- Many brands that look American are actually European-owned: for example, Holiday Inn, Burger King, Amoco, Mazola Oil, Vaseline, Dr. Pepper, Slim-Fast, and Pepsodent are all owned by companies in the United Kingdom.

- Boeing's commercial jetliner—the 787 Dreamliner—is assembled in Seattle from parts produced in factories of Boeing's corporate partners in Japan and Italy.

- Merrill Lynch has a diverse global workforce with 740 offices in 37 countries.

- Halliburton Corporation has operations in 70 countries, including the tax havens of Bahrain, Barbados, Bermuda, British Virgin Islands, Cayman Islands, Jersey, Liechtenstein, Mauritius, Netherlands Antilles, Nevis, Panama, and St. Lucia.

- Many U.S. companies outsource call centers to countries such as India. The savings are considerable since the average annual salary of a U.S. call center agent is $29,000 compared to $2,667 in India.

TRANSNATIONAL FINANCE

- In 2005, almost half of U.S. Treasury bonds were owned in Asia.

- In 2006, the average daily turnover in global currency markets was $1.5 trillion.

- Mexican immigrants in the United States send about $20 billion annually to relatives in their native country.

- Foreign direct investment amounts to about $400 billion yearly. But 80 percent of it goes to only 10 countries while 100 countries average just $100 million in FDI each year.

CULTURAL GLOBALIZATION

- There are approximately 27,000 Christian missionaries (one out of every two is American) in Islamic countries. In 1900, Africa had just 10 million Christians, about 9 percent of the population. In 2002, the Christian total was 46 percent of the population.

- U.S. Catholics, short on priests, are outsourcing prayer requests via e-mail to Indian priests for an average cost of $5 per prayer.

- An estimated 97 billion e-mails are sent worldwide daily.

- Two-thirds of the world's children are raised as bi-lingual speakers.

- Most of the world's books, newspapers, and e-mails are written in English.

- Four of the main sponsors of the Australian Open tennis tournament in 2004 were a South Korean car manufacturer (Kia), a Dutch beer (Heineken), a Swiss watch company (Rado), and American Express.

- The U.S. movie industry lost $2.3 billion in revenue to Internet piracy and $3.8 billion to bootlegged DVDs and other "hard goods" piracy in 2005. The leak of a single film in Wharton, N.J., in 2006 spread within weeks to 30,408 people on six continents.

GLOBALIZATION OF DISEASE

- HIV-AIDS has infected some 60 million people worldwide since crossing over from chimpanzees in the 1960s. Some 30 new diseases have cropped up since the mid-1970s, causing tens of millions of deaths, including SARS (severe acute respiratory syndrome) and Ebola hemorrhagic fever.

- In 2004, a U.S. corporation (Chiron) was contracted to supply half of the flu vaccine for the U.S. population. Its plant in the United Kingdom produced the vaccine but it was deemed unsafe, causing the United States to seek to replace some of this shortfall from Canada.

- The potentially deadly West Nile virus arrived in the United States from Africa in agricultural products transported in containers or by travelers. It has since spread to two-thirds of the states by migrating birds and mosquitoes.

- Global warming is causing tropical diseases such as malaria to migrate northward.

ENDNOTES

The structure of this listing came, in part, from "Globalization at a Glance," *International Socialist Review* 19 (August/September 2001), p. 4.

Sources: Philip Martin and Elizabeth Midgley, "Immigration: Shaping and Reshaping America," 2nd ed. *Population Bulletin* 61 (December, 2006); Joseph A. McFalls, Jr., "Population: A Lively Introduction," 5th ed. *Population Bulletin* 62 (March 2007); Brad Knickerbocker, "In the Coming Decades, the Effects of Global Warming are Likely to Turn Millions into Refugees," *Christian Science Monitor* (June 21, 2007). Zakaria, Fareed, "True or False: We are Losing the War Against

Radical Islam," *Newsweek* (July 2, 2007), pp. 38–41; Steve DiMeglio, "Fire Within Trailblazer Pak Successfully Rekindled," *USA Today* (June 7, 2007), p. 9C; Lou Dobbs, "Dangerously Dependent," *U.S. News & World Report*, (February 9, 2004); Peter Landesman, "The Girls Next Door," *New York Times Magazine* (January 25, 2004); Leah Platt, "Regulating the Brothel," *The American Prospect* (Summer 2001); Tim Weiner, "Low-Wage Costa Ricans Make Baseballs for Millionaires," *New York Times* (January 25, 2004); Martin Crutsinger, "Loss of U.S. Manufacturing Jobs Accelerating," Associated Press (April 30, 2007); Lou Dobbs, "Coming Up Empty," *U. S. News & World Report* (January 26, 2004); Michael Mandel, "Globalization vs. Immigration Reform," *BusinessWeek* (June 4, 2007), p. 40; Jyoti Thottam, "The Growing Dangers of China Trade," *Time* (July 9, 2007), pp. 28–31; Elizabeth Stanton, "Foreign U. S. Notes Rise to 80 Percent," Bloomberg.com (July 12, 2007); Steve Ettlinger, *Twinkie, Deconstructed* (New York: Hudson Street Press, 2007); Edward Iwata, "Companies Find Gold Inside Melting Pot," *USA Today* (July 9, 2007), 1B–2B; Dianne Solis, Alfredo Corchado, and Patricia Estrado, "Sustenance for Their Homeland," *Dallas Morning News* (September 29, 2003); Elizabeth Weise, "Buying American? It's Not in the Bag," *USA Today* (July 11, 2007), D1–D2; "Geographica," *National Geographic* 205 (April 2004); James Wallace, "Boeing Dreamliner 'Coming to Life,'" *Seattle Post-Intelligencer* (June 27, 2006): http://seattlepi.nwsource.com/business/275465japan27.html; Kerry Miller, "Hello India? Er, Des Moines?" *BusinessWeek* (June 25, 2007), p. 14. David Van Biema, "Missionaries Under Cover," *Time* (June 30, 2003); *World Watch*, "Unspoken Words," Volume 14 (May/June 2001); Payal Sampat, "Last Words," *World Watch* 14 (May/June, 2001); Michael Riley, "Families Without Borders" *Denver Post* (August 18, 2003); Ian Fletcher, "Troubled Waters," *Rocky Mountain News* (May 1, 2004); Rachel Guevera, "The Political Economy of A Narco-Terror State," *Z Magazine* 15 (October 2002); Geoffrey Cowley, "How Progress Makes Us Sick," *Newsweek* (May 5, 2003); T. R. Reid, "Buying American?" *Washington Post National Weekly Edition* (May 27/June 2, 2002); Julie Snider, "Where the Kids are From," *USA Today* (June 16, 2004); L. Jon Werthem, "The Whole World is Watching," *Sports Illustrated* (June 14, 2004); Chris Hawley, "More of the Earth is Turning to Dust," *Associated Press* (June 16, 2004); Phyllis Berman, "The Cosmopolitan Touch," *Forbes* (June 21, 2004); Steve Caulk, "Passage to India," *Rocky Mountain News* (July 12, 2003); Michael Scherer, "The World According to Halliburton," *Mother Jones* 28 (July/August 2003); Philip Jenkins, "The Next Christianity," *The Atlantic Monthly* (October 2002); Saritha Rai, "Short on Priests, U. S. Catholics Outsource Prayers to Indian Clergy," *New York Times* (June 13, 2004).

REFLECTION QUESTIONS FOR CHAPTER 2

1. How do the various dimensions of globalization affect your life? Consider, for example, job opportunities, stock market fluctuations, diseases (AIDS, SARS, West Nile virus), music, food, and fashion.

2. How has professional sport in the United States, especially hockey, baseball, and basketball, changed with the influx of foreign-born athletes?

3. Can the United States, as the greatest power in the world, go it alone or must it cooperate in multinational coalitions to deal effectively with the various dimensions of globalization?

4. What additional evidence can you provide for Friedman's "flat world" thesis, especially in the non-Western world?

3

Debating Globalization

INTRODUCTION

While globalization is not a new phenomenon, it is now much greater in scope and accelerating more rapidly than at any time in history. The world's people are interconnected as never before. The readings in this book examine the implications of this hugely important process. Scholars disagree on its effects. These are some of the major debates surrounding globalization: Have transnational corporations superseded nation-states? Does globalization enhance or undermine democracy? Is the culture of the West, spread through movies, television, advertising, consumerism, and the English language, making the world more homogeneous? With globalization, is the local no longer relevant? Each of these questions is subsumed under this fundamental question: Is globalization a good or a bad thing? Or more precisely, Who benefits and who does not from globalization? The answer to that question involves many questions: Will the world's inhabitants be more secure or insecure because of globalization? Is the environment more secure because of globalization or more likely to deteriorate further? Will the world's poor be uplifted by globalization or will their condition worsen further? Will the workers of the world benefit from technological advances and economic development or will wages be decreased as jobs go to the lowest-wage nations in a race to the bottom?

This section addresses a few of these issues. The first essay, by Pankaj Ghemawat, challenges Thomas Friedman's thesis that new technologies level the playing field of global competitiveness. He argues that even in a globalized world, geographical boundaries define our movements and constrain cross-border integration.

The next reading is a speech given at the University of London by distinguished economist Murray Weidenbaum. His argument provides a useful introduction to the issues that arise in this book by addressing the question, Is globalization a wonder land or a waste land? He provides arguments for both sides and identifies some "common ground on which people of good will on both sides of the heated controversies on globalization might agree."

Economists Christian Weller and Adam Hersh examine the relationship between free markets and poverty. They address the fundamental assumption of "The Washington Consensus," or neoliberal ideology that free markets (that is, the free flow of capital and commerce across political borders) benefits everyone. Free trade, in this view, alleviates poverty by raising overall growth rates and by bringing modern capitalism to the world's poorest. Weller and Hersh, to the contrary, argue that free trade results in slower growth, more vulnerability for poor countries, and a greater income disparity among individuals.

The final essay, provided by labor expert Jeff Faux, looks at one effort at free trade in North America—NAFTA. The question: After ten years, did NAFTA (the North American Free Trade Agreement) help workers and consumers in Canada, the United States, and Mexico? What has happened to wages? To the distribution of income and wealth? To the environment? To the economic integration of the three countries? And, what is the future of NAFTA?

6

Why the World Isn't Flat

PANKAJ GHEMAWAT

Globalization has bound people, countries, and markets closer than ever, rendering national borders relics of a bygone era—or so we're told. But a close look at the data reveals a world that's just a fraction as integrated as the one we thought we knew. In fact, more than 90 percent of all phone calls, Web traffic, and investment is local. What's more, even this small level of globalization could still slip away.

I deas will spread faster, leaping borders. Poor countries will have immediate access to information that was once restricted to the industrial world and traveled only slowly, if at all, beyond it. Entire electorates will learn things that once only a few bureaucrats knew. Small companies will offer services that previously only giants could provide. In all these ways, the communications revolution is profoundly democratic and liberating, leveling the imbalance between large and small, rich and poor. The global vision that Frances Cairncross predicted in her *Death of Distance* appears to be upon us. We seem to live in a world that is no longer a collection of isolated, "local" nations, effectively separated by high tariff walls, poor communications networks, and mutual suspicion. It's a world that, if you believe the most prominent proponents of globalization, is increasingly wired, informed, and, well, "flat."

It's an attractive idea. And if publishing trends are any indication, globalization is more than just a powerful economic and political transformation; it's a booming cottage industry. According to the U.S. Library of Congress's catalog, in the 1990s, about 500 books were published on globalization. Between 2000 and 2004, there were more than 4,000. In fact, between the mid-1990s and 2003, the rate of increase in globalization-related titles more than doubled every 18 months.

Amid all this clutter, several books on the subject have managed to attract significant attention. During a recent TV interview, the first question I was asked—quite earnestly—was why I still thought the world was round. The interviewer was referring of course to the thesis of *New York Times* columnist Thomas L. Friedman's bestselling book *The World Is Flat*. Friedman asserts that 10 forces—most of which enable connectivity and collaboration at a distance—are "flattening" the Earth and

SOURCE: "Why the World Isn't Flat," by Pankaj Ghemawat, *Foreign Policy*, March/ April 2007, pp. 54–60.

leveling a playing field of global competitiveness, the like of which the world has never before seen.

It sounds compelling enough. But Friedman's assertions are simply the latest in a series of exaggerated visions that also include the "end of history" and the "convergence of tastes." Some writers in this vein view globalization as a good thing—an escape from the ancient tribal rifts that have divided humans, or an opportunity to sell the same thing to everyone on Earth. Others lament its cancerous spread, a process at the end of which everyone will be eating the same fast food. Their arguments are mostly characterized by emotional rather than cerebral appeals, a reliance on prophecy, semiotic arousal (that is, treating everthing as a sign), a focus on technology as the driver of change, an emphasis on education that creates "new" people, and perhaps above all, a clamor for attention. But they all have one thing in common. They're wrong.

In truth, the world is not nearly as connected as these writers would have us believe. Despite talk of a new, wired world where information, ideas, money, and people can move around the planet faster than ever before, just a fraction of what we consider globalization actually exists. The portrait that emerges from a hard look at the way companies, people, and states interact is a world that's only beginning to realize the potential of true global integration. And what these trend's backers won't tell you is that globalization's future is more fragile than you know.

THE 10 PERCENT PRESUMPTION

The few cities that dominate international financial activity—Frankfurt, Hong Kong, London, New York—are at the height of modern global integration; which is to say, they are all relatively well connected with one another. But when you examine the numbers, the picture is one of extreme connectivity at the local level, not a flat world. What do such statistics reveal? Most types of economic activity that could be conducted either within or across borders turn out to still be quite domestically concentrated.

One favorite mantra from globalization champions is how "investment knows no boundaries." But how much of all the capital being invested around the world is conducted by companies outside of their home countries? The fact is, the total amount of the world's capital formation that is generated from foreign direct investment (FDI) has been less than 10 percent for the last three years for which data are available (2003–05). In other words, more than 90 percent of the fixed investment around the world is still domestic. And though merger waves can push the ratio higher, it has never reached 20 percent. In a thoroughly globalized environment, one would expect this number to be much higher—about 90 percent, by my calculation. And FDI isn't an odd or unrepresentative example.

The levels of internationalization associated with cross-border migration, telephone calls, management research and education, private charitable giving, patenting, stock investment, and trade, as a fraction of gross domestic product (GDP); all stand much closer to 10 percent than 100 percent. The biggest exception in absolute terms—the trade-to-GDP ratio—recedes most of the way back down toward 20 percent if you adjust for certain kinds of double-counting. So if someone asked me to guess the internationalization level of some activity about which I had no particular information, I would guess it to be much closer to 10 percent—the average for the nine categories of data in the chart—than to 100 percent. I call this the "10 Percent Presumption."

More broadly, these and other data on cross-border integration suggest a semiglobalized world, in which neither the bridges nor the barriers between countries can be ignored. From this perspective, the most astonishing aspect of various writings on globalization is the extent of exaggeration involved. In short, the levels of internationalization in the world today are roughly an order of magnitude lower than those implied by globalization proponents.

A STRONG NATIONAL DEFENSE

If you buy into the more extreme views of the globalization triumphalists, you would expect to see a world where national borders are irrelevant, and where citizens increasingly view themselves as members of ever broader political entities. True, communications technologies have improved dramatically during the past 100 years. The cost of a three-minute telephone call from New York to London fell from $350 in 1930 to about 40 cents in 1999, and is now approaching zero for voice-over-Internet telephony. And the Internet itself is just one of many newer forms of connectivity that have progressed several times faster than plain old telephone service. This pace of improvement has inspired excited proclamations about the pace of global integration. But it's a huge leap to go from predicting such changes to asserting that declining communication costs will obliterate the effects of distance. Although the barriers at borders have declined significantly, they haven't disappeared.

To see why, consider the Indian software industry—a favorite of Friedman and others. Friedman cites Nandan Nilekani, the CEO of the second-largest such firm, Infosys, as his muse for the notion of a flat world. But what Nilekani has pointed out privately is that while Indian software programmers can now serve the United States from India, access is assured, in part, by U.S. capital being invested—quite literally—in that outcome. In other words, the success of the Indian IT industry is not exempt from political and geographic constraints. The country of origin matters—even for capital, which is often considered stateless.

Or consider the largest Indian software firm, Tata Consultancy Services (TCS). Friedman has written at least two columns in the *New York Times* on TCS's Latin American operations: "[I]n today's world, having an Indian company led by a Hungarian-Uruguayan servicing American banks with Montevidean engineers managed by Indian technologists who have learned to eat Uruguayan veggie is just the new normal," Friedman writes. Perhaps. But the real question is why the company established those operations in the first place. Having worked as a strategy advisor to TCS since 2000, I can testify that reasons related to the tyranny of time zones, language, and the need for proximity to clients' local operations loomed large in that decision. This is a far cry from globalization proponents' oft-cited world in which geography, language, and distance don't matter.

Trade flows certainly bear that theory out. Consider Canadian-U.S. trade, the largest bilateral relationship of its kind in the world. In 1988, before the North American Free Trade Agreement (NAFTA) took effect, merchandise trade levels between Canadian provinces—that is, within the country—were estimated to be 20 times as large as their trade with similarly sized and similarly distant U.S. states. In other words, there was a built-in "home bias." Although NAFTA helped reduce this ratio of domestic to international trade—the home bias—to 10 to 1 by the mid-1990s, it still exceeds 5 to 1 today. And these ratios are just for merchandise; for services, the ratio is still several times larger. Clearly, the borders in our seemingly "borderless world" still matter to most people.

Geographical boundaries are so pervasive, they even extend to cyberspace. If there were one realm in which borders should be rendered meaningless and the globalization proponents should be correct in their overly optimistic models, it should be the Internet. Yet Web traffic within countries and regions has increased far faster than traffic between them. Just as in the real world, Internet links decay with distance. People across the world may be getting more connected, but they aren't connecting with each other. The average South Korean Web user may be spending several hours a day online—connected to the rest of the world in theory—but he is probably chatting with friends across town and e-mailing family across the country rather than meeting a fellow surfer in Los Angeles. We're more wired, but no more "global."

Just look at Google, which boasts of supporting more than 100 languages and, partly as a result, has recently been rated the most globalized Web site. But Google's operation in Russia (cofounder Sergey Brin's native country) reaches only 28 percent of the market there, versus 64 percent for the Russian market leader in search services, Yandex, and 53 percent for Rambler.

Indeed, these two local competitors account for 91 percent of the Russian market for online ads linked to Web searches. What has stymied Google's expansion into the Russian market? The biggest reason is the difficulty of designing a search engine to handle the linguistic complexities of the Russian language. In addition, these local competitors are more in tune with the Russian market, for example, developing payment methods through traditional banks to compensate for the dearth of credit cards. And, though Google has doubled its reach since 2003, it's had to set up a Moscow office in Russia and hire Russian software engineers, underlining the continued importance of physical location. Even now, borders between countries define—and constrain—our movements more than globalization breaks them down.

TURNING BACK THE CLOCK

If globalization is an inadequate term for the current state of integration, there's an obvious rejoinder: Even if the world isn't quite flat today, it will be tomorrow. To respond, we have to look at trends, rather than levels of integration at one point in time. The results are telling. Along a few dimensions, integration reached its all-time high many years ago. For example, rough calculations suggest that the number of long-term international migrants amounted to 3 percent of the world's population in 1900—the high-water mark of an earlier era of migration—versus 2.9 percent in 2005.

Along other dimensions, it's true that new records are being set. But this growth has happened only relatively recently, and only after long periods of stagnation and reversal. For example, FDI stocks divided by GDP peaked before World War I and didn't return to that level until the 1990s. Several economists have argued that the most remarkable development over the long term was the declining level of internationalization between the two World Wars. And despite the records being set, the current level of trade intensity falls far short of completeness, as the Canadian-U.S. trade data suggest. In fact, when trade economists look at these figures, they are amazed not at how much trade there is, but how little.

It's also useful to examine the considerable momentum that globalization proponents attribute to the constellation of policy changes that led many countries—particularly China, India, and the former Soviet Union—to engage more extensively with the international economy. One of the better-researched descriptions of these policy changes and their implications is provided by economists Jeffrey Sachs and Andrew Warner:

> The years between 1970 and 1995, and especially the last decade, have witnessed the most remarkable institutional harmonization and economic integration among nations in world history. While economic integration was increasing throughout the 1970s and 1980s, the extent of integration has come sharply into focus only since the collapse of communism in 1989. In 1995, one dominant global economic system is emerging.

Yes, such policy openings are important. But to paint them as a sea change is inaccurate at best. Remember the 10 Percent Presumption, and that integration is only beginning. The policies that we fickle humans enact are surprisingly reversible. Thus, Francis Fukuyama's *The End of History*, in which liberal democracy and technologically driven capitalism were supposed to have triumphed over other ideologies, seems quite quaint today. In the wake of Sept. 11, 2001, Samuel Huntington's *Clash of Civilizations* looks at least a bit more prescient. But even if you stay on the economic plane, as Sachs and Warner mostly do, you quickly see counterevidence to the supposed decisiveness of policy openings. The so-called Washington Consensus around market-friendly policies ran up against the 1997 Asian currency crisis and has since frayed substantially—for example, in the swing toward neopopulism across much of Latin America. In terms of economic outcomes, the number of countries—in Latin America, coastal Africa, and the former Soviet Union—that have dropped out of the "convergence club" (defined in terms of narrowing productivity and structural gaps vis-à-vis the advanced industrialized countries) is at least as impressive as the number of countries that have joined the club. At a multilateral level, the suspension of the Doha round of trade talks in the summer of 2006— prompting *The Economist* to run a cover titled "The Future of Globalization" and depicting a beached wreck—is no promising omen. In addition, the recent wave of cross-border mergers and acquisitions seems to be encountering more protectionism, in a broader range of countries, than did the previous wave in the late 1990s.

Of course, given that sentiments in these respect have shifted in the past 10 years or so, there is a fair chance that they may shift yet again in the next decade. The point is, it's not only possible to turn back the clock on globalization-friendly policies, it's relatively easy to imagine it happening. Specifically, we have to entertain the possibility that deep international economic integration may be inherently incompatible with national sovereignty—especially given the tendency of voters in many countries, including advanced ones, to support more protectionism, rather than less. As Jeff Immelt, CEO of GE, put it in late 2006, "If you put globalization to a popular vote in the U.S., it would lose." And even if cross-border integration continues on its upward path, the road from here to there is unlikely to be either smooth or straight. There will be shocks and cycles, in all likelihood, and maybe even another period of stagnation or reversal that will endure for decades. It wouldn't be unprecedented.

The champions of globalization are describing a world that doesn't exist. It's a fine strategy to sell books and even describe a potential environment that may someday exist. Because such episodes of mass delusion tend to be relatively short-lived even when they do achieve broad currency, one might simply be tempted to wait this one out as well. But the stakes are far too high for that. Governments that buy into the flat world are likely to pay too much attention to the "golden straitjacket" that Friedman emphasized in his earlier book, *The Lexus and the Olive Tree*, which is supposed to ensure that economics matters more and more and politics less and less. Buying into this version of an integrated world—or worse, using it as a basis for policymaking—is not only unproductive. It is dangerous.

7

Globalization

Wonder Land or Waste Land?

MURRAY WEIDENBAUM

It is a special pleasure to offer this talk on globalization in honor of T. S. Eliot. That is so because, when I listen to the critics of globalization, I quickly conjure up a dark vision of Eliot's term "waste land." The arguments in favor of globalization, in contrast, seem to describe a very different and upbeat world, a bright wonderland. Today, I will try to navigate a position in between these two polar alternatives. As you may suspect, this will be a difficult undertaking and I ask for your indulgence.

I confess that it is difficult to take a neutral position. As an economist, I often have weighed in on the pro-globalization side. But I have just returned from a one-year sabbatical and have had ample opportunity to immerse myself in the views of the various participants in the globalization debate.

I have learned to sidestep the semantic issue of defining precisely the very word "globalization." As one wit describes it, globalization is one of the great vacuum words of our time. It sucks up any meaning anyone wishes to ascribe to it. In my talk today, I focus on the array of impacts that arise from the increasing tendency for national borders to be crossed by people, goods, services, information, and ideas.

To put it candidly, I find truth, exaggeration, and error on all sides of the debate on globalization. Very few look at both the bright side and the dark side. Most economists and business leaders focus on the benefits of globalization. The litany is familiar. A greater flow of international trade and investment stimulates economic growth. That rising output requires more employment and income payments and thus generates a higher living standard for consumers. Rising living standards in turn increase the willingness of the society to devote resources to the environment and other important social goals.

Global competition also keeps domestic businesses on their toes, it forces them to innovate and improve product quality and industrial productivity. After all, if competition is good, spreading it out internationally must be even better. More fundamentally, rapidly developing economies tend to generate a

SOURCE: Murray Weidenbaum, "Globalization: Wonder Land or Waste Land?" *Vital Speeches of the Day (2002)*. Reprinted with permission from the author.

new middle class, and that is the bulwark of support for personal liberty as well as economic freedom.

Finally, we are told that economic isolationism does not work. The most striking case was sixteenth-century China, where one misguided emperor abruptly cut off trade and commercial intercourse with other nations. China had been the wealthiest, most technologically advanced, and arguably the most powerful nation on the face of the globe. Yet it promptly went into a decline from which it has yet to fully emerge. Frankly, my real disagreement with this line of thinking is not that I believe the facts or analysis to be wrong but that the entire approach is sadly inadequate in terms of responding to the concerns of the critics.

The other voices in the globalization debate emphasize the dark side. Workers feel threatened by unfair competition from low-cost sweatshops overseas. Citizens generally worry about the conditions in those foreign sweatshops, especially the presence of children in the workplace. People who care about the environment see the pollution caused by the long distance movement of goods as well as the shift of production to overseas locations with low or no environmental standards. All of us witness global financial crises and widespread recession and, far worse, mass starvation amidst the collapse of whole societies in Africa.

Meanwhile, we read about the growing inequality of income around the world. Apparently, the poor are getting poorer while the rich are getting richer. Globalization may be good for the compilers of economic statistics. But according to this viewpoint, it is the antithesis of justice and fairness. At the same time, government officials fear the loss of sovereignty, while we all face the rising power of international crime syndicates, spreading epidemics, and most dramatically the audacious attacks by global terrorist groups.

Well, which is it? Is globalization the bright sun or the dark side of the moon? Let us note that historians say that, measured by trade and investment flows, the world economy may have been more integrated in the nineteenth century than it is today. For example, before passports were so generally required for crossing borders, people were far freer to travel and to migrate than is now the case.

The extent of economic interdependence across national boundaries—globalization did not decline in the early [twentieth] century because of mass protests or a bad press. The causes were far more fundamental—World War I, the worldwide depression of the 1930s, and the subsequent separation of the major nations into democratic and totalitarian camps that culminated in World War II. That long period was a time of rising isolationism, both political and economic.

It is possible that the world is approaching a somewhat similar turning point. The increasingly negative public reaction to the rapid and pervasive changes generated by globalization is beginning to overtake the positive aspects—at least in terms of perception. The instinctive response by economists is to correct the substantial amount of misinformation that has fueled the backlash to opening markets and to expanding the reach of competition. Frankly, no amount of technical brilliance is going to convince the people who are genuinely concerned over the dark side of globalization.

And surely, the education and information approach does not help with the newest negative forces. I am referring to the rise of international terrorist networks. The sad fact is that the combination of terrorism and our strong response to it is making international trade more difficult and costly and international investment more hazardous and financially more risky. To sum up the most recent trends, the costs of globalization are rising while the benefits are declining.

It is not clear how those of us seriously concerned with world commerce should proceed under these circumstances. Obviously, the terrorists are not going to be helpful. It is their diabolical use of modern technology that has been pacing the rise of globalization, which has enabled them to be so effective in the devastation that they generate. As an economist, I have to reluctantly add, their efforts have been very cost-effective.

It is the two more conventional sides of the globalization debate—those who favor freer markets and those who have other peaceful priorities—that I turn to. However, even here I do not see the degree of trust and openmindedness necessary for the educational approach to succeed. I leave that important task to other occasions. I have to confess that it is tempting to respond to some of the more naive protesters in Seattle. My favorite sign was the one that blithely proclaimed, "Food is for people, not for export." I shudder to think of the malnutrition and worse that would follow from a ban on the export of food.

Instead, I will try to identify some common ground on which people of good will on both sides of the heated controversies on globalization might possibly agree. The focus is on useful but undramatic policy changes, what we can call the "nuts and bolts" of problem solving. I note in passing that the greatest opposition to globalization comes from those who believe that they have no stake in it.

So here are my five suggestions:

1. REFORM THE WORLD TRADE ORGANIZATION

Many of the criticisms of the WTO are on target. They deserve a positive response. The agency has become too close and too bureaucratic. But I see no value in trying to shut it down. Its fundamental notion of advancing the rule of law on an international front is an appealing idea. An American in England is deeply aware of our great debt to your nation for teaching us the importance of the rule of law and of the need to establish the institutions to carry it out effectively.

It does not diminish my adherence to free and open trade to state that the WTO has become too inbred and too rigid in its operations. For starters, the general sessions of trade negotiations should be open to the public, like our Congress is and, I believe, your Parliament. So should the hearings at which the various interest groups present their views. Yet, like our legislative committees, the members should be expected to go into closed executive sessions when they begin to do the actual negotiations and drafting of trade agreements. That is a common sense distinction which experience teaches us is practical and workable.

Similarly, the WTO's critical process for settling trade disputes should be opened up so that the critics can see for themselves the nitty gritty of the workings of the WTO rules. Specifically, the dispute settlement process should be expanded to include the submission of amicus briefs by interested public and private groups. Even our Supreme Court allows for such submissions by outsiders, so-called "friends of the court." Also the hearings of the dispute panels should be open to government and private observers as courtroom proceedings typically are. This does not require that the deliberations of the panels be public events.

2. HELP THE PEOPLE HURT BY GLOBALIZATION

Every significant economic change generates winners and losers. It does not satisfy the people hurt by globalization to tell them that far more people benefit from international trade and investment. Although I believe that response is accurate, it is so cavalier that it is bound to infuriate those concerned with the dark side of globalization.

A two-prong approach is needed. In the advanced industrialized nations, we must do a better job of helping the people who lose their jobs due to imports or the movement of factories to overseas locations. Simultaneously, we must grapple with the issue of the labor and environmental standards that are followed in poor (and thus usually low cost) countries by the companies that provide products for export to developed nations. Many of these overseas factories are either owned by companies in the developed world or they sell the bulk of their output to those Western companies. Let us take up each of these two issues in turn.

In the developed nations, such as our own, the most effective adjustment policy to help those who lose their jobs due to globalization—or for other reasons, notably technological advance—is to achieve a growing economy that generates a goodly supply of new jobs. In the absence of a successful macroeconomic policy, no adjustment programs will work well. Nevertheless, some more specific and constructive actions can be taken to improve the adjustment process.

Often, laid-off workers need just a modest bit of help, but they need it quickly. I am referring to the people who went straight from school to work and never had to conduct a serious job hunt. For them, the most effective assistance is modest but essential: help them locate a new job, show them how to prepare for a job interview, and how to fill out a job application. That may seem too obviously elementary to any college graduate, but we are talking about people who feel they have been treated badly by an economic system they do not really understand.

Many other unemployed people find that their job skills are obsolete or that much of their knowledge is only useful to their previous employer. They may be long on [what] we can call institutional information, but extremely short on the math and the language capabilities required for many new and well-paying jobs. These people could benefit from some pertinent education and training. Such "trade adjustment" programs have existed in the United States for four decades, but their track record is not very inspiring.

Those public sector adjustment assistance programs need to be adjusted. They should be made more user friendly. "One stop" registration should replace the current uncoordinated array of assistance. It is disheartening for a newly unemployed worker to feel like a ping-pong ball being tossed from bureau to bureau, rarely encountering a government official who seems to have any real interest in his or her situation.

At least in the United States, there is a type of educational and training institution that can be geared to serving unemployed blue-collar and white-collar workers. It is the network of community and junior colleges who serve a very different group of people than do the more prestigious senior colleges and universities.

Older workers present an especially difficult challenge. They have limited motivation to undertake training programs that, at best, will prepare them for positions that pay much less than their customary wages—and in a labor market where they will compete against youngsters half their age.

Some innovations are needed. One example of fresh thinking is the idea of providing "wage insurance" to pay a major portion of the difference in earnings between the new job and the previous position. The idea is to give the older workers the incentive to get back to work quickly before their skills become rusty. To the extent that such older workers demonstrate to their new employers their greater worth in terms of seasoned judgment and good work habits, they may find the wage gap between the old and new jobs narrowing and thus minimizing the need to draw on the wage insurance plan.

An even more contentious area is the issue of establishing labor and environmental standards for overseas locations that make products for export to the developed nations. There are several contending groups involved, each with its own goals and objectives.

The labor unions in the industrialized countries resent the competition from workers in countries with lower costs of production and hence lower working standards. Certainly compared with pay and factory conditions in the United States and the United Kingdom, it is easy and sometimes accurate to label these places as "sweatshops."

As would be expected, employers have a somewhat different attitude toward the matter. They view low-cost production sites overseas as necessary to meet competition. Many Western firms report that the factories they own or buy from in developing countries pay their workers substantially above locally prevailing wages. They also claim to maintain above-average working conditions.

There is a third force in this debate, which really complicates the issue. It consists of the governments of the developing countries, such as India and Brazil. They openly resent what they describe as the newly formed concern on the part of Westerners with the working conditions in their countries. They see that interest as a poorly disguised form of protectionism designed to keep their products out of the markets of the advanced economies.

A somewhat similar lineup of interest groups occurs in the case of environmental standards in developing countries. On the international trade front, unions and environmental groups tend to join forces. On domestic issues, however, they often go their separate ways, especially when the question seems to involve jobs versus the environment.

In terms of action on globalization matters, most unions and many environmental organizations insist on making labor and environmental standards a part of any new international trade agreement. Products produced in violation of the standards would be barred from entering other nations. The opposition to that approach is hardly limited to teachers of international trade theory. The most vehement opponents are business interests in the developed nations and governments in developing countries. The result is a standoff almost ensuring the failure of any new round of trade liberalization. Let us, however, continue our effort to develop alternative positions and thus find some common ground.

3. STRENGTHEN THE INTERNATIONAL LABOR ORGANIZATION

One global organization that warrants more attention is ILO. It is the only international agency in which labor is fully represented. Yet unions are reluctant to use the ILO to enforce international labor standards—and for good reason. When it comes to ensuring compliance with the enlightened standards it adopts, the ILO has been a paper tiger.

Worse yet, the U.S. Congress has not gotten around to approv[ing] all of the four "core" labor standards the ILO has promulgated—the right to form unions, ridding the workplace of discrimination in employment, and eliminating child and forced labor. Ironically, compared to our own detailed and pervasive labor laws and regulations, the core ILO standards are basic but far more limited. Yes, I am skipping over the problems arising from the bureaucratic language that found its way into the ILO process.

The U.S. Congress should quickly endorse all four of the ILO core labor standards. To follow up, the United States should take the lead in urging the United Kingdom and other industrialized nations to join us in providing adequate resources and support to the ILO. I would not focus on expanding some of the ILO staff. Large bureaucracies are notoriously unresponsive.

I do mean increasing the funds for such enlightened activities as the special program which provides financial assistance to very poor families in developing nations whose children are taken off factory employment. The idea is to enable those youngsters to stay in school. In so many cases, the families cannot afford to pay for such education on their own—or even to forgo the money the children had been earning.

4. GIVE PEOPLE A VOICE VIA THE INTERNET

There is a way of promoting adherence to the ILO labor standards without resorting to trade sanctions or other forms of compulsion. The ILO should post on the Internet the names of the countries that are not complying with the core labor standards. Such a "seal of disapproval" should be widely publicized. This information should be made available to the media worldwide. Consumers would then be encouraged not to buy goods made in those nations.

Did I say boycott? Well, yes and no. No, I am not advocating that governments use force to keep out the products from the offending countries. But, yes, I do advocate giving each individual consumer the ability to back up concerns with personal action—with that action being based on knowledge.

This approach does not provide the entertainment of puppet-parading protesters. However, it may be more effective in the long run. To be successful, this approach requires a citizenry that takes the pains to inform itself and then acts voluntarily on an individual basis. Given the widespread access to the Internet, such a consumer effort could be powerfully effective. Personally, I like the idea of consumers making up their own minds rather than relying on the compulsion of government or even the intimidation of group pressure.

If I show some enthusiasm for voluntary compliance with labor standards it is because in the last few years I have been a member of a team of independent outsiders who check the compliance of an American toy company (Mattel) with the high standards it has voluntarily set for domestic and overseas production. I do not claim perfection for the voluntary approach, but the genuine progress that has been made in local working conditions appeals to my sense of realism. While governments and non-governmental organizations continue to debate and disagree, Mattel and other U.S. private enterprises are succeeding in improving the work environment in their factories in developing countries. Based on my own on-the-ground inspections, it is clear that Western-owned or managed factories are at the top of the scale, often setting the pace for local firms.

Why are these Western-based companies so altruistic? Certainly, a good portion of their motivation is to avoid adverse publicity as well as pressure from customers and shareholders. But I can also attest to the fact that you can successfully appeal to a profit-maximizing business executive who gets a special pleasure from selling a toy that makes a child laugh.

5. REDUCE THE FREQUENCY OF GLOBAL ENCOUNTERS

Finally, I would like to present a really offbeat idea.

Virtually every meeting of global leaders generates a predictable response: [P]rotesters mass in an effort to close it down. This arouses my sense of irony. I believe the protesters are right, but for the wrong reason. The annual economic summit meeting of the heads of the major governments is not the occasion for the exercise of too much power, but too little. The phenomenon of highly orchestrated annual meetings of world leaders has degenerated into costly global photo opportunities at which presidents and prime ministers strut on the world stage—and accomplish little of substance.

One of the easiest forecasts to make in advance of any annual summit is that once again the joint communique will be disappointingly bland. Thus, the protesters are fundamentally misguided in massing at summit cities in an effort to influence a supposedly powerful decision making operation. Few decisions of importance are made there. Today's summit meeting at its best is merely a colorful social occasion. The leaders of the major nations get to meet each other.

Each government represented feels obliged to send a huge delegation to back up its national leader. After all, who knows what technical questions will be raised? In any event, that justifies the presence of a vast array of supporting officials and staff members. Given the millions of dollars directly spent on these taxpayer-financed junkets—in addition to the substantial indirect costs for security, etc.—it is easy to conclude that the summit meeting flunk the simplest cost/benefit test. Under the circumstances, I offer a modest proposal: declare a moratorium on global summits. At a minimum, cancel the meeting scheduled for the summer of 2002. Let the leaders use modern telecommunications to communicate. In plain language, if one prime minister wants to speak to another, he or she should pick up the phone to call. Use the money saved for some worthy endeavor such as providing emergency treatment for the sick.

I predict that, aside from folks who enjoy taking expensive international trips paid for by someone else, there will be little clamor to revive the expensive custom of holding frequent meetings of world leaders. Surely no serious government function will be adversely affected by not holding a global summit meeting later this year.

As the various sides in the debate on globalization continue to harden their positions, any movement to the high middle ground will become increasingly difficult. The development of a feeling of trust, or at least common understanding, is a badly needed precondition. In this presentation, I have admittedly only covered a small sample of the numerous contentious issues that eventually need to be dealt with.

Meanwhile, one modest change would help. The various participants in the often-heated discussions on globalization should consider moderating their vocabularies. So often the argument seems to be carried on between "greedy, profiteering monopolists" and "impractical free-trade theorists," on one side, and "environmental whackos" and "corrupt union bosses" on the other. The introduction of a bit of mutual good will would surely help.

In any event, the serious concerns generated by a more closely linked global marketplace must be faced. More of us need to understand and deal with both the dark side and the bright side of globalization. Some of us may believe that international commerce (i.e., globalization) is more of a wonder land than a waste land. But, as a practical matter, the status quo is unstable. Real improvement in government policies on international commerce will not take place until we respond constructively to the genuine concerns of the other voices in the globalization debate. Surely, we cannot take the position, "Stop the world, I want to get off."

8

Free Markets and Poverty

CHRISTIAN E. WELLER
ADAM HERSH

For better than two decades, the orthodox recipe for global growth has been embodied in the so-called Washington Consensus. This approach, advocated by the United States and enforced by the World Bank and the International Monetary Fund (IMF), holds that growth is maximized when barriers to the free flow of capital and commerce are dismantled and when individual economies are exposed to the discipline, consumer markets, and entrepreneurs of the world economic system. Proponents of this view have contended that the free-market approach to development will also alleviate poverty, both by raising overall growth rates and by bringing modern capitalism to the world's poorest.

Yet the actual experience since 1980 contradicts almost every one of these claims. Indeed, the free-trade/free-capital formula has led to slower growth and more vulnerability for poor countries—and to greater income disparity among individuals. In 1980 median income in the richest 10 percent of countries was 77 times greater than in the poorest 10 percent; by 1999 that gap had grown to 122 times. Progress in poverty reduction has been limited and geographically isolated. The number of poor people rose from 1987 to 1998; in many countries, the share of poor people increased (in 1998 close to half the population in many parts of the world were considered poor). In 1980 the world's poorest 10 percent, or 400 million people, lived on the equivalent of 72 cents a day or less. The same number of people had 79 cents per day in 1990 and 78 cents in 1999. The income of the world's poorest did not even keep up with inflation.

Why has the laissez-faire approach worsened both world growth and world income distribution? First, the IMF and the World Bank often commend austerity as an economic cure-all in order to reassure foreign investors of a sound fiscal and business climate—but austerity, not surprisingly, leads to slow growth. Second, slow growth itself can mean widening income inequality, since high growth and tight labor markets are what increase the bargaining power of the poor. (Economists estimate that poverty increases by 2 percent for every 1 percent of decline in growth.) Third, the hands-off approach to global development encourages foreign capital to seek regions and countries that offer the cheapest production costs—so even low-income countries must worry that some other,

SOURCE: Christian E. Weller and Adam Hersh, "Free Markets and Poverty," *The American Prospect* (Winter 2002), pp. A13–A15. Reprinted with permission from *The American Prospect*, Volume 13, Number 1: January 01, 2002. *The American Prospect*, 11 Beacon Street, Suite 1120, Boston, MA 02108. All rights reserved.

even more desperate workforce will do the same work for a lower wage. Finally, small and newly opened economies in the global free market are vulnerable to investment fads and speculative pressures from foreign investors—factors that result in instability and often overwhelm the putative benefits of greater openness. All of these upheavals disproportionately harm the poorest.

CAPITAL AND TRADE

Because capital controls were reduced or eliminated virtually everywhere over the past 20 years, the flow of capital to developing countries increased rapidly, from $1.9 billion in 1980 to $120.3 billion in 1997 (the last year before the global financial crisis). Even in 1998, in the wake of financial crisis, the flow of capital remained remarkably high at $56 billion (although a substantial share of this money consisted of short-term portfolio investments).

Unfortunately, faster capital mobility in a deregulated environment means an increase in speculative financing and, thus, greater financial instability. Under such conditions, the poor are unlikely to escape poverty through economic growth because they are ill equipped to weather the macroeconomic shocks.

Moreover, higher-income people can protect themselves more effectively from the fallout of a crisis. They have capital that they can move overseas. At the same time, in the IMF/World Bank formula, a crisis invariably calls for a reduction in public spending—at precisely a moment when the poor are more dependent on social safety nets. So on both counts, laissez-faire widens the gap between rich and poor.

Trade liberalization—the complement to deregulated capital markets in the global deregulation agenda—also plays a significant role in expanding inequality and limiting efforts to reduce poverty. It induces rapid structural change as well as a decline in real wages, working conditions, and living standards. It also gives teeth to employers' threats to close plants or to relocate or "outsource" production abroad, where labor regulations are less stringent and more difficult to enforce—thus undermining workers' attempts to organize and bargain for improved wages and working conditions. This trend fuels a race to the bottom in which governments vie for needed international investment by scrambling to offer employers the cheapest body of laborers.

The connection between rapid trade liberalization and inequality is reflected in downward wage pressures and rising inequality in industrializing as well as industrialized economies. A 1997 report by the United Nations Conference on Trade and Development, for instance, found that trade liberalization in Latin America led to widening wage gaps, falling real wages for unskilled workers (often more than 90 percent of the labor force in developing countries), and rising unemployment.

Evidence is overwhelming that income inequality is rising in industrializing countries. But there is also a broad consensus—even among laissez-faire cheerleaders—that income inequality has risen in developed nations as well since 1980. In a 1997 paper for the *Journal of Economic Literature*, Peter Gottschalk and Timothy M.

Smeeding found that "almost all industrial economies experienced some increase in wage inequality among prime-aged males" in the 1980s and early 1990s. Further, data from the widely respected Luxembourg Income Study show that among 24 such countries, 18 experienced a rise in income inequality, only 5 experienced a decline in inequality (Denmark, Luxembourg, the Netherlands, Spain, and Switzerland), and [one] (France) saw no change.

While a widening gap between the rich and the poor within countries is not universal, it appears to have occurred in most countries and is affecting most of the world's population.

PROBLEMATIC POSTER CHILDREN

The World Bank's conclusion that the lot of the poor has improved during the era of increasing trade- and capital-liberalization relies substantially on data from China and India. But both countries are anomalies. In reality, the facts in India and China undermine the case for a connection between greater deregulation and falling poverty and inequality. While in China the percentage who are poor has fallen, there has nonetheless been a rapid rise in inequality—most notably, from 1985 to 1995, between rural and urban areas and between provinces with urban centers and those without them. Also, a large number of China's workers labor under abhorrent, and possibly worsening, slave- or prison-labor conditions. This not only means that many workers are left out of China's economic growth; it also makes China an unappealing development model for the rest of the world. Thus, improvements in China are not universally shared and leave many workers behind, often in deplorable conditions.

Using India to illustrate the benefits of unregulated globalization is equally problematic, since the country achieved its progress while remaining relatively closed off to the global economy. Total "goods trade" (exports plus imports) was about 20 percent of India's gross domestic product in 1998, or 10 percentage points less than in China and only about one-fifth the level of such export-oriented countries as Korea. Moreover, the IMF views India as something of a laggard in deregulating its economy. IMF reports regularly recommend further liberalization of India's trade and capital flows—the only large developing economy for which this is the case.

More broadly, to use India and China as poster children for the IMF/World Bank brand of liberalization is laughable. Both nations have sheltered their currencies from global speculative pressures (a serious sin, according to the IMF). Both have been highly protectionist (India has been a leader of the bloc of developing nations resisting WTO pressures for laissez-faire openness). And both have relied heavily on state-led development and have opened to foreign capital only with negotiated conditions. The Heritage Foundation, in its annual *Index of Economic Freedom*, ranks India and China as tied for spot 121—among the least economically open nations in the world. Yet by letting in foreign capital in a limited and negotiated way, India and China have benefited from investment without totally sacrificing economic sovereignty. There may be a larger lesson here.

A BROADER PERSPECTIVE

The World Bank's assertion that "between countries, globalization is mostly re-ducing inequality" seems to contradict the IMF's assessment that "the relative gap between the richest and the poorest countries has continued to widen" in the 1990s. Given this confusion, it is useful to take a global perspective that looks at the distribution of world income across all countries and across all people.

Distribution among countries unambiguously worsened in the 1980s and 1990s. In other words, rich nations have gotten richer and poor ones have gotten poorer. The median per capita income of the world's richest 10 percent of countries was 77 times that of the poorest 10 percent in 1980, 120 times greater in 1990, and 122 times greater in 1999. The ratio of the average per capita incomes shows an even more dramatic increase.

World-income distribution across people (rather than countries) witnessed equitable improvement to some extent in the late 1990s, after a dramatic rise in inequality during the previous years. While the richest 10 percent of the world's population had, on average, incomes that were 79 times higher than those of the poorest 10 percent in 1980, their incomes were 120 times higher in 1990. That ratio dropped to 118 in 1999. The improvement in equality was somewhat more pronounced in terms of median incomes; yet even under this measure, income distribution was remarkably less equitable in 1999 than in 1980.

The few gains in the 1990s come solely from rising incomes in China. If China is excluded, there is an unambiguous trend toward growing income in-equality across the remaining world population in the 1980s and 1990s. But since income distribution in the People's Republic has become substantially less equi-table in the 1990s, the inclusion of China's per capita GDP in the distribution of

DISTRIBUTION OF WORLD INCOME
Ratio of Top 10 Percent to Bottom 10 Percent

	1980	1990	1999
By countries			
Ratio of average incomes	86.2%	125.9%	148.8%
Ratio of median incomes	76.8	119.6	121.8
By population			
Ratio of average incomes	78.9	119.7	117.7
Ratio of median incomes	69.6	121.5	100.8
By population, excluding China			
Ratio of average incomes	90.3	135.5	154.4
Ratio of median incomes	81.1	131.2	153.2

NOTE: Distributions are based on per capita GDP in current U.S. dollars (IMF data).
SOURCE: Authors' calculations based on IMF data.

world income across all people exaggerates improvements in the world's income distribution in the 1990s. Put simply, inequality is a bigger problem at the end of the nearly 20-year experiment with unregulated global capitalism than it was before deregulation became the rule.

Despite official claims to the contrary, the evidence clearly shows that the laissez-faire era has been one of slower growth and greater inequality. And the apparent improvement of that trend in the 1990s is the result solely of rising per capita income in China, where the enormous population tends to distort world averages. Even so, income inequality within countries is also growing. Success in reducing poverty has been limited.

The promises of more-equal income distribution and reduced poverty around the world have failed to materialize under the current form of unregulated globalization. It is time for multinational institutions and other international policy makers to develop a different set of strategies and programs to provide real benefits to the poor.

9

NAFTA at 10

Where Do We Go from Here?

JEFF FAUX

Ten years ago, the North American Free Trade Agreement was sold to the people of the United States, Mexico and Canada as a simple treaty eliminating tariffs on goods crossing the three countries' borders. But NAFTA is much more: It is the constitution of an emerging continental economy that recognizes one citizen—the business corporation. It gives corporations extraordinary protections from government policies that might limit future profits, and extraordinary rights to force the privatization of virtually all civilian public services. Disputes are settled by secret tribunals of experts, many of whom are employed privately as corporate lawyers and consultants. At the same time, NAFTA excludes protections for workers, the environment and the public that are part of the social contract established through long political struggle in each of the countries.

As Jorge Castañeda, Mexico's recent foreign secretary, observed, NAFTA was "an accord among magnates and potentates: an agreement for the rich and powerful . . . effectively excluding ordinary people in all three societies." Thus was NAFTA a model for the neoliberal governance of the global economy.

The business-backed politicians who pushed the agreement through the three legislatures promised that NAFTA would generate prosperity that would more than compensate "ordinary" people for its lack of social protections. Foreign investors would make Mexico an economic tiger, turning its poor workers into middle-class consumers who would then buy U.S. and Canadian goods, creating more jobs in the high-wage countries.

But as soon as the ink was dry on NAFTA, U.S. factories began to shift production to maquiladora factories along the border, where the Mexican government assures a docile labor force and virtually no environmental restrictions. The U.S. trade surplus with Mexico quickly turned into a deficit, and since then at least a half-million jobs have been lost, many of them in small towns and rural areas where there are no job alternatives.

Meanwhile, Mexico's overall growth rate has been half of what it needs to be just to generate enough jobs for its growing labor force. The NAFTA-

SOURCE: Jeff Faux, "NAFTA at 10: Where Do We Go From Here?" Reprinted with permission from the February 2, 2004 issue of *The Nation*. For subscription information, call 1-800-333-8536. Portions of each week's *Nation* magazine can be accessed at http://www.thenation.com.

inspired strategy of export-led growth undermined Mexican industries that sold to the domestic markets as well as the sixty-year-old social bargain in which workers and peasant farmers shared the benefits of growth in exchange for their support for a privileged oligarchy. NAFTA provided the oligarchs with new partners—the multinational corporations—allowing them to abandon their obligations to their fellow Mexicans. Average real wages in Mexican manufacturing are actually lower than they were ten years ago. Two and a half million farmers and their families have been driven out of their local markets and off their land by heavily subsidized U.S. and Canadian agribusiness. For most Mexicans, half of whom live in poverty, basic food has gotten even more expensive: Today the Mexican minimum wage buys less than half the tortillas it bought in 1994. As a result, hundreds of thousands of Mexicans continue to risk their lives crossing the border to get low-wage jobs in the United States.

Canada, which since 1989 has had a similar trade agreement with the United States, and which does much less business with Mexico, was less directly affected. But NAFTA strengthened Canadian corporations' ability to threaten workers and governments with moving south, helping undermine the country's traditionally strong labor and social standards.

In all three countries NAFTA has worsened the distribution of income and wealth. While ordinary people paid the costs, the benefits went to the continent's "rich and powerful." Canadian and U.S. corporate investors got guaranteed access to Mexico's cheap labor as well as its privatized public assets. Mexican elites brokered the deals. In one example, well-connected Mexicans bought the country's second-largest commercial bank from the government for $3.3 billion and sold it to Citigroup for $12.5 billion.

Yet despite its failures, NAFTA set in motion the economic integration of Canada, Mexico and the United States, which cannot now be stopped. Every day, more intracontinental connections in finance, marketing, production and other business networks are being hard-wired for a consolidated North American market. Ford pickup trucks are assembled in Mexico with engines from Ontario and transmissions from Ohio and Michigan. Canadian, Mexican and U.S. investors have created a labyrinth of interconnected corporate assets. After a temporary post-9/11 slowdown, the cross-border movement of people—unskilled workers, educated professionals, retirees—continued.

Expanded markets require expanded rules. Out of public sight, the rule-books are being filled in by NAFTA tribunals, trigovernmental commissions, administrative judges. Business-supported academic centers are humming with new proposals, ranging from guestworker programs, to the privatization of Canadian water and Mexican oil, to continental business tax policies. As a former Canadian ambassador to the United States recently commented, "Few days go by without new ideas for deepening NAFTA."

But while corporate business and its political clients are organized continentally, progressives are not. One reason is that the opposition to NAFTA in all three countries was in large part rooted in economic and political nationalism. The political heat that almost defeated the agreement in the U.S. Congress was fueled by the specter of American jobs moving to Mexico. The Canadian opposition painted NAFTA as a threat to Americanize Canadian culture. In Mexico, opposition was rooted in its people's historic mistrust of Yankee imperialism.

Once the fight over NAFTA was settled, opposition groups moved back to domestic issues or moved on to defend against neoliberalism in other global settings, such as the proposed Free Trade Area of the Americas and the new round of World Trade Organization negotiations. These are important battles, but the capacity of North American activists to influence these negotiations is marginal. For example, if the FTAA is permanently derailed, it will not be over a lack of social protections but because Latin American and U.S. business interests cannot make a deal.

Back home, however, North American opponents of neoliberalism—because they can be a force in the domestic politics of all three nations—have more leverage to develop a socially responsive model of economic integration between rich and poor economies. Indeed, given the influence of the United States in setting the rules for the global economy, a visible, sustained challenge to the NAFTA model here may be the most important contribution progressives on this continent can make to the building of a more just global economic system.

A continental progressive movement would build on its existing infrastructure in each nation—labor, environmentalists, human rights activists, progressive churches and populist legislators—and the fact that the majority of ordinary citizens in all three nations want a market system with social protections.

One initial organizing step might be to connect existing demands to rewrite NAFTA. For example, over the past year Mexican farmers demonstrated throughout the country—including breaking down the door to the Mexican Congress—demanding that NAFTA's agricultural provisions be changed. Had U.S. and Canadian small farmers, labor unions and environmentalists joined them with their own demands, the Mexican government would not have been able to isolate the farmers with the argument that changing NAFTA is politically impossible.

A new continental agreement could include financial assistance from the United States and Canada to Mexico for building the economic and social infrastructure it needs for growth, just as the European community has redistributed funds to its poorest members in order to create a stronger and more balanced economy. Continentwide enforceable labor, human rights and environmental protections ought to be established to prevent the erosion of living standards in Canada and the United States, and to insure that Mexican workers share in the benefits of rising productivity. Provisions of NAFTA that erode the ability of the local public sectors in all three countries to promote the welfare of their citizens should be stricken.

Progressive legislators in all three countries could begin working out proposals covering issues such as corporate governance, public health and safety, and investment in education that could be simultaneously introduced in all three capitals. A continental labor organizing campaign against a single employer could have an electrifying effect—demonstrating that workers in Canada, Mexico and the United States have more in common with one another than with the CEOs who may share their formal nationality.

Creating a continental political consciousness does not mean forming one nation. Few are ready for that—particularly the majority of Mexicans and Canadians appalled by the U.S. governing class's current imperial obsessions. But despite all the obvious difficulties, if progressives do not want to see a continental society built on NAFTA's reactionary template, they have little choice but to grasp hands across the borders and work together to build an economy that serves the continent's "ordinary" people.

REFLECTION QUESTIONS FOR CHAPTER 3

1. The readings in this section focus on the consequences of globalization for societies, institutions within societies (democracy), and opportunities for individuals (jobs, wages). Ghemawat disagrees with Friedman's view of information technology in the global age as profoundly liberating. What do you think of the reasons and evidence he gives to make the argument that the world's playing field is not leveling?

2. Are the arguments in this section more persuasive for "globalization is good" or for "globalization is bad"? Why? Continually revisit this question as you read the other chapters in this book.

3. Weidenbaum suggests five policy changes that will minimize the problems with globalization. Examine each, asking, Is it on the right track? Does it go far enough? Are there other policy changes that Weidenbaum has missed? Again, keep his policy suggestions in mind as you traverse this book. Your critique should become more sophisticated as you engage the arguments of other commentators.

4. Three of the essays in this section address the question, Is free trade the panacea for the world's economic problems? Is it? If so why? If not, why not?

4

Economic Globalization

At the heart of the globalization phenomenon is the complex web of economic transnational interconnections through the flow of goods and services, and capital. While trade between countries has existed for centuries, the current era "is the first in which the international economic system has become truly *interdependent*."[1] The goods we consume and the prices we pay for them are determined by transnational corporations and by the wages paid to workers who produce these products. Jobs move across borders to the low-wage economies. Workers move across these same borders to find more profitable work. Trillions of dollars move across borders instantaneously. The effects of a financial crisis in one country or region ripple across the globe causing havoc in stock and bond prices, commodity prices, the default of huge loans, and the devaluation of currencies. Most significant is the incredible power of transnational corporations. Barbara Ehrenreich says:

> There are 193 nations in the world, many of them ostensibly democratic, but most of them are dwarfed by the corporations that alone decide what will be produced, and where, and how much people will be paid to do the work. In effect, these multinational enterprises have become a kind of covert world government—motivated solely by profit and unaccountable to any citizenry. Only a small group of humans on the planet, roughly overlapping the world's 475-member billionaire's club, rule the global economy. And wherever globalization impinges, inequality deepens.[2]

The first selection in this chapter, by the *Dollars and Sense* collective, provides an overview of the global economy, with an emphasis on the international institutions supporting free trade and capitalism—the World Bank, the International Monetary Fund, the World Trade Organization, and the International Standards

Organization. The bias of these organizations against the world's working class and poor and toward corporations is made clear in the discussion.

Labor expert David Moberg presents a case study of a U.S. corporation (Maytag) moving from Galesburg, Illinois, to Mexico, and its consequences for Galesburg and the larger implications for other U.S. and Mexican workers as well.

Jesse Gordon and Knickerbocker Designs provide a visual of the labor history of a Gap sweatshirt. The process begins with workers in Uzbekistan earning 2 cents a pound harvesting cotton. The raw cotton is processed in South Korea, where workers make $4 an hour. Then on to Russia, where seamstresses make the sweatshirts for $39 to $69 a month. Then the product is shipped, stored, and trucked to distribution centers in the United States, where a nonunion entry-level worker at the Gap sells the sweatshirts for $48, while making $6 an hour.

Although globalization has transformed social relations around the world, it has not erased old forms of inequality. Today, gender and race remain central in global strategies of capitalism. Anibel Ferus-Comelo shows how gender is a key factor in electronics manufacturing. She describes the gendered processes that recruit, employ, and exploit migrant women in the electronics workforce around the world. She then outlines the ways in which women struggle for workplace justice in the face of corporate hostility.

In the next reading, William Robinson makes a case for the centrality of race in global capitalism. Not only are racial inequalities used in the new systems of production, but globalization creates and requires pools of immigrant labor that global elites and transnational capital shape for their own purposes. Latinos in the United States are now the superexploited sector. The immigrant's rights movement is the leading edge of the social struggle against oppressive class relations in global capitalism.

In the final article, Joseph E. Stiglitz, Nobel Laureate and former chief economist for the World Bank, offers an important counter to ongoing practices of corporate abuse in the global economy. He offers a five-pronged agenda for simultaneously making globalization "work" and achieving corporate social responsibility.

ENDNOTES

1. O'Meara, Patrick, Howard D. Mehlinger, and Matthew Krain, *Globalization and the Challenges of a New Century* (Bloomington, University of Indiana Press, 2000), p. 215.
2. Ehrenreich, Barbara, foreword to Sarah Anderson and John Cavanagh, *Field Guide to the Global Economy* (New York: The New Press, 2000), p. x.

10

The ABCs of the Global Economy

THE *DOLLARS AND SENSE* COLLECTIVE

In the 1960s, U.S corporations changed the way they went after profits in the international economy. Instead of producing goods in the U.S. to export, they moved more and more toward producing goods overseas to sell to consumers in those countries and at home. They had done some of this in the 1950s, but really sped up the process in the '60s.

Before the mid–1960s, free trade probably helped workers and consumers in the United States while hurting workers in poorer countries. Exporters invested their profits at home in the United States, creating new jobs and boosting incomes. The AFL-CIO thought this was a good deal and backed free trade.

But when corporations changed strategies, they changed the alliances. By the late 1960s, the AFL-CIO began opposing free trade as they watched jobs go overseas. But unionists did not see that they had to start building alliances internationally. The union federation continued to take money secretly from the U.S. government to help break up red unions abroad, not a good tactic for producing solidarity. It took until the 1990s for the AFL-CIO to reduce (though not eliminate) its alliance with the U.S. State Department. In the 1990s, unions also forged their alliance with the environmental movement to oppose free trade.

But corporations were not standing still; in the 1980s and 1990s they were working to shift the architecture of international institutions created after World War II to work more effectively in the new global economy they were creating. More and more of their profits were coming from overseas—by the 1990s, 30% of U.S. corporate profits came from their direct investments overseas, up from 13% in the 1960s. This includes money made from the operations of their subsidiaries abroad. But the share of corporate profits earned overseas is even higher than that because the 30% figure doesn't include the interest companies earn on money they loan abroad. And the financial sector is an increasingly important player in the global economy.

Financial institutions and other global corporations without national ties now use governments to dissolve any national restraints on their activities. They are global, so they want their government to be global too. And while trade used to be taken care of through its own organization (GATT) and money vaguely managed through another organization (the International Monetary

SOURCE: The *Dollars and Sense* Collective, "The ABCs of the Global Economy,"
March/April 2000. Reprinted by permission of *Dollars & Sense*, a progressive economics
magazine www.dollarsandsense.org.

Fund), the new World Trade Organization erases the divide between trade and investment in its efforts to deregulate investment worldwide.

In helping design some of the global institutions after World War II, John Maynard Keynes assumed companies and economics would operate within national bounds, with the IMF and others regulating exchanges across those borders. The instability created by ruptured borders is made worse by the deregulation sought by corporations, and especially, the financial sector. The most powerful governments of the world seem oblivious to this threat in giving them what they want.

This is a world-historical moment in which it is possible to stop the corporate offensive, a moment when the ruling partnership composed of the United States, Europe and to a lesser extent Japan is fracturing, as the European Union reaches its limit on the amount of deregulation it will take and Japan's economy is in turmoil. This may allow those opposing the ruling bloc—Third World governments (which may be conservative), labor, and environmentalists worldwide—to build alliances of convenience with sympathetic elements within the EU to guide the reshaping of the global institutions in a liberatory manner.

What follows is a primer on the most important of those institutions. We hope in the near future to publish primers on other aspects of the global economy: regional trade agreements and alternative visions of how to regulate it. Stay tuned.

—*Abby Scher*

THE WORLD BANK AND INTERNATIONAL MONETARY FUND

Where Did They Come From?

The basic institutions of the postwar international capitalist economy were framed, in 1944, at an international conference in the town of Bretton Woods, New Hampshire. Among the institutions coming out of the conference were the World Bank and the International Monetary Fund (IMF). These two are often discussed together because they were founded together, because countries must be members of the IMF before they can become members of the World Bank, and because both practice what is known as "structural adjustment" (where borrower countries unable to obtain credit from other sources must change government policies before loans are released).

At both the World Bank and IMF, the number of votes a country receives is based on how much capital it gives the institution, so rich countries like the United States enjoy disproportionate voting power. In both, five powerful countries (the United States, Great Britain, France, Germany, and Japan) get to appoint their own representatives to the institution's executive board (with 19 other directors elected by the rest of the 150-odd member countries). The president of the World Bank is elected by the Board of Executive Directors, and traditionally nominated by the U.S. representative. The managing director of the IMF, meanwhile, is traditionally a European. The governments of a few rich countries, obviously, call the shots in both institutions.

Why Should You Care?

Just after World War II, the World Bank mostly loaned money to Western European governments to help rebuild their countries. It was during the long tenure (1968–1981) of former U.S. Defense Secretary Robert S. McNamara as president that the bank turned towards "development" loans to Third World countries. McNamara brought the same philosophy to "development" that he had used in war—more is better. Ever since, the Bank's approach has drawn persistent criticism for favoring large, expensive projects regardless of their appropriateness to local conditions. Critics have argued that the bank pays little heed to the social and environmental impact of the projects it finances, and that it often works through dictatorial elites that channel benefits to themselves rather than those who need them (and leave the poor to foot the bill later).

The most important function of the IMF is as a "lender of last resort" to member countries that cannot borrow money from other sources. The loans are usually given to prevent a country from defaulting on previous loans from private banks. Funds are available from the IMF, on the condition that the country implement what is formally known as a "structural adjustment program" (SAP), but more often referred to as an "austerity plan." Typically, a government is told to eliminate price controls or subsidies, devalue its currency or eliminate labor regulations like minimum wage laws—all actions whose costs are born by the working class and the poor whose incomes are cut.

The conditions imposed by the IMF and the World Bank, which places similar conditions on "structural adjustment" loans, are motivated by an extraordinary devotion to the free-market model. As Colin Stoneman, an expert on Zimbabwe, put it, the World Bank's prescriptions for that country during the 1980s were "exactly those which someone with no knowledge of Zimbabwe, but familiarity with the World Bank, would have predicted."

The IMF and World Bank wield power disproportionate to the size of the loans they give out because private lenders take their lead in deciding which countries are credit-worthy. Both institutions have taken advantage of this leverage, and of debt crises in Latin America, Africa, and now Asia, to impose their cookie-cutter model (against varying levels of resistance from governments and people) on poor countries around the world.

—Alejandro Reuss

THE MULTILATERAL AGREEMENT ON INVESTMENT (MAI), TRADE-RELATED INVESTMENT MEASURES (TRIMS), AND THE INTERNATIONAL MOVEMENT OF CAPITAL

Where Did They Come From?

You're probably not the sort of person who would own a chemical plant or luxury hotel, but imagine you were. Imagine you built a chemical plant or luxury hotel in a foreign country, only to see a labor-friendly government take power and threaten your profits. This is the scenario which makes the CEOs of footloose global corporations wake up in the middle of the night in a cold sweat. To avert such threats, ministers of the richest countries met secretly at the Organization for Economic Cooperation and Development (OECD) in Paris in 1997 and tried to hammer out a bill of rights for international investors, the Multilateral Agreement on Investment (MAI).

When protests against the MAI broke out in the streets and the halls of government alike in 1998 and 1999, scuttling the agreement in that form, corporations turned to the World Trade Organization to achieve their goal. (See "Rage Against the Machine" by Chantell Taylor, *Dollars & Sense*, September/October 1998.)

What Are They Up To?

Both the MAI and Trade Related Investment Measures (or TRIMs, the name of the WTO version) would force governments to compensate companies for any losses (or reductions in profits) they might suffer because of changes in public policy. Governments would be compelled to tax, regulate, and subsidize foreign business exactly as they do local businesses. Policies designed to protect fledgling national industries (a staple of industrial development strategies from the United States and Germany in the 19th century to Japan and Korea in the 20th) would be ruled out.

TRIMs would also be a crowning blow to the control of governments over the movement of capital into or out of their countries. Until fairly recently, most governments imposed controls on the buying and selling of their currencies for purposes other than trade. Known as capital controls, these curbs significantly impeded the mobility of capital. By simply outlawing conversion, governments could trap investors into keeping their holdings in the local currency. But since the 1980s, the IMF and the U.S. Treasury have pressured governments to lift these controls so that international companies can more easily move money around the globe. Corporations and wealthy individuals can now credibly threaten to pull liquid capital out of any country whose policies displease them.

Malaysia successfully imposed controls during the Asian crisis of 1997 and 1998, spurring broad interest among developing countries. The United States wants to establish a new international discussion group—the Group of 20 (G-20), consisting of ministers from 20 developing countries handpicked by the U.S.—to consider reforms. Meanwhile, it continues to push for the MAI-style liberation of capital from any control whatsoever.

Why Should You Care?

It is sometimes said that the widening chasm between the rich and poor is due to the fact that capital is so easily shifted around the globe while labor, bound to family and place, is not. But there is nothing natural in this. Human beings, after all, have wandered the earth for millennia—traversing oceans and continents, in search of food, land, and adventure—whereas a factory, shipyard, or office building, once built, is almost impossible to move in a cost-effective way. Even liquid capital (money) is less mobile than it seems. To be sure, a Mexican can fill a suitcase with pesos, hop a plane and fly to California, but once she disembarks, who's to say what the pesos will be worth, or whether they'll be worth anything at all? For most of this century, however, capitalist governments have curbed labor's natural mobility through passports, migration laws, border checkpoints, and armed border patrols, while capital has been rendered movable by treaties and laws that harmonize the treatment of wealth around the world. The past two decades especially have seen a vast expansion in the legal rights of capital across borders. In other words, labor fights with the cuffs on, while capital takes the gloves off.

WORLD INTELLECTUAL PROPERTY ORGANIZATION (WIPO) AND TRADE-RELATED ASPECTS OF INTELLECTUAL PROPERTY RIGHTS (TRIPS)

What Are They Up To?

One of the less familiar members of the "alphabet soup" of international economic institutions, the World Intellectual Property Organization (WIPO) has governed "intellectual property" issues since its founding in 1970 (though it oversees treaties and conventions dating from as early as 1883). Companies are finding it harder to control intellectual property in two new fields—computer software and biotechnology—because it is so cheap and easy to reproduce electronic information and genetic material in virtually unlimited quantities. This is what makes software, music and video "piracy" widespread.

In the old days, "intellectual property" only covered property rights over inventions, industrial designs, trademarks, and artistic and literary works. Now it covers computer programs, electronic images and recordings, and even biological processes and genetic codes.

WIPO has been busy staking out a brave new world of property rights in the electronic domain. A 1996 WIPO treaty, which now faces ratification battles around the world, would outlaw the "circumvention" of electronic security measures. It would be illegal, for example, to sidestep the security measures on a website (such as those requiring that users register or send payment in exchange for access). The treaty, if ratified, would also prevent programmers from cracking open commercial software to view the underlying code. This could prevent programmers from crafting their own programs so that they are compatible with existing software, and prevent innovation in the form of "reengineering"—drawing on one design as the basis of another. Reengineering has been at the heart of many [a] country's economic development—not just Taiwan but also the United States. Lowell, Massachusetts, textile manufacturers built their looms based on English designs.

WIPO now faces a turf war over the intellectual property issue with none other than the World Trade Organization (WTO). Wealthy countries are attempting an end run around WIPO because it lacks enforcement power and less developed countries have resisted its agenda. But the mass-media, information-technology, and biotechnology industries in wealthy countries stand to lose the most from "piracy" and to gain the most in fees and royalties if given more extensive property rights. So they introduced, under the name "Trade-Related Aspects of Intellectual Property Rights" (TRIPS), extensive provisions on intellectual property into the most recent round of WTO negotiations.

TRIPs would put the muscle of trade sanctions behind intellectual property rights. It would also stake out new intellectual property rights over plant, animal, and even human genetic codes. The governments of some developing countries have objected, warning that private companies based in rich countries will declare ownership over the genetic codes of plants long used for healing or crops within their countries. By manipulating just one gene of a living organism, a company can be declared the sole owner of an entire plant variety.

Why Should You Care?

These proposals may seem like a new frontier of property rights, but except for the defense of ownership over life forms, TRIPS are actually a defense of the old regime of property rights. It is because current computer- and bio-technology make virtually unlimited production and free distribution possible that the fight for private property has become so extreme. By extending private property to previously unimagined horizons, we are reminded of the form of power used to defend it.

—Alejandro Reuss

THE WORLD TRADE ORGANIZATION (WTO)

Where Did It Come From?

Since the 1950s, government officials from around the world have met irregularly to hammer out the rules of a global trading system. Known as the General Agreements on Trade and Tariffs (GATT), these negotiations covered, in excruciating detail, such matters as what level of taxation Japan would impose on foreign rice, how many American automobiles Brazil would allow into its market, and how large a subsidy France could give its vineyards. Every clause was carefully crafted, with constant input from business representatives who hoped to profit from expanded international trade.

The GATT process however, was slow, cumbersome and difficult to monitor. As corporations expanded more rapidly into global markets they pushed governments to create a more powerful and permanent international body that could speed up trade negotiations as well as oversee and enforce provisions of the GATT. The result is the World Trade Organization, formed out of the ashes of GATT in 1994.

What Is It Up To?

The WTO functions as a sort of international court for adjudicating trade disputes. Each of its 135 member countries has one representative, who participates in negotiations over trade rules. The heart of the WTO, however, is not its delegates, but its dispute resolution system. With the establishment of the WTO, corporations now have a place to complain to when they want trade barriers— or domestic regulations that limit their freedom to buy and sell—overturned.

Though corporations have no standing in the WTO—the organization is, officially, open only to its member countries—the numerous advisory bodies that provide technical expertise to delegates are overflowing with corporate representation. The delegates themselves are drawn from trade ministries and confer regularly with the corporate lobbyists and advisors who swarm the streets and offices of Geneva, where the organization is headquartered. As a result, the WTO has become, as an anonymous delegate told the *Financial Times,* "a place where governments can collude against their citizens."

Lori Wallach and Michelle Sforza, in their new book *The WTO: Five Years of Reasons to Resist Corporate Globalization*, point out that large corporations are essentially "renting" governments to bring cases before the WTO, and in this way, to win in the WTO battles they have lost in the political arena at home. Large shrimping corporations, for example, got India to dispute the U.S. ban on shrimp catches that were not sea-turtle safe. Once such a case is raised, the resolution process violates most democratic notions of due process and openness. Cases are heard before a tribunal of "trade experts," generally lawyers, who, under WTO rules, are required to make their ruling with a presumption in favor of free trade. The WTO puts the burden squarely on governments to justify any restriction of what it considers the natural order of things. There are no amicus briefs (statements or legal opinion filed with a court by outside parties), no observers, and no public record of the deliberations.

The WTO's rule is not restricted to such matters as tariff barriers. When the organization was formed, environmental and labor groups warned that the WTO would soon be rendering decisions on essential matters of public policy. This has proven absolutely correct. Currently, the WTO is considering whether "selective purchasing" laws—like a Massachusetts law barring state agencies and local governments from buying products made in Burma and intended to withdraw an economic lifeline to that country's dictatorship—are a violation of "free trade." It is feared that the WTO will rule out these kinds of political motives from government policy making. The organization has already ruled against Europe for banning hormone-treated beef and against Japan for prohibiting pesticide-laden apples.

Why Should You Care?

At stake is a fundamental issue of popular sovereignty—the rights of the people to regulate economic life, whether at the level of the city, state, or nation. Certainly, the current structure of institutions like the WTO allows for little if any expression of the popular will. Can a city, state, or country insist that goods sold in its markets meet labor and environmental standards determined in a democratic forum by its citizens? What if the U.S., for example, insisted that clothing manufactured for the Gap by child laborers not be permitted for sale here? The U.S. does not allow businesses operating within its borders to produce goods with child labor, so why should we allow those same businesses—Disney, Gap, or Walmart—to produce their goods with child labor in Haiti and sell the goods here?

—*Ellen Frank*

INTERNATIONAL STANDARDS
ORGANIZATION (ISO)

There's at least one global institution shaping commerce that corporations control completely, with no pretense of public involvement. That is the International Standards Organization (ISO).

It was founded in 1947 (around the same time as the International Monetary Fund, World Bank and GATT), with the aim of easing trade by standardizing the dimensions of industrial products. Most famously, it set the dimensions of screw threads so that an auto manufacturer in the United States can be confident that screws it buys in China can be used in its cars. More recently, the ISO trumpets its success in standardizing ATM and credit card dimensions so they can be used in machines worldwide.

Without set standards, buyers cannot roam the world in search of the cheapest deal; the dissimilar products thus act as a "technical barrier to trade." Not surprisingly, the ISO, although privately run, is intimately linked to the World Trade Organization with whom it says it is creating "a strategic partnership."

"The political agreements reached within the framework of the WTO require underpinning by technical agreements" devised by the ISO, according to the ISO.

"From an environmental perspective, the ISO isn't ideal because it's captured by industry," says trade lawyer Stephen Porter of the Washington, D.C. Center for International and Environmental law. Companies send their expert reps to national standards organizations, that in turn send reps to the ISO.

That might not be a problem if the ISO stuck to screws, but in the 1990s it expanded its scope to setting environmental standards, including the process used for producing organic agricultural products.

"The part that's most troublesome is when an ISO standard becomes a default standard under the WTO rules," says Porter. "Does it become impossible to go beyond that in a practical matter if Austria wants to set an environmental standard that is 130% of the ISO standard?" And once ISO standards become part of the WTO, what was a voluntary system receives the force of law, without public involvement.

—Abby Scher

THE INTERNATIONAL LABOR ORGANIZATION (ILO)

Every year it is becoming more obvious that the global economy needs global regulation to protect the interests of workers and their communities. This was a central demand of some WTO protesters in Seattle. But who can regulate at a global level, and how can this regulation be made democratically accountable? There are no easy answers to these questions, but we can learn a lot by studying the successes and shortcomings of the International Labor Organization.

Where Did It Come From?

The ILO was established in 1919 in the wake of World War I, the Bolshevik revolution in Russia, and the founding of the Third (Communist) International, a world federation of revolutionary socialist political parties. Idealistic motives mingled with the goal of business and political elites to offer workers an alternative to revolution, and the result was an international treaty organization (established by agreement between governments) whose main job was to promulgate codes of practice in work and employment.

After World War II the ILO was grafted onto the UN structure, and it now serves a wide range of purposes: drafting conventions on labor standards (182 so far), monitoring their implementation, publishing analyses of labor conditions around the world, and providing technical assistance to national governments.

Why Should You Care?

The ILO's conventions set high standards in such areas as health and safety, freedom to organize unions, social insurance, and ending abuses like workplace discrimination and child labor. It convenes panels to investigate whether countries are upholding their legal commitment to enforce these standards, and by general agreement their reports are accurate and fair. ILO publications, like its flagship journal, *The International Labour Review,* its World Labor and Employment Reports, and its special studies, are of very high quality. Its staff, which is head-quartered in Geneva and numbers 1,900, has many talented and idealistic members. The ILO's technical assistance program is minuscule in comparison to the need, but it has changed the lives of many workers. (You can find out more about the ILO at its website: www.ilo.org.)

As a rule, international organizations are reflections of the policies of their member governments, particularly the ones with the most clout, such as the United States. Since governments are almost always biased toward business and against labor, we shouldn't expect to see much pro-labor activism in official circles. The ILO provides a partial exception to this rule, and it is worth considering why. There are probably four main reasons:

- The ILO's mission explicitly calls for improvements in the conditions of work, and the organization attracts people who believe in this cause. Compare this to the mission of the IMF (to promote the ability of countries to repay their international debts) or the WTO (to expand trade), for instance.

- Governments send their labor ministers (in the U.S., the Secretary of Labor) to represent them at the ILO. Labor ministers usually specialize in social protection issues and often serve as liaisons to labor unions. A roomful of labor ministers will generally be more progressive than a similar gaggle of finance (IMF) or trade (WTO) ministers.

- The ILO's governing body is based on tripartite principles: representatives from unions, employers, and government all have a seat at the table. By institutionalizing a role for nongovernmental organizations, the ILO achieves a greater degree of openness and accountability.

- Cynics would add that the ILO can afford to be progressive because it is largely powerless. It has no enforcement mechanism for its conventions, and some of the countries that are quickest to ratify have the worst records of living up to them.

On Balance?

The ILO has significant shortcomings as an organization. Perhaps the most important is its cumbersome, bureaucratic nature: it can take forever for the apparatus to make a decision and carry it out. (Of course, that beats the IMF's approach: decisive, reactionary, and authoritarian.) The experience of the ILO tells us that creating a force capable of governing the global economy will be extremely difficult, and that there are hard tradeoffs between democracy, power, and administrative effectiveness. But it also demonstrates that reforming international organizations—changing their missions and governance systems—is worth the effort, especially if it brings nongovernmental activists into the picture.

—Peter Dorman

Resources: Arthur MacEwan, "Markets Unbound: The Heavy Price of Globalization," *Real World International* (*Dollars and Sense*, 1999); David Mermelstein, ed., *The Economic Crisis Reader* (Vintage, 1975); Susan George and Fabrizio Sabelli, *Faith and Credit: The World Bank's Secular Empire* (Penguin Books, 1994); Hans-Albrecht Schraepler, *Directory of International Economic Organizations* (Georgetown University Press, 1997); Jayati Ghosh, Lectures on the history of the world economy, Tufts University, 1995; S.W. Black, "International Monetary Institutions," *The New Palgrave: A Dictionary of Economics*, John Eatwell, Murray Milgate, and Peter Newman, eds. (The Macmillan Press Limited, 1987).

11

Maytag Moves to Mexico

DAVID MOBERG

Galesburg, Illinois—Many Americans dream of getting rich. Aaron Kemp had more modest ambitions. "I wanted to work at a decent job and earn a decent wage, with decent benefits, so I can raise my kids, give them a decent education and maybe take them out to Pizza Hut on a Friday night. I don't need a Mercedes, just a ho-hum existence, and now," he says, with sadness and anger in his voice, "it seems hard to even do that."

Eight years ago, Kemp began working at the factory of Maytag Corporation, the largest employer in Galesburg, a western Illinois town of 34,000 and the birthplace of poet Carl Sandburg. In September, Maytag finally closed the plant, after sending a large part of the work that 1,600 people had recently been performing to a new Maytag factory in Reynosa, Mexico; another large part to Daewoo, a Korean multinational subcontractor that is expected to build a plant in Mexico; and a few dozen jobs to a plant in Iowa. Now Kemp, a 31-year-old union safety and education official with a muscular build and a small goatee, has a temporary job as a counselor to laid-off workers at two-thirds his old pay.

The local Machinists union fought the shutdown, taking their case to the streets, to the press, to politicians and to Maytag shareholders, even winning national attention when Senator-elect Barack Obama mentioned their cause in his Democratic convention keynote speech. But the union could not stop the Maytag jobs from being added to the tally of 2.7 million manufacturing jobs lost since 2000. Those several million jobs were eliminated for many reasons—including declining demand, rising efficiency and increased imports—but a significant portion are the result of U.S. multinational corporations, like Maytag, moving production out of the country.

Although the U.S Bureau of Labor Statistics concluded that during the first three months of this year only 4,633 workers lost jobs because of investment shifts overseas, a study for the U.S-China Economic and Security Review Commission by Kate Bronfenbrenner of Cornell University and Stephanie Luce of the University of Massachusetts found that at least more than five times that number of jobs were lost in the same period. They also estimate that in 2004 more than 400,000 jobs will be shifted from the United States to other

SOURCE: David Moberg, "Maytag Moves to Mexico," from *In These Times* (January 17, 2005), pp. 22–23, 28. Reprinted with permission from *In These Times*.

countries. That's nearly twice the rate in 2001, and it represents about one-fourth of all mass layoffs in 2004.

Despite the trend toward outsourcing white-collar jobs, Bronfenbrenner and Luce found that more than four-fifths of job shifts were still in manufacturing industries and more than one-third of the estimated 400,000 jobs shifted went to Mexico. But China is in second place, and rapidly rising in popularity. They also found that companies disproportionately target unionized jobs, which represent 39 percent of all jobs shifted out of the United States but only 8.2 percent of the private workforce. The Midwest has been hardest hit, most of all Illinois, which in the first three months of 2004 lost at least 7,555 jobs—almost all to Mexico.

LOCAL LOSSES CUT DEEP

The loss of 1,600 jobs with the Maytag closing is hard on Galesburg, where 5 percent of the town's workforce lost jobs, as well as the small surrounding towns. But the ripple effects—from lost jobs at nearby suppliers (including a workshop for the disabled that employed 100 people working on Maytag subassemblies) to indirect effects of declining consumption and reduced tax revenues—will raise the total job loss in the region to roughly 4,166, according to a Western Illinois University study.

That's only a part of the region's woes. In January, the new Australian owners of Butler Manufacturing, which makes steel buildings, will close their Galesburg plant—dumping both 270 manufacturing employees and the only unionized Butler facility. In the past few years, other area factories have closed or greatly cut back on their workforce, including a rubber hose manufacturer, a ceramics manufacturer, and several small industrial parts and equipment makers.

Some, but not all, of these other job losses involve shifts out of the country. They become part of the national problem posed by the growing trade deficit that may approach a record $600 billion this year. As more governments and financial market players have perceived this deficit—and the federal budget deficit—as unsustainable, the value of the dollar has fallen. The deficit increase partly reflects rising oil prices and a growing trade imbalance with China, whose currency, the yuan, is pegged to the dollar and, according to critics, undervalued. But the deficit is also a result of the shift in jobs manufacturing tradable goods.

A declining dollar should reduce this trade deficit. But changes in the American economy may blunt its effect. With the decline in its manufacturing base, the United States has fewer producers of tradable goods for export and relies more on imports for essential goods, even if their price in dollars rises sharply. The United States even runs deficits in agricultural commodities and advanced technology, while the small trade surplus in services has been shrinking. The surge in offshoring of white-collar work undercuts the traditional expectation that the United States would simply shift to theoretically higher skilled jobs as it lost manufacturing.

The attention focused on offshoring call-center or software jobs has reinforced the assumption, at least in elite political circles, that manufacturing is a lost cause, especially if the product can be made in China.

MAYTAG WORKERS ARGUE
FOR QUALITY, MORALITY

But Maytag workers had a strategy for saving their jobs. David Bevard, the articulate and thoughtful local union president, wanted Maytag to continue to position itself as a high-quality, premium-priced, Made-in-America classic; he argued that the company was damaging itself by undermining workers at the Galesburg plant who wanted to maintain high standards of quality and by accepting "junk" from offshore suppliers. Union members also wanted their protests to make other employers think twice about shifting jobs overseas. And they saw themselves in a global battle for justice.

Workers losing their $15 an hour jobs in Galesburg have a surprising empathy for the Mexican maquiladora workers who would be doing the same work for roughly one-sixth the wage. "The only people being done more a disservice than the people in Galesburg are the people who are going to have our jobs," Kemp says, sitting around the union hall before the shutdown occurred. "They're the only ones more exploited. It shouldn't be American workers against Chinese or Mexican workers, but working people against greed."

"We represent 1,600 in the Galesburg plant, but as a union representative, I feel I'm representing all workers everywhere and try to speak for all those workers," union vice-president Doug Dennison says. "This is so much bigger than a union issue. It's almost accepted what's happening in Galesburg is OK, that it's OK to do that."

"It's exploitation of the many for the benefit of the few," says Kemp. "Sometimes there's a fine line between what's legal and what's right."

"Morality," Dennison adds. They clearly think that is missing, as well as their power to do much about their situation. While most workers blamed "corporate greed" for the plant closing, they also blamed the government for enabling or encouraging that greed. And among an otherwise strongly Democratic crowd, people remember that it was Bill Clinton who pushed through NAFTA. "People in both parties are allowing this to happen," Toby Ladendorf laments on closing day. "Who's going to defend us?"

CONCESSIONS CAN'T COMPETE
WITH BOTTOM LINE

Over the decades, Galesburg workers had grown accustomed both to the security of the Maytag jobs and to intimations of insecurity, especially as the industry consolidated into a handful of domestic appliance makers. When Maytag bought the plant in 1986, workers were encouraged by its reputation for quality. But by 1992, as a precondition to making an investment of $180 million, Maytag was demanding concessions from the union and public assistance to keep the plant open, including $7.5 million in state grants and loans, a $3 million city grant paid through increased sales taxes, and local tax abatements through 2004 worth about $4 million. (After the closing, the state passed new legislation to make expected public benefits of such aid clear and to recover money if the goals are not met. And the Knox County state's attorney is trying to recover excess tax abatements.)

The union tried to cooperate to increase productivity, says Bevard, but management was only interested in cutting jobs. Union business agent Mike Patrick suggested that management adopt the "high performance work organization" model that worked well at companies like Harley-Davidson, giving workers responsibility and authority to use their knowledge at work. "Maytag had no intention of giving employees any control," Patrick says. "They wanted to stay with the command and control model." Indeed, Maytag tried to tighten control further and force more concessions, provoking workers to the brink of a strike in 2002.

Then on October 12, 2002, Maytag announced that the plant would close beginning in 2003. Managers told the union that the plant was "not competitively viable.'"

Maytag was profitable, but revenue and profits have been stagnant or declining and the company's stock price has dropped. Big box retailers like Home Depot were taking a larger share of the market and demanding lower prices from manufacturers. Also, other refrigerator makers had begun producing in Mexico, and Maytag already had subassembly operations in Reynosa. About three hours of direct labor are needed to build the cheaper refrigerators, and with cheaper Mexican labor that can make a difference of $50 on a $350 refrigerator, not counting the savings accrued from lower social and environmental regulations. Maytag will save money eventually, but there was speculation in Galesburg that Maytag was simply following the crowd offshore or trying to please Wall Street to boost its stock price.

GALESBURG STRUGGLES TO RETOOL

In October, the unemployment rate in Galesburg was 9.1 percent. Knox County is on the state's youth poverty warning list. Galesburg recovered from major workplace closings in the 1980s partly through expansion of factories like Maytag, as well as accepting a state prison that residents previously opposed.

Now, to survive, laid-off workers must retrain as welders, nurses, office managers and computer technicians. But even in these growing occupations, there are far more trainees than available local jobs. Many look to long commutes or relocations in order to find jobs, or they prepare to compete with their kids for $7 to $8 Wal-Mart jobs. Meanwhile, economic development officials try to attract investment but rarely mention manufacturing, except to convert the region's abundant corn and soybeans into marketable products. The town has a new logistics park, entrepreneurial centers, and business incubators, and there's some talk about Galesburg becoming an education laboratory, a tourist center or an "agurb" retirement center for upscale refugees from cities like Chicago, a three-and-a-half-hour drive away.

The town is playing up its historic—and rebounding—strength as a railroad center and its interstate highway connections in the search for warehouses and distribution facilities. Last summer a delegation went to China, looking for investors and Chinese companies seeking distribution centers for the kinds of goods once manufactured in towns like Galesburg. It was a sign, local citizens thought, of how globalized the town was becoming.

"Globalization is such a fraud," says Bevard. "It's just a rush to the bottom for cheap labor. Instead of reducing the United States to the Third World, we should be elevating the standards of those countries." Then, perhaps, the Aaron Kemps of this country could hope once again for a ho-hum but decent life for themselves and their kids.

12

The Sweat Behind the Shirt

The Labor History of a Gap Sweatshirt

JESSE GORDON
KNICKERBOCKER DESIGNS

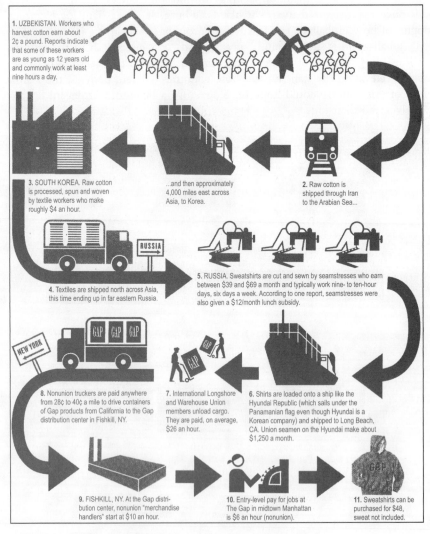

1. UZBEKISTAN. Workers who harvest cotton earn about 2¢ a pound. Reports indicate that some of these workers are as young as 12 years old and commonly work at least nine hours a day.

3. SOUTH KOREA. Raw cotton is processed, spun and woven by textile workers who make roughly $4 an hour.

...and then approximately 4,000 miles east across Asia, to Korea.

2. Raw cotton is shipped through Iran to the Arabian Sea...

4. Textiles are shipped north across Asia, this time ending up in far eastern Russia.

5. RUSSIA. Sweatshirts are cut and sewn by seamstresses who earn between $39 and $69 a month and typically work nine- to ten-hour days, six days a week. According to one report, seamstresses were also given a $12/month lunch subsidy.

8. Nonunion truckers are paid anywhere from 28¢ to 40¢ a mile to drive containers of Gap products from California to the Gap distribution center in Fishkill, NY.

7. International Longshore and Warehouse Union members unload cargo. They are paid, on average, $26 an hour.

6. Shirts are loaded onto a ship like the Hyundai Republic (which sails under the Panamanian flag even though Hyundai is a Korean company) and shipped to Long Beach, CA. Union seamen on the Hyundai make about $1,250 a month.

9. FISHKILL, NY. At the Gap distribution center, nonunion "merchandise handlers" start at $10 an hour.

10. Entry-level pay for jobs at The Gap in midtown Manhattan is $6 an hour (nonunion).

11. Sweatshirts can be purchased for $48, sweat not included.

SOURCE: Jesse Gordon and Knickerbocker Designs, "The Sweat Behind the Shirt." Reprinted with permission from the March 10, 2001 issue of *The Nation*. For subscription information, call 1–800–333–8536. Portions of each week's *The Nation* magazine can be accessed at http://www.thenation.com.

13

Double Jeopardy

Gender and Migration in Electronics Manufacturing

ANIBEL FERUS-COMELO

On November 19, 1998, more than 6,300 workers from fourteen production units of a US$1 billion Indian multinational company went on an indefinite strike in Bangalore, the "Silicon Valley of India."[1] About 80 percent of these striking workers were women between the ages of 18 and 25. This was the bitterest high-profile labor dispute that Bangalore had experienced in decades. At the heart of the dispute was an apprenticeship program that the company had abused to retain shop-floor workers in tenuous employment on subminimum wages for many years. Besides a lack of job security, the women had to stand all day with no fans or chairs for relief from the heat and fatigue. They had no breaks, needed permission to use the toilet, and were forced to work overtime without extra pay.

Both the conditions that these workers faced and their courage to change them are part of a prevailing reality at the bottom of the labor market in the global electronics industry. Many production workers in key industrial clusters throughout the world are women and either internal or transnational migrants. Yet, with the exception of a brief burst of studies during the 1980s, the electronics industry has been conspicuously absent in discussions related to the politics of race, gender, migration, and employment. In contrast, there has been a steady growth in efforts to analyze the participation of migrant women and the nature of employment in the global garments industry. This disparity is all the more remarkable given the striking similarities between the two industries—their growth as a pillar of national development in many countries; the staggering power, wealth, and influence of the leading multinationals; the pyramid-shaped corporate structure; the use of subcontracting; and the intolerable exploitation of workers. The antisweatshop movement that has taken root in North America and Western Europe now presents an opportunity to draw attention to the proletarians of the "IT [information technology] revolution" and to demand corporate responsibility on a global scale.

Note: References for this reading are not included here. Interested readers should consult the original.
SOURCE: Anibel Ferus-Comelo, "Double Jeopardy: Gender and Migration in Electronics Manufacturing." In *Challenging the Chip*, Ted Smith, David A. Sonnenfeld, and David Naguib (eds.). Philadelphia: Temple University Press, 2006, pp. 42–54.

This chapter has three parts. The first explores reasons for the apparent preferential employment of young, migrant women in electronics manufacturing, arguing that this is mainly due to the corporate search for ever-greater flexibility and productivity. The second section shows that this emphasis on flexible production, which has proven to be an effective corporate strategy, in practice frequently violates workers' fundamental rights. The third section presents some examples of the response from workers and labor organizations. This paper draws from reports of management practices and working conditions in the electronics industry throughout the world supplemented by primary research in two important nodes of the electronics industry—Silicon Valley, California; and Bangalore, India.

FEMINIZATION AND MIGRATION
FOR FLEXIBLE PRODUCTION

Between 1985 and 2000, women represented more than half of the electronics production workforce in Hong Kong, Macau, Singapore, Taiwan, the Czech Republic, Malaysia, Indonesia, Puerto Rico, Slovenia, Cuba, the Philippines, Thailand, and Sri Lanka. The latest figures available show that the proportion of female employment in this industry was highest in Hong Kong (79 percent), Macau (69 percent), and Malaysia (68 percent; see Table 4.1).

Women's employment in the electronics industry broadly reflects the geographical specialization in industrial segments. A survey of forty-four countries between 1990 and 2000 showed that women formed the majority of the work-

T A B L E 4.1   Percentage of Female Employment in the Electronics Industry in Select Countries

Country	1985	1990	1995	1999
Hungary	47.4	45.0	n/a	n/a
Japan	45.6	n/a	39.1	37.0
Korea	50.9	49.1	45.6	41.2
Macedonia	30.0	27.3	22.2	n/a
Malaysia	73.7	75.4	66.5	67.8
Manaus, Brazil	n/a	54.3	49.5	42.3
Portugal	46.7	33.7	45.5	n/a
Puerto Rico	n/a	61.0	53.4	52.8
Taiwan	58.8	54.5	52.7	n/a
USA	41.9	42.7	41.8	n/a

SOURCE: Elaborated from the United Nations Development Organization (UNIDO 2003); Departamento Intersindical de Estatistica e Estudos Socio-Economicos (DIEESE 2001), for Brazil; and ILO (1998), for Macedonia.

force in a variety of countries: in Slovakia, Bulgaria, and the Czech Republic in office and computer equipment manufacturing; in Bulgaria, the Czech Republic, Latvia, Mexico, Portugal, and Slovakia in radio, television, and telecommunications equipment production; in Cyprus, the Czech Republic, Portugal, Slovakia, and Slovenia in electronic components production; and in Argentina, Bermuda, the Czech Republic, Ireland, the Philippines, Portugal, and Slovenia in medical, precision, and optical equipment manufacturing (ILO 2002).

Since the 1960s, the global labor market in electronics manufacturing also has drawn workers across national borders. Immigrant workers in electronics production tend to be women, and there is a growing trend toward the "feminization of migration," as women migrate from such countries as the Philippines, Bangladesh, and Sri Lanka to high-tech hot spots in Singapore, Hong Kong, Malaysia, and Taiwan (David 1996, 39). In some countries, electronics manufacturing draws internally displaced workers. The industry predominates in Export Processing Zones (EPZs), Free Trade Zones (FTZs), and other specially designated industrial zones, which have mushroomed throughout the world as a result of export-oriented development policies. Nearly all of the workers in these zones, such as the Batamindo Industrial estate in Indonesia, are migrant women from other, poorer regions of the country (David 1996; Holdcroft 2003; Kelly 2002; Rosa 1994).

Some commentators have suggested that women workers are typically young and single (Fuentes and Ehrenreich 1983). However, other studies have shown some geographical variation in the age, marital status, and level of formal education of the women employed in electronics manufacturing, especially in comparison to garment workers (Chhachhi 1997a, 1999; Fernández-Kelly 1983; Goldstein 1989; Rosa 1994). Generally, women and migrants work in machine operation, assembly, packaging, and quality control in a variety of workplaces, from the sprawling factories that employ thousands of production workers, as in Mexico, Thailand, and China (Leong and Pandita 2006) to units of 10 to 15 employees, or even in their own homes, as uncovered in Silicon Valley, California. As the following sections show, electronics firms achieve flexible and efficient production by employing women and migrants.

Labor Control and Productivity

Several studies on electronics employers' recruitment practices have attributed an apparent preference for women workers, particularly in Asia, to gendered cultural and biological stereotypes. The "nimble fingers" thesis argues that managers perceive women as more suitable than men for assembly work because they are seen to have "natural" traits, such as manual dexterity ("fast-fingered women"), patience, and a tendency to be meticulous (Elson and Pearson 1980; Lim 1978). These allegedly innate feminine characteristics are required to manipulate intricate wires and repeat the same finite number of tasks all day. Managers tend to undervalue the mental concentration and "aptitude and alertness" that an assembler in fact applies to ensure high product quality.[2] Chhachhi's (1999) study of the electronics industry in Delhi, India, found that women's jobs are often downgraded as "unskilled" and casualized, regardless of the actual skills the job entails (see also Standing 1999). Such an appraisal of who is capable of boring, monotonous, manual work also has racial dimensions in Silicon Valley (Hossfeld 1990).

Youth implies that workers have little or no experience with waged employment; thus they have more modest expectations than do older women workers, especially those who were formerly employed in unionized workplaces (Chhachhi 1999; Rosa 1994). Companies also deliberately recruit young women because they are in better physical health than older women, with sharper eyesight and quicker reflexes (Elson and Pearson 1989; Hossfeld 1991a). Women in Thailand's EPZs typically work for a couple of years before returning to their places of origin to set up their own small businesses or to start families (Theobald 1996). In some segments of the industry that require a trained, skilled workforce and in places where there is a relative shortage of qualified workers, retaining workers is a high priority. However, in parts of such countries as China and Malaysia, where the labor market is fairly loose for jobs in which little training is involved, labor turnover is not a serious concern, because the workers' physical productivity declines over time, mainly due to occupational health hazards and injuries (see CAFOD 2004; Standing 1999). Sometimes management actively promotes a high labor turnover so that they can minimize the nonwage costs that accrue over time.

At a domestic firm in Bangalore, managers exert great pressure on workers to force them to resign voluntarily (Ferus-Comelo 2005). For example, once women workers get married, line supervisors aggravate their working conditions by assigning them compulsory overtime without extra pay, so that family members will pressure them to quit their jobs. If they become pregnant, they are assigned to the most arduous tasks. They are forced to stand all day, to work in heated areas of the shop floor with no breaks, or to load and unload heavy consignment, to make their jobs as unbearable as possible. Only a few manage to endure this treatment and are retained, whereas most are replaced by a fresh group of women.

Labor control is further extended to the workers' reproductive roles. In many countries, such as Costa Rica, Honduras, Mexico, Thailand, and China, women are subjected to pregnancy tests as a routine part of the job screening process and sometimes even throughout their job tenure (CAFOD 2004; David 1996; Wong 2002). Women who are pregnant are either denied jobs or fired. If a migrant worker in Taiwan gets pregnant, not only can her contract be terminated, but also, despite a change in the law that allows pregnant workers to stay, she faces the choice of abortion or deportation (Stein 2003). To avoid losing their jobs, women in India and Thailand have frequent, unsafe abortions (CLIST 2003; Rajalakshmi 1999). By monitoring women's bodies this way, companies try to avoid any legally mandated social costs of reproduction, such as higher health care bills, maternity leave, and child care.

Numerical Flexibility

Although workforce characteristics and related social attributes consolidate labor control, the nature of electronics production and the industry's structure equally create a niche for migrant women workers in precarious employment (Hossfeld 1991a). To achieve flexibility in order to respond to fluctuating market demands, managers give priority to the unencumbered expansion and contraction of the labor force. They increasingly rely on contract labor, that is, workers who are recruited on a casual basis through temporary hiring agencies for short-term "on-call" assembly jobs. These workers, who are frequently foreign or internal migrants, are not considered direct employees of the companies for whom they work.

Such two-tiered systems are practiced in mature high-tech sites like Silicon Valley, as well as in new industrial nodes, such as Bangalore (Ferus-Comelo 2005). According to an executive manager of the largest employment agency in Silicon Valley, 10 to 35 percent of a workforce of 6,000 placed in jobs weekly where "you might find the United Nations, at least 20 to 30 different countries represented," are in electronics firms and a majority do assembly work. Contract workers in Bangalore are hired at two levels—"skilled" and "unskilled"—regardless of their technical training and other qualifications. Casual workers are forced to take a day off each month so that they do not have continuous, six-month employment, which would legally qualify them for permanent status.

In Penang, Malaysia, overseas contract workers are recruited by agencies in Indonesia and Bangladesh to provide managers numerical flexibility during production slowdowns when their contracts are not renewed. At the same time, however, such workers are a source of labor-market stability, as their visas are conditional on employment in a specific company, which prevents them from job hopping (Kelly 2002). At MMI Precision, a factory producing electronic parts and components in Thailand, 700 of the 2,500 workers are casual employees recruited by 11 different agencies (Wong 2002). Similarly, of the 7,000 workers at an IBM plant in Guadalajara, Mexico, only 500 were hired directly by IBM or its contract manufacturer, Sanmina SCI. The remaining 6,500 were hired through different recruitment agencies (CAFOD 2004).

Differences in workers' employment status can be traced along gender lines, as women are often overrepresented among the workers in part-time casual employment (ILO 1998). A study of Nokia's employment practices at its plant in Manaus, Brazil, showed that of the 265 workers hired indirectly through labor agencies, 212 (80 percent) were women and 53 (20 percent) were men. An additional four women were designated as "trainees," thus raising the total number of women working in the plant to 750 in comparison to 606 men (SO 2002).

Insecure employment for women workers extends to home working, as home assembly is one way that electronics manufacturers can attain the speed of delivery required by their customers. Numerous reports have documented the employment of immigrant women in home-based electronics assembly in such places as New York, Madrid, and Silicon Valley (Dangler 1989; Ewell and Ha 1999a; Portes, Castells, and Benton 1989). However, Chhachhi (1999, 20) argued that, despite the few cases of home-based work in Delhi, India, there is "no trend toward home-based work" because of the pressure for quality control and the financial risk of releasing expensive imported components outside the factory and strict oversight. This was confirmed in an analysis of the supply chain strategy of electronics production implemented in Bangalore (Ferus-Comelo 2005), suggesting that there are spatial variations in the industry's reliance on home-based work.

THE HUMAN TOLL OF HIGH-TECH

Women and migrants' working conditions in the electronics industry need to be understood within the broader context of economic necessity and their limited opportunities for safe, secure, well-paying jobs (Dangler 1989; Kabeer 2000). Migrant workers tend to compare the conditions of employment in the electronics industry to former employment or to the bleak economic and political situations they have left behind in their places of origin (Beech 2003; Hossfeld 1991b). Women workers, particularly in countries with strict gender-based cultural restrictions, also rate highly the noneconomic benefits of their jobs, such as personal independence, the importance of which cannot be underestimated (Balakrishnan 2002; Ong 1987). This does not, however, mean that working conditions in the electronics industry are ideal. Due to the nature of electronics manufacturing and the organizational structure adopted by corporate management, young, migrant women face a number of labor rights violations.

The most common problem is compulsory overtime with no extra pay. A survey conducted among 200 workers in 20 electronics companies in Penang, Malaysia, found that 60 percent of workers had to do overtime, and half did at least five hours of overtime per day (Wangel 2001). In China, very long, compulsory overtime is the norm, and workers typically clock an average of 100 to 120 hours of overtime per month, with a seven-day work week (CAFOD 2004; Leong and Pandita 2006). They often must put in off-the-record, illegal overtime to achieve the minimum hourly wage. In some places, such as Malaysia and Indonesia, and in Sri Lanka's EPZs, migrant women lead highly regimented lives in hostels, which allows plant managers to control and monitor workers' availability for overtime and leaves workers no other viable alternative (Kelly 2002; Rosa 1994). Migrant workers in South Korea, who are limited to a single three-year stay, are considered trainees their first two years and are thereby exempt from most of the country's labor laws, especially those related to minimum wage and overtime (Stein 2003).

An ILO survey of wages in electronics production showed that women throughout the global industry typically earned less than men. The wage differential between men and women was greatest in Japan and Malaysia, worsening over time in Japan and improving in Malaysia. Women's earnings were less than 70 percent of men's in Thailand, Singapore, Cyprus, and Korea (ILO 1998). Lower wages often are justified by the inaccurate perception that women workers are secondary (to male) wage earners in their households and therefore can afford to work for less (Hossfeld 1990; Mitter 1986). A gender difference in wages also may be due to the occupational segregation that characterizes this industry and the higher proportion of women working part time or on a temporary basis (ILO 1998).

The management practice of outsourcing production to contract manufacturers and a wide, global network of suppliers increases the pressure on firms at the lower end of the chain to deliver cost savings (Lüthje 2006). Consequently, the subcontractors seek ways to minimize their labor costs and thereby gain or retain production orders. This form of restructuring has serious negative implications for workers at the tail end of subcontracting chains, which is where most migrant women work.

An expansion of jobs in small units or the informal economy, including home-based work, allows women the flexibility to combine employment with domestic commitments (Balakrishnan 2002; Dangler 1989). However, conditions in these second and third tiers of employment are frequently far worse than those in large companies, which come under the purview of labor legislation.[3] Small enterprises employing a hundred employees or fewer tend to elude regulatory control and are hosts to a variety of labor problems, including low wages, late or nonpayment of wages, work targets, and lack of employment security. Home-based workers, a majority of whom are women and immigrants, are paid a piece-rate that falls far below the local minimum wage. To increase payments and meet deadlines set by companies or middle-men, they often involve their underage children and family members in assembly work.[4]

Production workers in electronics manufacturing are subjected to a range of acute and chronic forms of damage to their bodies (see LaDou; Hawes and Pellow 2006; see also Fox 1991). These industrial health hazards are distinctly gendered due to the labor processes predominately undertaken by women and the high-pressure environment in which they work (Theobald 2002). Resigning from hazardous employment is not a viable option for most workers, particularly migrant women. Their marginalization in the local labor market underscores their dependence on jobs that are harmful to their health, even if they are not aware of the risks.

The health and safety of migrant workers, especially those holding an irregular status, is a major concern, as they often avoid seeking medical treatment because of prohibitive costs, an inability to take time off work, problems with transportation, or the fear of drawing attention to themselves and losing their jobs or being deported (ILO 2004). Sahabat Wanita, a national feminist workers' organization in Malaysia that supports the rights of electronics workers in EPZs, has documented cases of migrant workers who are sent home to die without any explanation or compensation for their medical conditions (Papachan n.d.). In addition to corporate negligence, linguistic barriers, poor access to health care, and a lack of familiarity with local health care systems compound migrants' work-related risks (ILO 2004).

FIGHTING AGAINST THE ODDS

Although the global electronics industry thrives on the labor of young, migrant women workers, these workers' rights are consistently violated with impunity, as the contemporary global electronics manufacturing workforce remains largely unorganized. In Bangalore, for instance, unionization is limited to the public-sector plants that were established in India's post-Independence era,[5] whereas forming a national union for electronics workers in Malaysia remains virtually prohibited (Wangel 2001). In 2000, the electronics industry had only eight enterprise-level unions, representing just 5 percent of the country's 150,000 workers—80 percent of whom are women—and leaving the rest unprotected (ICFTU 2001).

Recent union-organizing efforts in the high-tech industry have been limited to the male-dominated, professional occupations. For example, the Union Network International (UNI), a Global Union Federation, has launched a campaign to organize software programmers and financial, executive, technical, and consultancy workers in Bangalore and Hyderabad in India.[6] The Communications Workers of America (CWA) has been involved in organizing contingent workers, including long-term temporary workers (or "perma-temps") in white-collar career occupations at IBM in Alliance@IBM and WashTech, the Washington Alliance of Technology Workers.

Because organizing migrant workers is tremendously difficult due to the temporary nature of their employment and the ways in which managers use ethnicity and place of origin to pit workers against each other and against unions (see Hossfeld 1990; Kelly 2002), unions are being forced to find new ways to extend collective bargaining rights to migrants. Although some countries reserve the right to organize for nationals (ILO 2004), there are additional structural challenges to overcome. In the Guangdong Province of China, contrary to the national trend, women outnumber men among the 10 million migrant workers (estimated at 60 percent) who are drawn to the labor-intensive industries, including electronics (*China Labor Bulletin* 2004). Here, the Guangdong Federation of Trade Unions (GFTU) rejected the Ruian Migrants' Management Association (RMMA), which was established in 2002 to support the rights of migrants, particularly temporary workers, in Ruian City, and, considering it a rival workers' organization, declared it illegal (*China Labor Bulletin* 2002).

Many unions, both nationally and internationally, support women's participation and have clear policies to make organizing women workers a high priority. However, it is unclear to what extent these unions have been able to move beyond rhetoric to accommodate women workers' specific employment conditions and issues in their action agendas. The Iron and Steel Trades Confederation (ISTC) in Scotland and the General Confederation of Italian Labor (CGIL) in Italy are examples of positive union responses at different levels to the changing structure of employment and unfair managerial practices.[7] Recognizing the increases in home working and self-employment, the CGIL set up a new union organization to organize and represent these workers, whereas the ISTC doubled its membership in two years and has the highest female membership of any ISTC branch due to its active organizing agenda (May 1999; Rigby 1999).

Reports from various parts of the world, such as Japan, Italy, Thailand, and Taiwan, suggest that, once organized, women workers' interests are not given high enough priority within the male-dominated culture and leadership of union structures (see Hossfeld 1991b; WWW 1991, for examples). Without women's participation in critical decision-making processes and in leadership positions, unions risk a male bias in organizing priorities and strategies. As long as they are seen as "special interests" or as being divisive, migrant women workers' issues will remain unaddressed by labor unions.

Forms of Resistance

Contrary to the stereotypes of women workers as meek and docile, they continue to struggle for justice in the workplace, often in the face of tremendous corporate hostility (Fuentes and Ehrenreich 1983; Mitter 1986). As a result, they are frequently fired, demoted, assigned to night shifts, or transferred to factories in other locations that have worse working conditions (CAFOD 2004; Ferus-Comelo 2005; WWW 1991).

Women workers also have been physically attacked for their trade union activities.[8] In the walkout described at the beginning of this chapter, management harassment of workers started immediately after the first open union meeting. Managers were able to mobilize the state machinery against workers. On the second day of the strike, the police beat the workers so viciously that some could hardly move; one woman miscarried. Throughout the strike, hundreds were arrested on trumped-up charges. A union leader elected during the strike was threatened about having acid thrown on her face and having her legs broken if she testified on behalf of the injured workers in hospital. She was illegally fired immediately after her election. Police harassed other women activists at home during the night, causing them a great deal of anguish due to the social stigma they suffered as a result.

Despite such virulent opposition, workers deploy a range of strategies to challenge and subvert management. Some women are finding ways to deal with unfair management practices without directly confronting their bosses or risking their jobs. For example, coworkers of a six-month-pregnant woman in Mumbai's Santa Cruz Electronics Export Processing Zone (SEEPZ) surreptitiously rotated places with her because she was forced to stand for hours at a stretch without a break and was refused light work by the manager (Cecilia 1992). Similarly, Asian women workers in Silicon Valley circumvented the manager's divide-and-rule tactics when a Mexican coworker was assigned to their workstation to "learn speed" by deliberately slowing down their work and setting a lower average quota than usual (Hossfeld 1990, 174). In general, workers draw inspirational tactics from their cultures to respond to hegemonic relations of accumulation and domination in the workplace (Ong 1987; Sargeson 2001).

Women also have waged militant campaigns using hunger strikes and slowdowns of production as tools. For example, the striking workers in Bangalore lay down in front of the company gates to prevent scab workers from being bussed into work. Small groups of four or five women took turns on a hunger strike to protest against the false arrests and police brutality (Ferus-Comelo 2005). Similarly, women workers in Silicon Valley, who were predominately immigrants from Mexico, held a hunger strike to protest against poverty wages, a lack of benefits, and unsafe working conditions at Versatronex (Bacon 1993). Hundreds of 20- to 23-year-old women workers in a South Korean electronics firm occupied the factory to prevent relocation when the company threatened to move (Hamilton 1991, 31–32).

Recourse to the law is another way workers seek justice. About 890 migrant workers organized and filed a grievance against Seagate Technology in Shenzhen, China, for cheating them of their pensions worth thousands of dollars each (Schoenberger 2002). However, a few landmark cases demonstrate the bitter defeats that many workers face. For instance, production worker Mayuree Taeviya made history in 1994 when she accused the Japanese-owned Electro-Ceramics Ltd. in Thailand of poisoning its employees. After a four-year legal campaign, which resulted in her being socially ostracized, almost cost her husband his job and sapped her not only of all her savings but also of her will to live, Mayuree lost her case (H. M. Chiu 2003; Foran and Sonnenfeld 2006).

Organizational Support

A number of organizations have emerged to support women workers who do not have access to union protection. Women's associations, church groups, and university-based labor advocates attempt to bridge the divide between the workers and the general public and to provide workers with much-needed moral support (WWW 1991). Through various service-based projects, such as legal advice forums, these organizations strive to build long-term relationships with workers and foster a sense of trust and collectivity that may result in further action. For example, women members of the Malaysian Trade Union Congress (MTUC) run two hostels for women who work in the EPZs in the suburbs of Kuala Lumpur (David 1996; Rosa 1994). Besides providing workers with safe, decent, and affordable housing, these hostels create a space within which women can discuss their problems and raise their awareness of the benefits of unions. Not only has this initiative been replicated elsewhere, but also a small group of women who lived in these hostels have helped create a new union in the electronics industry.

Similarly, the work of community-based organizations in different parts of the world has been effective in the absence of local union representation. The Center of Reflection and Action in Labor Issues (CEREAL) is involved in popular education among electronics workers in Guadalajara, Mexico (CAFOD 2004). The Hong Kong–based Christian Industrial Committee (CIC), an independent, church-supported workers' support center, has collaborated with the Electronics Industry Employees General Union to organize workers at Digital Equipment Corporation (DEC), who were mostly young women with children.

These workers succeeded in convincing the company to provide either transportation or severance pay to workers when it relocated its assembly plant to a different site (Wilson 1986). Workers in Philips Hong Kong also have managed to make significant improvements in their workplace environment with the assistance of the Neighborhood and Workers' Service Center (WWW 1991, 73–75). In Thailand, the Centre for the Advancement of Lanna Women (CALW)[9] at Chiang Mai University organized study groups and seminars to encourage electronics workers to discuss their concerns and strategize about how to tackle them (Theobald 1996). CALW also played an important role as a workers' advocate by circulating a petition to demand that the government reopen an investigation into the deaths of electronics workers in the Northern Regional Industrial Estate (NRIE) in Lamphun in 1993 (Foran and Sonnenfeld 2006).

CONCLUSION

This chapter provides some explanations for the prevalence of women, especially foreign and internal migrant workers, in global hotspots for electronics production. It is the industry's drive toward flexible production that makes the control and replacement of workers, as well as the expansion and contraction of the workforce, high managerial priorities. Although migrant workers' prospects and terms of employment may be better than those in their places of origin, they often face conditions far inferior to those available to local nationals. Electronics manufacturing not only provides women with jobs that may pay relatively higher wages than some other occupations but also fosters economic insecurity. Migrant women workers typically experience wage discrimination, job casualization, and considerable health problems. Despite existing international labor standards to protect them, their rights as workers are too often undermined. Combined with their weak labor market position, the limited collective bargaining strength of electronics production workers worldwide translates into a double jeopardy for migrant women workers.

Examples of labor action illustrate workers' belief in collective action and their ability to unite against unjust managerial practices. Although women and migrant workers fight hard for more control over their work lives, it would be wrong to romanticize their efforts for workplace improvements. Despite their courage, determination, and solidarity, production workers' actions often end in defeat. Under the threat of production relocation and the termination of production contracts, workers have been forced to either compromise or surrender. In most cases, the fight for justice is a long, drawn-out process that demands tenacity and endurance, but holds no promises (Ku 2006; McCourt 2006). Organizing workers, especially women and migrants, in electronics manufacturing is a critical task whose time is long overdue.

ENDNOTES

I am grateful for comments and encouragement from Matt Griffith, Sean Gogarty, Paul Bailey, Jenny Holdcroft, Molly Kenyon, Dr. Jane Wills, the participants of the Wageningen editorial workshop, the editors and three anonymous reviewers; and for the financial support of the Department of Geography, Queen Mary, University of London; *Antipode: A Radical Journal of Geography*; the Developing Areas Research Group of The Royal Geographical Society; and the University of London Central Research Fund.

1. Information about this strike was compiled from interviews with strikers and union officers (see Ferus-Comelo 2005).

2. *Financial Times*, February 5, 1985, p. 22, cited in Elson and Pearson (1989), p. 2.

3. Working in large factories that are covered by labor legislation does not necessarily protect workers' rights, as working conditions are most often determined by managerial practice and not by state regulation.

4. See Ewell and Ha (1999a, 1999b) regarding an investigation into home-based work in Silicon Valley.

5. See Ferus-Comelo (2005) for a fuller account of the state of the local labor movement.

6. UNI refers to the sector as Industry, Business and Informational Technology Services Sector. See www.union-network.org/unisite/Sectors/IBITS/IBITS.html, for a current report. I am grateful to Andrew Bibby for this information.

7. The *Confederazione Generale Italiana del Lavoro (CGIL)* is a national federation of unions in Italy.

8. See Hamilton (1991) for an account of the violent resistance to unionization that women workers faced at an electronics assembly factory in South Korea.

9. *Lanna* means "Northern Thai."

14

Globalization and the Struggle for Immigrant Rights in the United States

WILLIAM ROBINSON

It is an honor and a privilege to be here with you today, with the leaders and organizers of one of the most vital, just and cutting edge struggles of our time. I am very grateful to Javier Rodriguez and the other conveners of this conference for inviting me to participate. I want to start by highlighting three things that are unprecedented and interconnected, three current "upsurges," and I am not referring to Bush's escalation in Iraq.

The first is an upsurge in Latino immigration to the United States. Officially, there are 34 million immigrants in the U.S., 12 to 15 million of them undocumented, although we know that these are underestimates. Migration levels in recent years have surpassed those of the turn of the 19th century. Of these 34 million, 18 to 20 million are from Latin America, the majority from Mexico, but also from Central America, the Dominican Republic, Peru, Ecuador, Colombia, Brazil, Argentina, and elsewhere.

The second is the upsurge of repression, racism, and discrimination against immigrants—the minutemen, the denial of drivers licenses, attacks, evictions, escalating raids, public segregation and anti-immigrant jim crow, and so on. We are witnessing the criminalization of immigrants and the militarization of their control by the state.

Third is an unprecedented mass immigrant rights movement. We saw last Spring the largest demonstrations in U.S. history. They had the powers that be quite frightened. This is what *poder popular* looks like; what "power of the people" means.

What is the larger context and backdrop of anti-immigrant politics and immigrant struggle? In an attempt to answer this question I would like to put forward 10 points for analysis and discussion.

1. These upsurges are situated in an age of globalization; this new system of global capitalism we face.

    This system entered a new stage that began in late 1970s and 1980s, a transnational phase. The capitalist system began dramatic expansion

SOURCE: William Robinson, "Globalization and the Struggle for Immigrant Rights in the United States." Keynote Presentation for "El Gran Paro Americano II" Immigrant Rights Conference, Feb 3–4 2007, Los Angeles.

worldwide, and that is the structural underpinnings of the immigrant issue and ultimately those are the structures we need to address. Capital has become truly transnational, with newfound global mobility, and new powers to reorganize the whole world. In this new phase, the system depends on new methods of control over workers worldwide, and relies much more heavily on migrant workers who can be denied their rights and superexploited. This denial of rights and superexploitation requires, in turn, new levels of control and repression.

Since the 1980s, global capital has been waging a worldwide offensive against global labor with the objective of capturing natural resources, markets, and labor pools around the world. There is a new global production and financial system into which every country is being integrated. There are new levels of global social and economic integration and new webs of interdependence. A few thousand of the most powerful transnational corporations dominate the process.

Capitalist globalization constitutes in the practice a war of global rich against global poor. There has been a massive transfer of wealth from the poor to the rich. In the United States, relative wages have declined steadily since 1973 and we are witness to unprecedented global inequalities. Currently just a little more than 10 percent of the world's population consumes 85 percent of the world's wealth, while the rest have to make do with just 15 percent of that wealth—wealth which the world's workers generate but do not receive. This is the new global social apartheid. Hundreds of millions of people worldwide have been displaced—turned into workers for global economy and thrown into a new global labor market that global elites and transnational capital have been able to shape for their own purposes.

2. In Latin America every country has been violently integrated into global capitalism through free trade agreements, privatizations, deregulation and neo-liberal social and economic policies.

In Mexico, this process began under De la Madrid in 1982 and accelerated under Salinas de Gortari starting in 1988. But it really took off when NAFTA went into effect in 1994. Under Calderon the process will continue and even deepen. The PAN is now the party of global capital in Mexico. But this process has taken place all over Latin America. Throughout the continent millions of campesinos have been displaced, indigenous communities have been broken up, whole countries have become deindustrialized, millions of public sector workers fired, small businesses forced out by the onslaught of transnational corporations, such as Wal-Mart in Mexico (there are over 700 Wal-Marts now in Mexico and they are the country's biggest retailer and employer), systematic and ongoing austerity, the dismantling of social welfare programs, and so on. Hundreds of millions have been thrown into poverty, unemployment and dispossession. As a result, political and military conflict has spread.

These policies of neo-liberalism, free trade and capitalist globalization have been imposed by global elites and their local allies, and especially by the U.S. government, against the wishes of the great majority. They have created social disaster of unprecedented magnitude in Latin America; generated a crisis of survival for hundreds of millions.

This is the backdrop for transnational migration.

In more academic terms, we can say that the transnational circulation of capital and the disruption and deprivation it causes, in turn, generates the transnational circulation of labor. In other words, global capitalism creates immigrant workers. The wave of outmigration from socially and economically devastated communities in Latin America began in the 1980s and accelerated in the 1990s and in the new century, coinciding with globalization and neo-liberalism. In a sense, this must be seen as a coerced or forced migration, since global capitalism exerts a structural violence over whole populations and makes it impossible for them to survive in their homeland.

Yet, while transnational capital is free to move about the world, to reshape the world in its interests—transnational labor is subject to ever tighter and more repressive controls. 9/11 gave the Bush regime the pretext to step up the war against immigrant rights, side by side with its war in Iraq, and to militarize society and create the beginnings of a police state that is wielded against immigrants. The war in Iraq reflects the war on immigrants.

We should acknowledge that borders are not in our interest. Borders are instruments of dominant groups, of powerful economic groups, of capital, not labor. They are functional to the system as mechanisms of transnational control.

3. In this system, the U.S. and the global economy are increasingly dependent on immigrant labor that can be super-exploited and super-controlled.

We have the following data for the percentage of the workforce in different categories in California that were immigrants in 1980 and in 1990:

T A B L E 4.2

	1980	1990
Construction	20	64
Janitor	26	49
Farm worker	58	91
Maid	34	76
Electronics	37	60
Child care	20	58
Restaurant	29	69
Gardener	37	66
Drywall install	9	48

Since 1990 there has been further tremendous increase in employer dependence on immigrant labor. The U.S. and global economy would grind to halt without immigrant labor. Employers don't want expensive labor, labor with citizenship rights. They are seeking cheap, super-exploited labor, super-controllable labor. Moreover, that 20 percent of the population that is affluent or well off want cheap "throw-away labor." They want to be able to maintain their privileges by drawing on an army of maids, nannies, gardeners, *jornaleros,* and so on. Elites, employers, and the more affluent wish to maintain a reserve army of immigrant labor.

4. Sustaining such an immigrant labor force means creating—and reproducing —the division of workers into immigrants and citizens.

   This is a new axis of inequality worldwide, between citizen and non-citizen. This axis is racialized. We are talking about racialized class relations worldwide, meaning that these are class relations of exploitation, but they are also racist relations of racial/ethnic oppression and discrimination against latinos/as and other immigrants.

5. The phenomenon of a super-exploited and super-controlled latino/a immigrant workforce is part of a larger global phenomenon, that of transnational migration flows worldwide and the creation of immigrant labor pools. We have:

   ▪ latinos/as and other immigrants in North America;

   ▪ Turkish, Eastern European, North African and Asian labor in Europe;

   ▪ Indian and Pakistani workers in Middle East oil producing countries;

   ▪ Central and Southern African immigrants in South Africa;

   ▪ Nicaraguans in Costa Rica, Peruvians in Chile, Bolivians in Argentina;

   ▪ Asian immigrants in Australia;

   ▪ Thai, Korean, and others in Japan, etc. . . .

   In all these cases, repressive state controls over transnational labor creates conditions for "immigrant labor" as a distinct category of workers in relation to capital. The creation of these distinct categories—"immigrant labor" groups all over the world—becomes central to the whole global capitalist economy. What we are seeing is the rise of a transnational capitalist class that draws on immigrant labor pools around the world for its own use. And alongside this transnational capitalist class we see the rise of a transnational, or global, working class, but one split between immigrant and native workers. In this situation, borders and nationality are used by capital, the powerful, and the privileged, to exploit, control, and dominate this global working class.

6. In all these cases—but returning the focus to the United States—the system needs this immigrant labor; it can't function without it, without this reserve army of immigrant labor.

   But—and this is the crux of the matter: the system needs it to remain just that—immigrant labor: vulnerable, undocumented, without citizenship and civil, political, and labor rights, deportable . . . in a word—controllable. The aim of powers that be is not to do away with latino/a (and other) immigrants, but to exercise repressive control over immigrants. It is this condition of deportable that they wish to create or preserve since that condition assures the ability to super-exploit with impunity, and dispose of without consequences should this labor organize, demand its rights and its dignity. Latino/a immigrant labor is the new super-exploited sector of the labor force. We need here to focus on black-brown unity. African Americans used to constitute the super-exploited segment of the working class outside of the Southwest (where chicanos/as played this role). But in the 1960s and 1970s African Americans organized to win their civil rights. They fought also for full social and labor rights, launched the movement for black liberation and also became the most militant group within the trade unions. Since they have citizenship rights they cannot be deported. So capital decided that African Americans were not desirable workers since they were not as vulnerable and were too militant and organized. Employers turned massively in the 1980s and on to shifting to immigrant workers and to structurally marginalizing black workers. So while African Americans are increasingly the structurally unemployed and marginalized sector of the working class—subject to hostility, neglect, and incarceration—latinos/as are now increasingly the super-exploited sector. Black-brown unity is crucial. We cannot let the system pit African Americans and immigrants against each other.

7. This is therefore a contradictory situation:

    From the viewpoint of dominant groups, this situation presents a dilemma: how to have their cake and eat it too? How to super-exploit a latino/a immigrant population, yet how to simultaneously assure it is super-controllable and super-controlled? Hence the dual emphasis on guest-worker programs alongside heightened criminalization, enforcement and militarization.

8. We need to deepen a working-class focus! The immigrant issue is a labor issue, one in which we see how race and class come together.

    Let us recall that this is about transnational immigrants as workers for global capitalism, and latino/a immigrants as immigrant workers. Transnational capital wants a class of workers and the twin instruments in this endeavor become:

    1: division of the working class into immigrant and citizen, and;
    2: racialization of the former.

    The struggle of immigrant labor is the struggle of all working and poor people. Any improvement of status of immigrant labor, any advance in immigrant rights, is in the interests of all workers.

    But there is a technical point we need to stress because it is so important. Global elites and dominant groups around the world have imposed new capital-labor relations on all workers based on oppressive new systems of labor control and cheapening of labor. This involves diverse contingent categories of devalued labor, including subcontracted, outsourced, and flexibilized work, deunionization, casualization, informalization, part-time, temp, and contract work replacing steady full-time jobs, the loss of benefits, the erosion of wages, longer hours, and so on. This is what we could call the "Walmartization" of labor. It is not just immigrant workers—but all workers, immigrant and citizen alike—who are increasingly subject to these new capital-labor relations, in the Unites States and all around the world.

Here's the key point: an immigrant workforce reflects these new global class relations; from the viewpoint of dominant groups, they are the perfect workforce for global capitalism. Latino/immigrant workers are reduced to nothing but a commodity, a flexibilized and expendable input into the global capitalist economy, a transnationally mobile commodity deployed when and where capital need them throughout North America and utterly dehumanized in the process.

9. Why increasing hostility and oppression against the latino/a community, not just from the state and the right wing, but in the mass media, among the general public, and so on?

This system needs latino/a immigrant labor, yet the presence of that labor scares both dominant groups and privileged—generally white/native—strata. Dominant groups and privileged strata fear a rising tide of latino immigrants will lead to a loss of cultural and political control, so the dynamic becomes racialized hostility towards latinos/as, and the problematic [sic] becomes how to control them. Thus we have a rising tide of xenophobia and nativism, escalating racism, the minutemen.

Really, what this amounts to is the beginnings of fascism, of a 21st century fascism. The neo-fascist movement is led and manipulated by elites, but its base is drawn from those displaced from previously privileged positions. White working and middle class sectors who face downward mobility and insecurities brought about by capitalist globalization are particularly prone to being organized into racist anti-immigrant politics by right-wing forces. The loss of caste privileges for these white sectors of the working class is problematic for political elites and the state, since legitimation and domination in the United States have historically been constructed through the white racial hegemonic bloc. Therefore, anti-immigrant forces try to draw in white workers with appeals to racial solidarity and to xenophobia, and by scapegoating immigrant communities.

10. Conclusions: Some have called this the "new civil rights movement." It is, but this is about much more than "civil rights." This is fundamentally about human rights, about what kind of a world we are going to live in. No one can be left out in this struggle. There is no room for compromise—full legalization for all. But also: freedom from all forms of repression and persecution, and full labor, social, cultural and human rights for all.

The immigrant rights movement is the leading edge of popular struggle in the United States. In the larger picture, beyond its immediate demands, the movement for immigrant rights challenges the oppressive and exploitative class relations that are at the very core of global capitalism. Bound up with immigrant debate in the United States is the entire political economy of global capitalism, with all its injustices and inequalities. This is the same political economy that is now being sharply contested throughout Latin America by the upsurge in mass, popular, and democratic struggles.

The movement for immigrant rights in the United States is part and parcel of this larger Latin American—and worldwide—struggle for social justice and human dignity. We are integral to that worldwide struggle, on its cutting edge. We need to see our struggle as part of a broader transnational movement, to develop transnational links with other immigrant movements around the world and also with social movements of the poor, indigenous, and workers in Latin America and elsewhere.

Please allow me by way of conclusion to humbly express my opinion on one matter:

The powers that be in the United States were terrified by the mass mobilizations of Spring 2006 and as we know they tried to intimidate movement by unleashing a wave of shameless repression that is still continuing. Some have pointed out that we don't have the organizational capacity to defend all the victims of this repression. That is very true. But it is also true that the only real defense from this repression is not to back off, but to push forward, to step up and intensify, the mass struggle. Unfortunately, there is no change without sacrifice. Backing down or holding back only make it easier for the state and the right wing to retake the initiative and carry out repression. When you have seized the initiative—as we did last Spring—when your enemy is on the defensive, it is not the time to back down or demobilize but to sustain and deepen the offensive.

Let me close by quot[ing] from Che Guevara, from something he said that is written on a poster that I saw on my way in here, and that is fitting for the moment: "Seamos realistas; sonamos lo imposible."

Thank you.

15

Making Globalization Work

The Multinational Corporation

JOSEPH E. STIGLITZ

It is easy to understand why multinational corporations have played such a central role in globalization: it takes organizations of enormous scope to span the globe, to bring together the markets, technology, and capital of the developed countries with the production capacities of the developing ones. The question is how to ensure that developing countries get more benefits—and face fewer of the costs. In the following pages, I set out a five-pronged agenda that, though it will not eliminate all instances of corporate abuse, will I believe lessen them. Underlying most of these reforms is a simple objective: to align private incentives with social costs and benefits.

CORPORATE SOCIAL RESPONSIBILITY

Though many corporations, especially in the United States, continue to argue that their sole responsibility is to shareholders, many do recognize that their responsibility goes further. There is an element of self-interest here: doing good can be good for business, and doing bad can subject companies to expensive lawsuits. Bad behavior also can harm a company's image: the negative publicity surrounding the U.S. shoe company Nike after its suppliers in Vietnam mistreated local workers and the furor after Ken Saro-Wiwa was killed in Nigeria amid accusations that the Anglo-Dutch oil company Shell supported the military junta that murdered him were wake-up calls. Executives realized that they could be blamed for problems thousands of miles away from headquarters. Events like these have led to a number of voluntary initiatives by companies to improve the lot of their workers and the communities where they do business.

While increasingly more corporations see business social responsibility (BSR) as a matter of good business (and some studies suggest that socially responsible firms have performed better in the stock market than others), for many firms, their executives and employees, social responsibility is as much a moral issue as an economic one. Companies can be thought of as communities, people work-

SOURCE: Excerpt from Joseph E. Stiglitz, *Making Globalization Work*, New York: W. W. Norton, 2006, pp. 197–209.

ing together in a common purpose—say, to produce a product or provide a service. And as they work together, they care about each other, the communities in which they work, and the broader community, the world, in which we all live. This means that a company may not fire a worker the moment he is no longer needed, or that it may spend more money to reduce pollution than it is absolutely required to do by law. These companies may gain, of course, not just by avoiding the negative publicity described earlier; they may benefit from the higher quality labor force that they attract and improved morale: their workers feel better about working for a company that is socially responsible.

The BSR movement has helped bring about a change in the mindset of many corporations and of the individuals who work for them. It has also worked hard to develop tools to ensure that companies live up to their ideals: accounting frameworks are being developed that track contributions to the community and environmental impact, and these are helping firms think more about the full consequences of their actions.

Regrettably, in a world of ruthless competition, incentives often work against even those with the best of intentions. A mining company that is willing to skimp on safety and environmental safeguards will be able to underbid one of comparable efficiency that pursues sound environmental policies. The oil company that is willing to engage in bribery to obtain oil at a lower price will show higher profits than a comparable company that does not. The bank that is willing to help its clients avoid or evade taxes may do better—at least if it's not caught—than the one that discourages them from doing so.

There is a further problem. Today, all companies, even the worst polluters and those with the worst labor records, have hired public relations firms to laud their sense of corporate responsibility and their concern for the environment and workers' rights. Corporations are becoming adept at image manipulation, and have learned to speak in favor of social responsibility even while they continue to evade it.

As a result, important as it is, the BSR movement is not enough. It must be supplemented by stronger regulations. Those who are really serious about higher standards should welcome regulations that support the codes of conduct they publicly endorse, for such regulations would protect them from unfair competition from those who do not adhere to the same standards. Regulations will help prevent a race to the bottom.

LIMITING THE POWER OF CORPORATIONS

Corporations strive for profits, and one of the surest ways of garnering sustainable profits is to restrict competition—buying up competitors, squashing competitors by driving them out of business, or colluding with competitors to raise prices. The problem of anti-competitive behavior has been evident since the birth of economics: as Adam Smith put it, "People of the same trade seldom meet together, even for merriment and diversion, but the conversation ends in a conspiracy against the public, or in some contrivance to raise prices."[1] When there is a lack of competition, the potential for abuses of multinationals grows much worse.

For more than a century, the advanced industrial countries have recognized the dangers of monopolies and anti-competitive behavior, enacting laws to break up the former and to punish the latter. Collaborating with supposed competitors to fix prices is a criminal act in most advanced industrial countries, with stiff penalties in both criminal and civil actions: in the United States, those who are convicted in a criminal action may go to jail and those who can show that they have paid higher prices as a result of monopolization receive triple damages (three times the amount overcharged by the monopolists).

With the advent of globalization and globally traded commodities, monopolies, and cartels—and the problems they create—often have become global in scope.[2] Globalization has unleashed a new potential for anti-competitive behavior that may be harder both to detect and to curtail.

The nature of global monopolies was revealed by a rash of global pricing cases uncovered in the early 1990s, including two involving U.S. giant Archer Daniels Midland (ADM). One case involved vitamins; another, lysine (an essential amino acid fed to pigs); a third, corn fructose. In the lysine case, the cartel fixed prices, allocated market share, and fixed quotas, managing to increase prices by 70 percent within three months. ADM was fined $100 million; Michael Andreas, the son of the CEO, and one other executive were sent to jail. In the corn fructose case, ADM faced damage claims of up to $2 billion and agreed to pay $400 million. In the vitamin case, criminal penalties imposed by the United States and the EU on the conspirators amounted to more than $1.7 billion; though the civil suits have not all been settled yet, almost $600 million has been paid out so far and there are further claims in excess of a billion dollars. Those outside the United States and the EU, however, have little prospect of receiving significant compensation.

This reflects a general problem: while the benefits to the monopolists are global, enforcement remains fragmented, with each jurisdiction looking after its own citizens—meaning in practice that no one looks after consumers in small and developing countries. Worse still, home nations frequently fight in favor of their own global monopolies. This is natural; harm done to consumers and firms abroad is not their concern. When, in July 2001, the EU found that a proposed merger between the two U.S. giants GE and Honeywell would significantly reduce competition, the U.S. government vociferously complained. But the EU was right, and it took courage for the EU competition commissioner, Mario Monti, to stand up to the United States, fulfilling his obligation to enforce EU competition laws. His decision effectively blocked the merger.

Perhaps worse are instances where governments actually help to create global cartels to advance the interests of their own national companies. This happened while I was serving in the White House. In the face of weakening aluminum prices, Paul O'Neill, later to be secretary of the Treasury under President George W. Bush but at the time head of Alcoa, the world's largest producer of aluminum, pleaded for a global aluminum cartel to stabilize the market and protect America against "destructive" competition from Russia, then making its transition to a market economy. In a dramatic meeting, with the Council of

Economic Advisers and the Department of Justice both strenuously opposing the proposal, the Clinton administration decided to take the lead in creating a global cartel—such a clear violation of competitive market principles that Assistant Attorney General Anne Bingaman announced as the meeting ended that she might have to subpoena those at the meeting for violating anti-trust laws. The cartel resulted, as O'Neill had hoped it would, in higher prices and profits for Alcoa—but also in higher prices for consumers.[3] Indeed, the cartel worked so well from O'Neill's perspective that after he became Treasury secretary he proposed another, for steel, to raise prices and restore profits in the U.S. steel industry. But with so many more countries and firms involved in steel production than in aluminum, the complexity of establishing and maintaining a global steel cartel was far greater, and the attempt failed.

Perhaps the most successful global monopoly is Microsoft, which has succeeded in gaining global market power not only in PC operating systems but in key applications such as browsers. A firm is said to monopolize a market if it has an overwhelming share; as of August 2005, Microsoft operating systems accounted for 87 percent of the total PC market and 89.6 percent of the Intel-based PC market. The personal computer, the Internet, word processing, and spreadsheets almost define the modern economy—and a single company has obtained dominance in these key areas. When Microsoft bundles a program such as Media Player with its operating system, it is effectively selling the program at a zero price. No company can compete with that. Courts in the United States as well as in Europe found not only that Microsoft had monopoly power but that it had abused this power. The only controversy was over the appropriate remedy. Microsoft has had to pay billions to settle anti-trust claims; as a result of a 2004 ruling in Europe, Microsoft must offer a version of its operating system there without Media Player included. Still, with Microsoft's monopoly so entrenched, it is unlikely that, without much stronger action, a competitive marketplace will be restored.

Microsoft's monopoly power leads not only to higher prices but to less innovation. Innovators saw what happened to Netscape, the first major Internet browser, as it was squashed by Microsoft—a powerful warning to anyone discovering a major innovation that might compete with or be integrated into Microsoft's operating system. One possible solution might involve limiting Microsoft's intellectual property protection for its operating system to, say, three years. That would provide strong incentives for it to provide innovations of the kind that users value and for which they would be willing to pay. If it failed to innovate, others could innovate off its *old* operating system—it would become a free platform, on top of which innovations in applications could be built.

The failure to develop a global approach to global cartels and monopolies is yet another instance of economic globalization outpacing political globalization. The current piecemeal approach, with each country looking after its own citizens, is costly and inefficient, and especially ineffective in protecting those in developing countries, whose resources, we have noted, are no match for those of large multinationals. Even if they dared to take on Microsoft, there is an imbalance of legal resources; and in the end, Microsoft might threaten to leave (as it did to South Korea)—and without Microsoft's operating system, they would lose interconnectivity with the rest of the world.

Globalization of monopolies requires a global competition law and a global competition authority to enforce it, allowing both criminal prosecution and civil action in any case in which anti-competitive behavior affects more than one jurisdiction. This does not require the dismantling of national competition authorities. The risks and costs of monopolization are sufficiently great, and the dangers of large firms using political influence wherever they can to suppress prosecution are sufficiently large, that there is a need for multiple oversight. Both the United States and the EU have kept in place multiple oversight—in the United States, at the level of both state and federal government; in the EU, at the level of the EU itself and national governments.

IMPROVING CORPORATE GOVERNANCE

A third set of reforms focuses on the laws governing corporations themselves. How do we make corporations, and their officers, act in ways that are consistent with the broader public interest? What reforms in the legal system can help align private incentives with social costs and benefits?

One step in the right direction would be to have companies take into account all stakeholders—employees and the communities in which they operate, not just their shareholders. It should not, for instance, be a violation of their fiduciary responsibility to their shareholders for them to pursue good environmental policies, even if profits are thereby hurt.[4]

Limited liability law was intended to limit the liability of investors, not to absolve employees, however senior, of responsibility. But, as we have seen, sometimes that is the result. Executives should be held personally responsible for more of their actions, making it more difficult for them to hide behind the veil of their corporations. Recently, there have been some moves in this direction, among them the agreement by the board of directors of WorldCom to provide some compensation to investors who lost as a result of WorldCom's misrepresentations. In publicly owned corporations, financial penalties typically have little effect on the incentives of managers. Even a large payout by the corporation as compensation for damages will have little direct effect on them, and with managers and boards of directors protected by insurance, even when fines are levied on them directly the costs are borne by others.

Just as the effective enforcement of competition policy has been found to require criminal sanctions—prison—so too is it necessary in other arenas. In 2002, following the corporate accounting scandals in the United States, the U.S. Congress passed the Sarbanes-Oxley Act, which makes the CEO responsible for the company accounts. Sarbanes-Oxley has been criticized for being excessively stringent and costly to comply with; there is often a danger of overreaction, and with experience the legislation may get fine-tuned. But the costs of the abuses—the misallocation of resources, the loss of confidence in the market economy—were also large, almost surely of an order of magnitude greater than the costs of the regulation. Moreover, many of the costs are start-up costs; once firms have adjusted to the new system, annual costs will be lower.

If there is a case for making corporate officers individually responsible in the area of accountability to shareholders and other stakeholders, then there is an even stronger case in other areas. It is no less a crime to ruin the environment (stealing the heritage of the entire community) than to cheat investors by manipulating the books. Environmental damage done by corporations is longer lasting, and those injured are innocent bystanders who were neither party to any agreement nor stood to gain from investment. When a company has egregiously violated a nation's environmental laws, the CEO and others who made the decisions and took the actions should be held criminally liable.

Another important step in achieving congruence between private and social interests is to make it easier for compensation to be obtained when damage has been done. Making firms pay for the damage they inflict—injury to workers or to the environment—provides firms with greater incentives to act more responsibly and to ensure that their employees do so. Of course, legal systems are imperfect. Large corporations can hire the best lawyers, against whom the lawyers that (often poor) injured parties can afford are no match. Sophisticated legal tactics often enable clearly culpable American firms to go free; until recently, few of the cigarette companies responsible for millions of deaths had been made to pay compensation. But, as we have already seen, the problems of making an American company pay for the consequences of its actions in a developing country are even greater. Even when the corporation is found guilty, it may be difficult to enforce the judgment. The company may well have protected itself by limiting its assets within the country, and attaching assets outside the country may be nearly impossible.

Several changes would go a long way toward repairing the system. The first is to allow those in other countries to sue in the home country of the offending corporation. The United States has allowed such suits since 1789 under the Alien Tort Claims Act, which allows those injured abroad to bring suit in the United States for any injury "committed in violation of the law of nations or a treaty of the United States." There have been attempts in recent years to bring actions in U.S. courts against multinational corporations, with some small measure of success. Of course, corporations would like to restrict such suits, but, if we are to make globalization work, there is a need to establish such legal provisions worldwide. This is the only way that there can be effective enforcement, especially when the offending corporation has few assets in the country where the damage occurred. A further advantage of these suits is that an American or European firm can no longer complain that it lost because the plaintiff had a home-court advantage.

A complementary reform would be to allow judgments made in foreign courts to be enforced by courts in the advanced industrial countries. If a court in, say, Brazil finds that an American mining company has done a billion dollars' worth of damage but does not have a billion dollars' worth of assets in Brazil, Brazil could use U.S. courts to help it collect damages. This is the case today in most international commercial arbitrations—but these are directed at protecting investors. Once again, there is an asymmetry: there is less concern about protecting countries against damage done by footloose international firms, who limit their assets within a country as a way of controlling their liability exposure.

Some firms are wary about being subject to foreign courts, claiming that the courts are stacked against them. This is simply one of the prices that one has to, and should, pay if one wants to do business in a country—including, in particular, extracting that country's natural resources. Alternatively, any firm claiming, as a defense against the enforcement of an adverse judgment, that a proceeding abroad was unfair could be automatically subject to suit in its own country's courts, to be judged according to the higher environmental and other regulatory standards of the two countries. This is not double jeopardy in the usual sense: the firm could have accepted the first judgment; it subjects itself to a second court only because it refuses to accept the findings of the first. The stipulation that the company should be judged by the environmental standards of the home country reflects a presumption increasingly recognized by the business social responsibility movement—that there should not be a double standard, with, say, lower environmental standards in developing countries than in the United States and the EU.

In the lore of America's West, bandits would cross the state line to seek a safe haven. For international environmental bandits, there should be no safe haven. Any country in which the corporation (or the substantial owners of the corporation) has assets should provide a venue in which suits can be brought or in which enforcement actions to ensure payment of liabilities can be undertaken. The corporation may incorporate where it wants, but this should not make it any less accountable for its actions in other jurisdictions.

To make this effective, it may be necessary to pierce the corporate veil. Mining companies, for example, often incorporate subsidiaries to run a particular mine, so that when the mine is exhausted—and all that remains are the costs of cleanup—the subsidiary goes bankrupt, leaving the parent unscathed. A simple rule would be that in certain classes of liabilities, such as those associated with environmental abuses, any entity owning more than, say, 20 percent of the shares of a company could be held liable even if the corporation itself went bankrupt. Limited liability should not be sacrosanct. Like property rights—including intellectual property—it is a creation of man, to provide appropriate incentives; when that artifice fails to fulfill its social function, it needs to be modified.

GLOBAL LAWS FOR A GLOBAL ECONOMY

Eventually, we should be working toward the creation of international legal frameworks and international courts—as necessary for the smooth functioning of the global economy as federal courts and national laws are for national economies.

When consumers within the United States and certain other countries are hurt by price-fixing, they can band together, file what is called a "class action" suit, and if they succeed, they receive an amount that is triple the damages they incurred. This provides a strong incentive for firms not to engage in price-fixing. With global price-fixing, the harm done has become global, so consumers around the world need to band together and perhaps sue in, say, American courts. A recent Supreme Court decision gives the perpetrators, however, an easy way out. Once they have paid off the Americans who are injured, which may be just a fraction of the global liability, the plaintiffs have to find another venue.[5] By the same token, a single injured individual—say, in Bhopal—cannot afford to bring a suit; the maximum he or she can collect would be too small to pay any but the poorest of lawyers. But by acting collectively, the injured have some hope of redress. Those injured in Bhopal may have received far too little, but that they got as much as they did was a result of class action.

Not surprisingly, defense lawyers try to stop class actions by saying that the injured parties are sufficiently different that their cases cannot be consolidated. Insisting on a large number of separate cases against the same corporation for the same injury obviously imposes an enormous—in many cases, an impossible —burden on the legal system.

When a large number of individuals have been injured in a similar way, they should be able to band together to bring a single suit. We need to make it easier to pursue global class action suits, either in newly established global courts, or in national courts. Justice is far better served by recognizing the common element, to establish culpability and a base level of compensation, which can be supplemented if necessary by separate trials focusing on adjustments for unusual situations. For instance, price-fixing raises costs for all those who buy the product. A class action suit would establish that there has been price-fixing and calculate the amount prices have been raised from what they otherwise would have been. Of course, the magnitude of the injury suffered by a large producer in a developed country and a small consumer in a developing country will be very different. Having determined, however, the cartel's liability for price-fixing and ascertained the magnitude by which prices were increased, it would be a relatively easy matter to determine how much each should receive (which might have to be done in a series of mini-trials).[6]

And just as we recognize that access to justice for the poor requires the government to finance legal aid, this should be the case internationally as well: advanced industrial countries should provide legal assistance to those in developing countries.

REDUCING THE SCOPE FOR CORRUPTION

There are several other actions that advanced industrial countries can undertake in order to make it more difficult for corporations to get away with the worst kinds of misdeeds. As we noted earlier, there is now widespread recognition of the corrosive effects of corruption and the need to attack it at both the supply and demand sides. The United States' passage of the Foreign Corrupt Practices Act in 1997 was a major step in the right direction. Every government needs to adopt a foreign corrupt practices act, and penalties should be imposed on governments that do not enact or enforce such laws. This is the kind of new issue that should have been introduced as part of the development round of trade negotiations; it was not even broached. Bribery should be viewed as an unfair competitive practice and, just like any other unfair competitive practice outlawed under WTO rules, be subject to sanctions.

Bank secrecy aggravates the problems of corruption, providing a safe haven for ill-gotten gains. In the aftermath of the East Asian crisis, there were calls from the IMF and the U.S. Treasury for greater transparency in the Asian financial markets. When the developing countries pointed out that one of the problems in tracing the flow of funds was bank secrecy in offshore Western banks, there was a decided change in tone. The money is in these so-called offshore accounts not because the climate in the Cayman Islands is more conducive to banking; money goes there precisely because of the opportunities it affords for avoiding taxes, laws, and regulations. The existence of these opportunities is not an accidental loophole. The secrecy of the offshore banking centers exists because it is in the interests of certain groups in the advanced industrial countries.

There was an accord among the advanced industrial countries to do something about bank secrecy, but in August 2001 the Bush administration vetoed it. Then, when it was discovered that bank secrecy had been used to finance the terrorists involved in the September 11 attacks, the United States changed its views—but only where fighting terrorism was involved. Other forms of bank secrecy, as corrosive as they are to societies around the world, as bad as they are for development, are evidently still permissible; after all, bank secrecy is another way by which corporations increase the after-tax profits that are enjoyed by corporation owners. The international community should quickly broaden the rules against bank secrecy to areas beyond terrorism. The G-8 could itself bring this about, simply by forbidding any of their banks to have dealings with the banks of any jurisdiction that did not comply. The United States has shown that collective action can work: it has been effective in stopping the use of banks for financing terrorism. The same resolve should be used against corruption, arms sales, drugs, and tax evasion.

ENDNOTES

1. *The Wealth of Nations* (New York: Modern Library, 1937), p. 128.
2. An additional level of complexity is added by international agreements that are supposed to deal with anti-competitive behavior. While the WTO allows countries to use dumping duties, dumping, as traditionally defined, has little to do with anti-competitive behavior. Moreover, while dumping is a concern with firms that charge too little, the WTO seems unconcerned about the much greater danger of monopolization, of firms charging too much. In one instance the United States did accuse Japan of anti-

competitive behavior in film (Fuji outsold Kodak two to one, whereas in the United States, the ratios are reversed). But the U.S. position was not sustained.

3. See Stiglitz, *Globalization and Its Discontents* (New York: W. W. Norton, 2003).

4. Some European countries have legal frameworks that recognize the obligation of corporations not only to shareholders but also to others affected by their policies.

5. The reason that America is the preferred venue is that it has traditionally had the strongest competition laws. The 2005 Supreme Court decision was in *F. Hoffman–LaRoche, Ltd.* (a Swiss-based multinational operating in more than 150 countries) *v. Empagran SA*, an Ecuadorean company injured by having to pay higher prices for vitamin C that it used in shrimp and fish farming. Hoffman-LaRoche and other producers of vitamin C had been found guilty of price-fixing, but they first settled claims by Americans who also had been injured. With American claimants out of the case, the Supreme Court ruled that Empagran and twenty other foreign companies could not seek redress in U.S. courts. I thought the principles involved were so important for the preservation of global competition that I filed an *amicus curiae* (friend of the court) brief, describing the risks of global monopoly and what should be done. While the Court found against Empagran, its ruling did suggest an awareness of the problems posed by global monopolies.

6. There are innumerable dimensions to making a global legal regime that is both fair to the injured and incentivizes corporations to act responsibly. A more fundamental legal reform would separate out the issues of punishment and deterrence from the problem of just compensation. A claims board could establish, for instance, the magnitude of the damage suffered by each individual and provide compensation on that basis. A separate tribunal could establish the extent of the corporation's culpability, whether it took actions which caused harm—say, as a result of inappropriate environmental policies—and then assess, using a statistical model, appropriate penalties. Additional punitive damages might be assessed to provide further deterrence or in response to particularly outrageous behavior.

REFLECTION QUESTIONS FOR CHAPTER 4

1. What is the evidence of the global economy penetrating your community? Who benefits? Who does not?

2. What are the consequences for the United States of corporations outsourcing to low-wage economies? Who is hurt by this "race to the bottom"?

3. How is economic globalization both gendered and racialized around the world? What are the human tolls of these forms of stratification? What solutions do you see for these inequalities?

4. Should there be constraints on the business activities of transnational corporations? If so, what organization(s) should monitor and, if necessary, sanction them?

5

Political Globalization

The world is divided into nation-states. Each of the states has a sovereign government, that is, it is the ultimate authority within its territory. Each state assumes responsibility for education, the economy, environmental protection, foreign policy, and military defense for its territory and people[1] (Lechner and Boli, 2000:196–197). Wars are fought between these nations. But the globalization of the present threatens the old political patterns based on the nation-state. Some examples:

- The war on terrorism is not a war against a nation or nations but rather against religious and ethnic groups that have organized not only within nations but also across national boundaries.

- Nation-states alone are often incapable of combating terrorism, pollution, narcotics trafficking, and transnational crime networks.

- Global warming and environmental degradation, the products of activities within states, have transnational consequences.

- The world economy is beyond the control of nations given the power of transnational corporations, the massive number and magnitude of transnational financial transactions, and the interconnected markets. Hence, a stock market crash or a natural disaster, or the devaluing of a currency or other event in one nation, has important ramifications for markets, jobs, and stability in many other countries.

- There are transnational political entities created to deal with global problems. Some of these are the World Court, the United Nations, and its associated agencies such as the World Health Organization and the International Labor Organization. There also are the World Trade Organization and the International Monetary Fund, created primarily to manage the world economy.

■ Recently the nations of Europe have consolidated in a new political unit—
the European Union—with a common currency and a parliament.

The primacy of the nation-state is not over, but globalization has reduced its
significance. In the first essay, Joseph E. Stiglitz argues that integration with the
global economy works fine when sovereign countries define the terms. It works
disastrously, though, when globalization is managed for them by the
International Monetary Fund and other international economic institutions.

Philosopher Peter Singer examines how activities and public policies within
nations have dire consequences for the world's environment. The issue is an ethical one. Should political leaders see their role narrowly to promote the interest
of their citizens or should they be concerned with the welfare of people everywhere? He states, "How well we come through the era of globalization will depend on how we respond ethically to the idea that we live in one world."

Author Tina Rosenberg emphasizes the government's role in forcing small
Mexican farmers off the land through free trade treaties, lowered import barriers,
and subsidized agriculture. Her brief essay underscores the devastating effects of
U.S. and Mexican policies for the 18 million Mexicans who live on small farms
and eventually flock to overcrowded cities already congested with the poor and
unemployed.

Civic engagement is the topic of the final reading in this section. Jonathan
Fox explores the Mexican migrant organizations that bring people from one area
in Mexico to another one in the United States, and foster civic engagement in
two societies. According to Fox, binational migrants are reshaping civic life as
they campaign for their rights to be heard in both Mexico and the United
States.

ENDNOTES

1. Lechner, Frank J., and John Boli (eds.). 2000. *The Globalization Reader.*
    Malden, MA: Blackwell Publishers.

16

Globalism's Discontents

JOSEPH E. STIGLITZ

Few subjects have polarized people throughout the world as much as globalization. Some see it as the way of the future, bringing unprecedented prosperity to everyone, everywhere. Others, symbolized by the Seattle protestors of December 1999, fault globalization as the source of untold problems, from the destruction of native cultures to increasing poverty and immiseration. In this article, I want to sort out the different meanings of globalization. In many countries, globalization has brought huge benefits to a few with few benefits to the many. But in the case of a few countries, it has brought enormous benefit to the many. Why have there been these huge differences in experiences? The answer is that globalization has meant different things in different places.

The countries that have managed globalization on their own, such as those in East Asia, have, by and large, ensured that they reaped huge benefits and that those benefits were equitably shared; they were able substantially to control the terms on which they engaged with the global economy. By contrast, the countries that have, by and large, had globalization managed for them by the International Monetary Fund and other international economic institutions have not done so well. The problem is thus not with globalization but with how it has been managed.

The international financial institutions have pushed a particular ideology—market fundamentalism—that is both bad economics and bad politics; it is based on premises concerning how markets work that do not hold even for developed countries, much less for developing countries. The IMF has pushed these economics policies without a broader vision of society or the role of economics within society. And it has pushed these policies in ways that have undermined emerging democracies.

More generally, globalization itself has been governed in ways that are undemocratic and have been disadvantageous to developing countries, especially the poor within those countries. The Seattle protestors pointed to the absence of democracy and of transparency, the governance of the international economic institutions by and for special corporate and financial interests, and the absence of countervailing democratic checks to ensure that these informal and *public*

SOURCE: Joseph E. Stiglitz, "Globalism's Discontents." *The American Prospect* (Winter 2002) pp. A16–A21. Vol. 13, Number 1. Reprinted with permission from *The American Prospect*, Volume 13, Number 1: January 1, 2002. *The American Prospect*, 11 Beacon Street, Suite 1120, Boston, MA 02108. All rights reserved.

institutions serve a general interest. In these complaints, there is more than a grain of truth.

BENEFICIAL GLOBALIZATION

Of the countries of the world, those in East Asia have grown the fastest and done most to reduce poverty. And they have done so, emphatically, via "globalization." Their growth has been based on exports—by taking advantage of the global market for exports and by closing the technology gap. It was not just gaps in capital and other resources that separated the developed from the less-developed countries but differences in knowledge. East Asian countries took advantage of the "globalization of knowledge" to reduce these disparities. But while some of the countries in the region grew by opening themselves up to multinational companies, others, such as Korea and Taiwan, grew by creating their own enterprises. Here is the key distinction: Each of the most successful globalizing countries determined its own pace of change; each made sure as it grew that the benefits were shared equitably; each rejected the basic tenets of the "Washington Consensus," which argued for a minimalist role for government and rapid privatization and liberalization.

In East Asia, government took an active role in managing the economy. The steel industry that the Korean government created was among the most efficient in the world—performing far better than its private-sector rivals in the United States (which, though private, are constantly turning to the government for protection and for subsidies). Financial markets were highly regulated. My research shows that those regulations promoted growth. It was only when these countries stripped away the regulations, under pressure from the U.S. Treasury and the IMF, that they encountered problems.

During the 1960s, 1970s, and 1980s, the East Asian economies not only grew rapidly but were remarkably stable. Two of the countries most touched by the 1997–1998 economic crisis had had in the preceding three decades not a single year of negative growth; two had only one year—a better performance than the United States or the other wealthy nations that make up the Organization for Economic Cooperation and Development (OECD). The single most important factor leading to the troubles that several of the East Asian countries encountered in the late 1990s—the East Asian crisis—was the rapid liberalization of financial and capital markets. In short, the countries of East Asia benefited from globalization because they made globalization work for them; it was when they succumbed to the pressures from the outside that they ran into problems that were beyond their own capacity to manage well.

Globalization can yield immense benefits. Elsewhere in the developing world, globalization of knowledge has brought improved health, with life spans increasing at a rapid pace. How can one put a price on these benefits of globalization? Globalization has brought still other benefits: Today there is the beginning of a globalized civil society that has begun to succeed with such reforms as the Mine Ban Treaty and debt forgiveness for the poorest highly indebted countries (the Jubilee movement). The globalization protest movement itself would not have been possible without globalization.

THE DARKER SIDE OF GLOBALIZATION

How then could a trend with the power to have so many benefits have produced such opposition? Simply because it has not only failed to live up to its potential but frequently has had very adverse effects. But this forces us to ask, why has it had such adverse effects? The answer can be seen by looking at each of the economic elements of globalization as pursued by the international financial institutions and especially by the IMF.

The most adverse effects have arisen from the liberalization of financial and capital markets—which has posed risks to developing countries without commensurate rewards. The liberalization has left them prey to hot money pouring into the country, an influx that has fueled speculative real-estate booms; just as suddenly, as investor sentiment changes, the money is pulled out, leaving in its wake economic devastation. Early on, the IMF said that these countries were being rightly punished for pursuing bad economic policies. But as the crisis spread from country to country, even those that the IMF had given high marks found themselves ravaged.

The IMF often speaks about the importance of the discipline provided by capital markets. In doing so, it exhibits a certain paternalism, a new form of the old colonial mentality: "We in the establishment, we in the North who run our capital markets, know best. Do what we tell you to do, and you will prosper." The arrogance is offensive, but the objection is more than just to style. The position is highly undemocratic: There is an implied assumption that democracy by itself does not provide sufficient discipline. But if one is to have an external disciplinarian, one should choose a good disciplinarian who knows what is good for growth, who shares one's values. One doesn't want an arbitrary and capricious taskmaster who one moment praises you for your virtues and the next screams at you for being rotten to the core. But capital markets are just such a fickle taskmaster; even ardent advocates talk about their bouts of irrational exuberance followed by equally irrational pessimism.

LESSONS OF CRISIS

Nowhere was the fickleness more evident than in the last global financial crisis. Historically, most of the disturbances in capital flows into and out of a country are not the result of factors inside the country. Major disturbances arise, rather, from influences outside the country. When Argentina suddenly faced high interest rates in 1998, it wasn't because of what Argentina did but because of what happened in Russia. Argentina cannot be blamed for Russia's crisis.

Small developing countries find it virtually impossible to withstand this volatility. I have described capital-market liberalization with a simple metaphor: Small countries are like small boats. Liberalizing capital markets is like setting them loose on a rough sea. Even if the boats are well captained, even if the boats are sound, they are likely to be hit broadside by a big wave and capsize. But the IMF pushed for the boats to set forth into the roughest parts of the sea before they were seaworthy, with untrained captains and crews, and without life vests. No wonder matters turned out so badly!

To see why it is important to choose a disciplinarian who shares one's values, consider a world in which there were free mobility of skilled labor. Skilled labor would then provide discipline. Today, a country that does not treat capital well will find capital quickly withdrawing; in a world of free labor mobility if a country did not treat skilled labor well, it too would withdraw. Workers would worry about the quality of their children's education and their family's health care, the quality of their environment and of their own wages and working conditions. They would say to the government: If you fail to provide these essentials, we will move elsewhere. That is a far cry from the kind of discipline that free-flowing capital provides.

The liberalization of capital markets has not brought growth. How can one build factories or create jobs with money that can come in and out of a country overnight? And it gets worse: Prudential behavior requires countries to set aside reserves equal to the amount of short-term lending; so if a firm in a poor country borrows $100 million at, say, 20 percent interest rates short-term from a bank in the United States, the government must set aside a corresponding amount. The reserves are typically held in U.S. Treasury bills—a safe, liquid asset. In effect, the country is borrowing $100 million from the United States and lending $100 million to the United States. But when it borrows, it pays a high interest rate, around 4 percent. This may be great for the United States, but it can hardly help the growth of the poor country. There is also a high *opportunity* cost of the reserves; the money could have been much better spent on building rural roads or constructing schools or health clinics. But instead, the country is, in effect, forced to lend money to the United States.

Thailand illustrates the true ironies of such policies: There, the free market led to investments in empty office buildings, starving other sectors—such as education and transportation—of badly needed resources. Until the IMF and the U.S. Treasury came along. Thailand had restricted bank lending for speculative real estate. The Thais had seen the record: Such lending is an essential part of the boom-bust cycle that has characterized capitalism for 200 years. It wanted to be sure that the scarce capital went to create jobs. But the IMF nixed this intervention in the free market. If the free market said, "Build empty office buildings," so be it! The market knew better than any government bureaucrat who mistakenly might have thought it wiser to build schools or factories.

THE COSTS OF VOLATILITY

Capital-market liberalization is inevitably accompanied by huge volatility, and this volatility impedes growth and increases poverty. It increases the risks of investing in the country, and thus investors demand a risk premium in the form of higher-than-normal profits. Not only is growth not enhanced but poverty is increased through several channels. The high volatility increases the likelihood of recessions—and the poor always bear the brunt of such downturns. Even in developed countries, safety nets are weak or nonexistent among the self-employed and in the rural sector. But these are the dominant sectors in developing countries. Without adequate safety nets, the recessions that follow from capital-market liberalization lead to impoverishment. In the name of imposing budget discipline and reassuring investors, the IMF invariably demands expenditure reductions, which almost inevitably result in cuts in outlays for safety nets that are already threadbare.

But matters are even worse—for under the doctrines of the "discipline of the capital markets," if countries try to tax capital, capital flees. Thus, the IMF doctrines inevitably lead to an increase in tax burdens on the poor and the middle classes. Thus, while IMF bailouts enable the rich to take their money out of the country at more favorable terms (at the overvalued exchange rates), the burden of repaying the loans lies with the workers who remain behind.

The reason that I emphasize capital-market liberalization is that the case against it—and against the IMF's stance in pushing it—is so compelling. It illustrates what can go wrong with globalization. Even economists like Jagdish Bhagwati, strong advocates of free trade, see the folly in liberalizing capital markets. Belatedly, so too has the IMF—at least in its official rhetoric, though less so in its policy stances—but too late for all those countries that have suffered so much from following the IMF's prescriptions.

But while the case for trade liberalization—when properly done—is quite compelling, the way it has been pushed by the IMF has been far more problematic. The basic logic is simple: Trade liberalization is supposed to result in resources moving from inefficient protected sectors to more efficient export sectors. The problem is not only that job destruction comes before the job creation —so that unemployment and poverty result—but that the IMF's "structural adjustment programs" (designed in ways that allegedly would reassure global investors) make job creation almost impossible. For these programs are often accompanied by high interest rates that are often justified by a single-minded focus on inflation. Sometimes that concern is deserved; often, though, it is carried to an extreme. In the United States, we worry that small increases in the interest rate will discourage investment. The IMF has pushed for far higher interest rates in countries with a far less hospitable investment environment. The high interest rates mean that new jobs and enterprises are not created. What happens is that trade liberalization, rather than moving workers from low-productivity jobs to high-productivity ones, moves them from low-productivity jobs to unemployment. Rather than enhanced growth, the effect is increased poverty. To make matters even worse, the unfair trade-liberalization agenda forces poor countries to compete with highly subsidized American and European agriculture.

THE GOVERNANCE OF GLOBALIZATION

As the market economy has matured within countries, there has been increasing recognition of the importance of having rules to govern it. One hundred fifty years ago, in many parts of the world, there was a domestic process that was in some ways analogous to globalization. In the United States, government promoted the formation of the national economy, the building of the railroads, and the development of the telegraph—all of which reduced transportation and communication costs within the United States. As that process occurred, the democratically elected national government provided oversight: supervising and regulating, balancing interests, tempering crisis, and limiting adverse consequences of this very large change in economic structure. So, for instance, in 1863 the U.S. government established the first financial-banking regulatory authority—the Officer of the Comptroller of Currency—because it was important to have strong national banks, and that requires strong regulation.

The United States, among the least statist of the industrial democracies, adopted other policies. Agriculture, the central industry of the United States in the mid-nineteenth century, was supported by the 1862 Morrill Act, which established research, extension, and teaching programs. That system worked extremely well and is widely credited with playing a central role in the enormous increases in agricultural productivity over the last century and a half. We established an industrial policy for other fledgling industries, including radio and civil aviation. The beginning of the telecommunications industry, with the first telegraph line between Baltimore and Washington, D.C., was funded by the federal government. And it is a tradition that has continued, with the U.S. government's founding of the Internet.

By contrast, in the current process of globalization we have a system of what I call global governance without global government. International institutions like the World Trade Organization, the IMF, the World Bank, and others provide an ad hoc system of global governance, but it is a far cry from global government and lacks democratic accountability. Although it is perhaps better than not having any system of global governance, the system is structured not to serve general interests or assure equitable results. This not only raises issues of whether broader values are given short shrift; it does not even promote growth as much as an alternative might.

GOVERNANCE THROUGH IDEOLOGY

Consider the contrast between how economic decisions are made inside the United States and how they are made in the international economic institutions.

In this country, economic decisions within the administration are undertaken largely by the National Economic Council, which includes the secretary of labor, the secretary of commerce, the chairman of the Council of Economic Advisers, the treasury secretary, the assistant attorney general for antitrust, and the U.S. trade representative. The Treasury is only one vote and often gets voted down. All of these officials, of course, are part of an administration that must face Congress and the democratic electorate. But in the international arena, only the voices of the financial community are heard. The IMF reports to the ministers of finance and the governors of the central banks, and one of the important items on its agenda is to make these central banks more independent—and less democratically accountable. It might make little difference if the IMF dealt only with matters of concern to the financial community, such as the clearance of checks; but in fact, its policies affect every aspect of life. It forces countries to have tight monetary and fiscal policies: It evaluates the trade-off between inflation and unemployment, and in that trade-off it always puts far more weight on inflation than on jobs.

The problem with having the rules of the game dictated by the IMF—and thus by the financial community—is not just a question of values (though that is important) but also a question of ideology. The financial community's view of the world predominates—even when there is little evidence in its support. Indeed, beliefs on key issues are held so strongly that theoretical and empirical support of the positions is viewed as hardly necessary.

Recall again the IMF's position on liberalizing capital markets. As noted, the IMF pushed a set of policies that exposed countries to serious risk. One might have thought, given the evidence of the costs, that the IMF could offer plenty of evidence that the policies also did some good. In fact, there was no such evidence; the evidence that was available suggested that there was little if any positive effect on growth. Ideology enabled IMF officials not only to ignore the absence of benefits but also to overlook the evidence of the huge costs imposed on countries.

AN UNFAIR TRADE AGENDA

The trade-liberalization agenda has been set by the North, or more accurately, by special interests in the North. Consequently, a disproportionate part of the gains has accrued to the advanced industrial countries, and in some cases the less-developed countries have actually been worse off. After the last round of trade negotiations, the Uruguay Round that ended in 1994, the World Bank calculated the gains and losses to each of the regions of the world. The United States and Europe gained enormously. But sub-Saharan Africa, the poorest region of the world, lost by about 2 percent because of terms-of-trade effects: The trade negotiations opened their markets to manufactured goods produced by the industrialized countries but did not open up the markets of Europe and the United States to the agricultural goods in which poor countries often have a comparative advantage. Nor did the trade agreements eliminate the subsidies to agriculture that make it so hard for the developing countries to compete.

The U.S. negotiations with China over its membership in the WTO displayed a double standard bordering on the surreal. The U.S. trade representative, the chief negotiator for the United States, began by insisting that China was a developed country. Under WTO rules, developing countries are allowed longer transition periods in which state subsidies and other departures from the WTO strictures are permitted. China certainly wishes it were a developed country, with Western-style per capita incomes. And since China has a lot of "capitals," it's possible to multiply a huge number of people by very small average incomes and conclude that the People's Republic is a big economy. . . . But China is not only a developing economy; it is a low-income developing country. Yet the United States insisted that China be treated like a developed country! China went along with the fiction; the negotiations dragged on so long that China got some extra time to adjust. But the true hypocrisy was shown when U.S. negotiators asked, in effect, for developing-country status for the United States to get extra time to shelter the American textile industry.

Trade negotiations in the service industries also illustrate the unlevel nature of the playing field. Which service industries did the United States say were *very* important? Financial services—industries in which Wall Street has a comparative advantage. Construction industries and maritime services were not on the agenda, because the developing countries would have a comparative advantage in these sectors.

Consider also intellectual-property rights, which are important if innovators are to have incentives to innovate (though many of the corporate advocates of intellectual property exaggerate its importance and fail to note that much of the most important research, as in basic science and mathematics, is not patentable). Intellectual-property rights, such as patents and trademarks, need to balance the interests of producers with those of users—not only users in developing countries, but researchers in developed countries. If we underprice the profitability of innovation to the inventor, we deter invention. If we overprice its cost to the research community and the end user, we retard its diffusion and beneficial effects on living standards.

In the final stages of the Uruguay negotiations, both the White House Office of Science and Technology Policy and the Council of Economic Advisers worried that we had not got the balance right—that the agreement put producers' interests over users'. We worried that, with this imbalance, the rate of progress and innovation might actually be impeded. After all, knowledge is the most important input into research, and overly strong intellectual-property rights can, in effect, increase the price of this input. We were also concerned about the consequences of denying lifesaving medicines to the poor. This issue subsequently gained international attention in the context of the provision of AIDS medicines in South Africa. . . . The international outrage forced the drug companies to back down—and it appears that, going forward, the most adverse consequences will be circumscribed. But it is worth noting that initially, even the Democratic U.S. administration supported the pharmaceutical companies.

What we were not fully aware of was another danger—what has come to be called "biopiracy," which involves international drug companies patenting traditional medicines. Not only do they seek to make money from "resources" and knowledge that rightfully belong to the developing countries, but in doing so they squelch domestic firms who long provided these traditional medicines. While it is not clear whether these patents would hold up in court if they were effectively challenged, it is clear that the less-developed countries may not have the legal and financial resources required to mount such a challenge. The issue has become the source of enormous emotional, and potentially economic, concern throughout the developing world. This fall, while I was in Ecuador visiting a village in the high Andes, the Indian mayor railed against how globalization has led to biopiracy.

GLOBALIZATION AND SEPTEMBER 11

September 11 brought home a still darker side of globalization—it provided a global arena for terrorists. But the ensuing events and discussions highlighted broader aspects of the globalization debate. It made clear how untenable American unilateralist positions were. President Bush, who had unilaterally rejected the international agreement to address one of the long-term global risks perceived by countries around the world—global warming, in which the United States is the largest culprit—called for a global alliance against terrorism. The administration realized that success would require concerted action by all.

One of the ways to fight terrorists, Washington soon discovered, was to cut off their sources of funding. Ever since the East Asian crisis, global attention had focused on the secretive offshore banking centers. Discussions following that crisis focused on the importance of good information—transparency, or openness—but this was intended for the developing countries. As international discussions turned to the lack of transparency shown by the IMF and the offshore banking centers, the U.S. Treasury changed its tune. It is not because these secretive banking havens provide better services than those provided by banks in New York or London that billions have been put there; the secrecy serves a variety of nefarious purposes—including avoiding taxation and money laundering. These institutions could be shut down overnight—or forced to comply with international norms—if the United States and the other leading countries wanted. They continue to exist because they serve the interests of the financial community and the wealthy. Their continuing existence is no accident. Indeed, the OECD drafted an agreement to limit their scope—and before September 11, the Bush administration unilaterally walked away from this agreement too. How foolish this looks now in retrospect! Had it been embraced, we would have been further along the road to controlling the flow of money into the hands of the terrorists.

There is one more aspect to the aftermath of September 11 worth noting here. The United States was already in recession, but the attack made matters worse. It used to be said that when the United States sneezed, Mexico caught a cold. With globalization, when the United States sneezes, much of the rest of the world risks catching pneumonia. And the United States now has a bad case of the flu. With globalization, mismanaged macroeconomic policy in the United States—the failure to design an effective stimulus package—has global consequences. But around the world, anger at the traditional IMF policies is growing. The developing countries are saying to the industrialized nations: "When you face a slowdown, you follow the precepts that we are all taught in our economic courses: You adopt expansionary monetary and fiscal policies. But when we face a slowdown, you insist on contractionary policies. For you, deficits are okay; for us, they are impermissible—even if we can raise the funds through 'selling forward,' say, some natural resources." A heightened sense of inequity prevails, partly because the consequences of maintaining contractionary policies are so great.

GLOBAL SOCIAL JUSTICE

Today, in much of the developing world, globalization is being questioned. For instance, in Latin America, after a short burst of growth in the early 1990s, stagnation and recession have set in. The growth was not sustained—some might say, was not sustainable. Indeed, at this juncture, the growth record of the so-called post-reform era looks no better, and in some countries much worse, than in the widely criticized import-substitution period of the 1950s and 1960s when Latin countries tried to industrialize by discouraging imports, Indeed, reform critics point out that the burst of growth in the early 1990s was little more than a "catch-up" that did not even make up for the lost decade of the 1980s.

Throughout the region, people are asking: "Has reform failed or has globalization failed?" The distinction is perhaps artificial, for globalization was at the center of the reforms. Even in those countries that have managed to grow, such as Mexico, the benefits have accrued largely to the upper 30 percent and have been even concentrated in the top 10 percent. Those at the bottom have gained little; many are even worse off. The reforms have exposed countries to greater risk, and the risks have been borne disproportionately by those least able to cope with them. Just as in many countries where the pacing and sequencing of reforms has resulted in job destruction outmatching job creation, so too has the exposure to risk outmatched the ability to create institutions for coping with risk, including effective safety nets.

In this bleak landscape, there are some positive signs. Those in the North have become more aware of the inequalities of the global economic architecture. The agreement at Doha to hold a new round of trade negotiations—the "Development Round"—promises to rectify some of the imbalances of the past. There has been a marked change in the rhetoric of the international economic institutions—at least they talk about poverty. At the World Bank, there have been some real reforms; there has been some progress in translating the rhetoric into reality—in ensuring that the voices of the poor are heard and the concerns of the developing countries are listened to. But elsewhere, there is often a gap between the rhetoric and the reality. Serious reforms in governance, in who makes decisions and how they are made, are not on the table. If one of the problems at the IMF has been that the ideology, interests, and perspectives of the financial community in the advanced industrialized countries have been given disproportionate weight (in matters whose effects go well beyond finance), then the prospects for success in the current discussions of reform, in which the same parties continue to predominate, are bleak. They are more likely to result in slight changes in the shape of the table, not changes in who is *at* the table or what is on the agenda.

September 11 has resulted in a global alliance against terrorism. What we now need is not just an alliance *against* evil, but an alliance *for* something positive—a global alliance for reducing poverty and for creating a better environment, an alliance for creating a global society with more social justice.

17

Navigating the Ethics of Globalization

PETER SINGER

Consider two aspects of globalization: first, planes exploding as they slam into the World Trade Center, and second, the emission of carbon dioxide from the exhaust of gas-guzzling sport-utility vehicles. One brought instant death and left unforgettable images that were watched on television screens all over the world; the other makes a contribution to climate change that can be detected only by scientific instruments. Yet both are indications of the way in which we are now one world, and the more subtle changes to which sport-utility–vehicle owners unintentionally contribute will almost certainly kill far more people than the more visible aspect of globalization. When people in rich nations switch to vehicles that consume more fuel than the cars they used to drive, they contribute to changes in the climate of Mozambique or Bangladesh—changes that may cause crops to fail, sea levels to rise, and tropical diseases to spread.

As scientists pile up the evidence that continuing greenhouse-gas emissions will imperil millions of lives, the leader of the nation that emits the largest share of those gases has said: "We will not do anything that harms our economy, because first things first are the people who live in America." President Bush's remarks were not an aberration, but an expression of an ethical view that he may have learned from his father. The first President George Bush had said much the same thing at the 1992 Earth Summit in Rio de Janeiro.

But it is not only the two Bush administrations that have put the interests of Americans first. When it came to the crunch in the Balkans, the Clinton-Gore administration made it very clear that it was not prepared to risk the life of a single American in order to reduce the number of civilian casualties. In the context of the debate over whether to intervene in Bosnia to stop Serb "ethnic cleansing" operations directed against Bosnian Muslims, Colin L. Powell, then chairman of the Joint Chiefs of Staff, quoted with approval the remark of the 19th-century German statesman Otto von Bismarck, that all the Balkans were not worth the bones of a single one of his soldiers. Bismarck, however, was not thinking of intervening in the Balkans to stop crimes against humanity. As chancellor of imperial Germany, he assumed that his country followed its national interest. To use his remark today as an argument against humanitarian intervention is to return to 19th-century power politics, ignoring both the bloody

SOURCE: Peter Singer, "Navigating the Ethics of Globalization." *The Chronicle of Higher Education* (October 11, 2002), pp. B7–B10. Reprinted by permission of the author.
© Peter Singer 2002.

wars that style of politics brought about in the first half of the 20th century, and the efforts of the second half of the 20th century to find a better foundation for peace and the prevention of crimes against humanity.

That forces us to consider a fundamental ethical issue. To what extent should political leaders see their role narrowly, in terms of promoting the interests of their citizens, and to what extent should they be concerned with the welfare of people everywhere?

There is a strong ethical case for saying that it is wrong for leaders to give absolute priority to the interests of their own citizens. The value of the life of an innocent human being does not vary according to nationality. But, it might be said, the abstract ethical idea that all humans are entitled to equal consideration cannot govern the duties of a political leader. Just as parents are expected to provide for the interests of their own children, rather than for the interests of strangers, so too in accepting the office of president of the United States, President Bush has taken on a specific role that makes it his duty to protect and further the interests of Americans. Other countries have their leaders, with similar roles in respect to the interests of their fellow citizens.

There is no world political community, and as long as that situation prevails, we must have nation-states, and the leaders of those nation-states must give preference to the interests of their citizens. Otherwise, unless electors were suddenly to turn into altruists of a kind never before seen on a large scale, democracy could not function. Our leaders feel that they must give some degree of priority to the interests of their own citizens, and they are, so this argument runs, right to do so. But what does "some degree of priority" amount to, in practice?

Related to that question about the duties of national leaders is another one: Is the division of the world's people into sovereign nations a dominant and unalterable fact of life? Here our thinking has been affected by the horrors of Bosnia, Rwanda, and Kosovo. In Rwanda, a United Nations inquiry took the view that 2,500 military personnel, given the proper training and mandate, might have saved 800,000 lives. Secretary General Kofi Annan, who, as under secretary general for peacekeeping operations at the time, must bear some responsibility for what the inquiry termed a "terrible and humiliating" paralysis, has learned from that situation. Now he urges that "the world cannot stand aside when gross and systematic violations of human rights are taking place." What we need, he has said, are "legitimate and universal principles" on which we can base intervention. That means a redefinition of state sovereignty, or more accurately, an abandonment of the absolute idea of state sovereignty that has prevailed in Europe since the Treaty of Westphalia in 1648.

The aftermath of the attacks on September 11 underlined in a very different way the extent to which our thinking about state sovereignty has changed over the past century. In the summer of 1914 another act of terrorism shocked the world: the assassination of the Austrian Crown Prince Franz Ferdinand and his wife in Sarajevo, by a Bosnian Serb nationalist. In the wake of that outrage Austria-Hungary presented an ultimatum to Serbia in which it laid out the evidence that the assassins were trained and armed by the Black Hand, a shadowy Serbian organization headed by the chief of Serbian military intelligence. The Black Hand was tolerated or supported by other Serbian government officials, and Serbian officials arranged safe passage across the border into Bosnia for the seven conspirators in the assassination plot. Accordingly, Austria-Hungary's ultimatum demanded that the Serbs bring those responsible to justice and allow Austro-Hungarian officials to inspect files to ensure that that had been done properly.

Despite the clear evidence of the involvement of Serbian officials in the crime—evidence that, historians agree, was substantially accurate—the ultimatum Austria-Hungary presented was widely condemned in Russia, France, Britain, and the United States. Many historians studying the origins of the First World War have condemned the Austro-Hungarian ultimatum as demanding more than one sovereign nation may properly ask of another. They have added that the Austro-Hungarian refusal to negotiate after the Serbian government accepted many, but not all, of its demands is further evidence that Austria-Hungary, together with its backer Germany, wanted an excuse to declare war on Serbia. Hence those two nations must bear the guilt for the outbreak of the war and the nine million deaths that followed.

Now consider the American response to the terrorist attacks of September 11. The demands made of the Taliban by the Bush administration were scarcely less stringent than those made by Austria-Hungary of Serbia in 1914. (The main difference is that the Austro-Hungarians insisted on the suppression of hostile nationalist propaganda. Freedom of speech was not so widely regarded, then, as a human right.) Moreover, the American demand that the Taliban hand over Osama bin Laden was made without presenting to the Taliban any evidence at all linking him to the attacks of September 11. Yet the American demands, far from being condemned as a mere pretext for aggressive war, were endorsed as reasonable and justifiable by a wide-ranging coalition of nations.

When President Bush said, in speeches and press conferences after September 11, that he would not draw a distinction between terrorists and regimes that harbor terrorists, no ambassadors, foreign ministers, or United Nations representatives denounced that as a "vicious" doctrine or a "tyrannical" demand on other sovereign nations, as the Austro-Hungarian demands had been denounced. The U.N. Security Council broadly endorsed it, in its resolution of September 28, 2001.

It seems that world leaders now accept that every nation has an obligation to every other nation of the world to suppress activities within its borders that might lead to terrorist attacks carried out in other countries, and that it is reasonable to go to war with a nation that does not do so. If Kaisers Franz Joseph I and Wilhelm II could see this, they might well feel that, since 1914, the world has come round to their view.

Terrorism has made our world an integrated community in a new and frightening way. Not merely the activities of our neighbors, but those of the inhabitants of the most remote mountain valleys of the farthest-flung countries of our planet have become our business. We need to extend the reach of the criminal law there and to have the means to bring terrorists to justice without declaring war on an entire country in order to do it. For that we need a sound global system of criminal justice, so justice does not become the victim of national differences of opinion.

We also need, though it will be far more difficult to achieve, a sense that we really are one community, that we are people who recognize not only the force of prohibitions against killing each other but also the pull of obligations to assist one another. That may not stop religious fanatics from carrying out suicide missions, but it will help to isolate them and reduce their support. It was not a coincidence that just two weeks after September 11, conservative members of the U.S. Congress abandoned their opposition to the payment of $585 million in back dues that the United States owed the United Nations. Now that America was calling for the world to come to its aid to stamp out terrorism, it was apparent that it could no longer flout the rules of the global community to the extent that it had been doing before September 11.

We have lived with the idea of sovereign states for so long that they have come to be part of the background not only of diplomacy and public policy but also of ethics. Implicit in the term "globalization" rather than the older "internationalization" is the idea that we are moving beyond the era of growing ties between nations and are beginning to contemplate something beyond the existing conceptions of the nation-state. But this change needs to be reflected in all levels of our thought, and especially in our thinking about ethics.

For most of the eons of human existence, people living only short distances apart might as well, for all the difference they made to each other's lives, have been living in separate worlds. A river, a mountain range, a stretch of forest or desert, a sea—those were enough to cut people off from each other. Over the past few centuries the isolation has dwindled, slowly at first, then with increasing rapidity. Now people living on opposite sides of the world are linked in ways previously unimaginable.

One hundred and fifty years ago, Karl Marx gave a one-sentence summary of his theory of history: "The hand mill gives you society with the feudal lord; the steam mill, society with the industrial capitalist." Today he could have added: "The jet plane, the telephone, and the Internet give you a global society with the transnational corporation and the World Economic Forum."

Technology changes everything—that was Marx's claim, and if it was a dangerous half-truth, it was still an illuminating one. As technology has overcome distance, economic globalization has followed. In London supermarkets, fresh vegetables flown in from Kenya are offered for sale alongside those from nearby Kent. Planes bring illegal immigrants seeking to better their own lives in a country they have long admired. In the wrong hands the same planes become lethal weapons that bring down tall buildings. Instant digital communication spreads the nature of international trade from actual goods to skilled services. At the end of a day's trading a bank based in New York may have its accounts balanced by clerks living in India. The increasing degree to which there is a single world economy is reflected in the development of new forms of global governance, the most controversial of which has been the World Trade Organization, but the WTO is not itself the creator of the global economy.

Global market forces provide incentives for every nation to put on what the foreign-affairs columnist Thomas L. Friedman has called a "Golden Straitjacket," a set of policies that involve freeing up the private sector of the economy, shrinking the bureaucracy, keeping inflation low, and removing restrictions on foreign investment. If a country refuses to wear the golden straitjacket, or tries to take it off, then the electronic herd—the currency traders, stock and bond traders, and those who make investment decisions for multinational corporations—could gallop off in a different direction, taking the investment capital that countries want to keep their economy growing. When capital is internationally mobile, to raise your tax rates is to risk triggering a flight of capital to other countries with comparable investment prospects and lower taxation.

The upshot is that as the economy grows and average incomes rise, the scope of politics may shrink—at least as long as no political party is prepared to challenge the assumption that global capitalism is the best economic system. When neither the government nor the opposition is prepared to take the risk of removing the golden straitjacket, the differences between the major political parties shrink to differences over minor ways in which the straitjacket might be adjusted. Thus even without the WTO, the growth of the global economy itself marks a decline in the power of the nation-state.

Marx argued that in the long run we never reject advances in the means by which we satisfy our material needs. Hence history is driven by the growth of productive forces. He would have been contemptuous of the suggestion that globalization is something foisted on the world by a conspiracy of corporate executives meeting in Switzerland, and he might have agreed with Friedman's remark that the most basic truth about globalization is, *"No one is in charge."* For Marx that is a statement that epitomizes humanity in a state of alienation, living in a world in which, instead of ruling ourselves, we are ruled by our own creation, the global economy. For Friedman, on the other hand, all that needs to be said about Marx's alternative—state control of the economy—is that it doesn't work. (Whether there are alternatives to both capitalism and centrally controlled socialism that could work is another question, but not one for here.)

Marx also believed that a society's ethic is a reflection of the economic structure to which its technology has given rise. Thus a feudal economy in which serfs are tied to their lord's land gives you the ethic of feudal chivalry based on the loyalty of knights and vassals to their lord, and the obligations of the lord to protect them in time of war. A capitalist economy requires a mobile labor force able to meet the needs of the market, so it breaks the tie between lord and vassal, substituting an ethic in which the right to buy and sell labor is paramount.

Our newly interdependent global society, with its remarkable possibilities for linking people around the planet, gives us the material basis for a new ethic. Marx would have thought that such an ethic would serve the interests of the ruling class, that is, the rich nations and the transnational corporations they have spawned. But perhaps our ethic is related to our technology in a looser, less deterministic way than Marx thought.

Ethics appear to have developed from the behavior and feelings of social mammals. They became distinct from anything we can observe in our closest nonhuman relatives when we started using our reasoning abilities to justify our behavior to other members of our group. If the group to which we must justify ourselves is the tribe, or the nation, then our morality is likely to be tribal, or nationalistic. If, however, the revolution in communications has created a global audience, then we might feel a need to justify our behavior to the whole world. As Clive Kessler argued recently in *Third World Quarterly*, that change creates the material basis for a new ethic that will serve the interests of all those who live on this planet in a way that, despite much rhetoric, no previous ethic has ever done.

If this appeal to our need for ethical justification appears to be based on too generous a view of human nature, there is another consideration of a very different kind that leads in the same direction. The great empires of the past, whether Persian, Roman, Chinese, or British, were, as long as their power lasted, able to keep their major cities safe from threatening barbarians on the frontiers of their far-flung realms. In the 21st century the greatest superpower in history was unable to keep the self-appointed warriors of a different worldview from attacking both its greatest city and its capital.

My thesis is that how well we come through the era of globalization (perhaps whether we come through it at all) will depend on how we respond ethically to the idea that we live in one world. For the rich nations not to take a global ethical viewpoint has long been seriously morally wrong. Now it is also, in the long term, a danger to their security.

There is one great obstacle to further progress in this direction. It has to be said, in cool but plain language, that in recent years the international effort to build a global community has been hampered by the repeated failure of the United States to play its part. Despite being the single largest polluter of the world's atmosphere, and on a per-capita basis the most profligate of the major nations, the United States has refused to join the 178 states that have accepted the Kyoto Protocol. Along with Libya and China, the United States voted against setting up an International Criminal Court to try people accused of genocide and crimes against humanity. Now that the court seems likely to go ahead, the U.S. government has said that it has no intention of participating. Though it is one of the world's wealthiest nations, with the world's strongest economy, the United States gives significantly less foreign aid, as a proportion of its gross national product, than any other developed nation.

When the world's most powerful state wraps itself in what, until September 11, it took to be the security of its military might, and arrogantly refuses to give up any of its own rights and privileges for the sake of the common good—even when other nations are giving up their rights and privileges—the prospects of finding solutions to global problems are dimmed. One can only hope that when the rest of the world nevertheless proceeds down the right path, as it did in resolving to go ahead with the Kyoto Protocol, and as it is now doing with the International Criminal Court, the United States will eventually be shamed into joining in. If it does not do so, it risks falling into a situation in which it is universally seen by everyone except its own self-satisfied citizens as the world's "rogue superpower." Even from a strictly self-interested perspective, if the United States wants the cooperation of other nations in matters that are largely its own concern—such as the struggle to eliminate terrorism—it cannot afford to be so regarded.

I have argued that as more and more issues increasingly demand global solutions, the extent to which any state can independently determine its future diminishes. We therefore need to strengthen institutions for global decision making and make them more responsible to the people they affect. That line of thought leads in the direction of a world community with its own directly elected legislature, perhaps slowly evolving along the lines of the European Union.

There is little political support for such ideas at present. Apart from the threat that the idea poses to the self-interest of the citizens of the rich nations, many would say it puts too much at risk, for gains that are too uncertain. It is widely believed that a world government would be, at best, an unchecked bureaucratic behemoth that makes the bureaucracy of the European Union look like a lean and efficient operation. At worst, it would become a global tyranny, unchecked and unchallengeable. Those thoughts have to be taken seriously. They present a challenge that should not be beyond the best minds in the fields of political science and public administration, once those people adjust to the new reality of the global community and turn their attention to issues of government beyond national boundaries.

We need to learn from the experience of other multinational organizations. The European Union is a federal body that has adopted the principle that decisions should always be taken at the lowest level capable of dealing with the problem. The application of that principle, known as subsidiarity, is still being tested. But if it works for Europe, it is not impossible that it might work for the world.

To rush into world federalism would be too risky, but we could accept the diminishing significance of national boundaries and take a pragmatic, step-by-step approach to greater global governance. There is a good case for global environmental and labor standards. The World Trade Organization has indicated its support for the International Labor Organization to develop core labor standards. If those standards are developed and accepted, they would not be much use without a global body to check that they are being adhered to, and to allow other countries to impose trade sanctions against goods that are not produced in conformity with the standards. Since the WTO seems eager to pass this task over to the ILO, we might see that organization significantly strengthened.

Something similar could happen with environmental standards. It is even possible to imagine a United Nations Economic and Social Security Council that would take charge of the task of eliminating global poverty, and would be voted the resources to do it. These and other specific proposals for stronger global institutions to accomplish a particular task should be considered on their merits.

The 15th and 16th centuries are celebrated for the voyages of discovery that proved that the world is round. The 18th century saw the first proclamations of universal human rights. The 20th century's conquest of space made it possible for a human being to look at our planet from a point not on it, and so to see it, literally, as one world. Now the 21st century faces the task of developing a suitable form of government for that single world. It is a daunting moral and intellectual challenge, but one we cannot refuse to take up. It is no exaggeration to say that the future of the world depends on how well we meet it.

18

Why Mexico's Small Corn Farmers Go Hungry

TINA ROSENBERG

Macario Hernández's grandfather grew corn in the hills of Puebla, Mexico. His father does the same. Mr. Hernández grows corn, too, but not for much longer. Around his village of Guadalupe Victoria, people farm the way they have for centuries, on tiny plots of land watered only by rain, their plows pulled by burros. Mr. Hernández, a thoughtful man of 30, is battling to bring his family and neighbors out of the Middle Ages. But these days modernity is less his goal than his enemy.

This is because he, like other small farmers in Mexico, competes with American products raised on megafarms that use satellite imagery to mete out fertilizer. These products are so heavily subsidized by the government that many are exported for less than it costs to grow them. According to the Institute for Agriculture and Trade Policy in Minneapolis, American corn sells in Mexico for 25 percent less than its cost. The prices Mr. Hernández and others receive are so low that they lose money with each acre they plant.

In January, campesinos from all over the country marched into Mexico City's central plaza to protest. Thousands of men in jeans and straw hats jammed the Zócalo, alongside horses and tractors. Farmers have staged smaller protests around Mexico for months. The protests have won campesino organizations a series of talks with the government. But they are unlikely to get what they want: a renegotiation of the North American Free Trade Agreement, or Nafta, protective temporary tariffs and a new policy that seeks to help small farmers instead of trying to force them off the land.

The problems of rural Mexicans are echoed around the world as countries lower their import barriers, required by free trade treaties and the rules of the World Trade Organization. When markets are open, agricultural products flood in from wealthy nations, which subsidize agriculture and allow agribusiness to export crops cheaply. European farmers get 35 percent of their income in government subsidies, American farmers 20 percent. American subsidies are at record levels, and last year, Washington passed a farm bill that included a $40 billion increase in subsidies to large grain and cotton farmers.

SOURCE: Tina Rosenberg, "Why Mexico's Small Corn Farmers Go Hungry." *New York Times* (March 3, 2003). Online: http://www.nytimes/2003/03/03/opinion/03Mon3.html.

It seems paradoxical to argue that cheap food hurts poor people. But three-quarters of the world's poor are rural. When subsidized imports undercut their products, they starve. Agricultural subsidies, which rob developing countries of the ability to export crops, have become the most important dispute at the W.T.O. Wealthy countries do far more harm to poor nations with these subsidies than they do good with foreign aid.

While such subsidies have been deadly for the 18 million Mexicans who live on small farms—nearly a fifth of the country—Mexico's near-complete neglect of the countryside is at fault, too. Mexican officials say openly that they long ago concluded that small agriculture was inefficient, and that the solution for farmers was to find other work. "The government's solution for the problems of the countryside is to get campesinos to stop being campesinos," says Victor Suárez, a leader of a coalition of small farmers.

But the government's determination not to invest in losers is a self-fulfilling prophecy. The small farmers I met in their fields in Puebla want to stop growing corn and move into fruit or organic vegetables. Two years ago Mr. Hernández, who works with a farming cooperative, brought in thousands of peach plants. But only a few farmers could buy them. Farm credit essentially does not exist in Mexico, as the government closed the rural bank, and other bankers do not want to lend to small farmers. "We are trying to get people to rethink and understand that the traditional doesn't work," says Mr. Hernández. "But the lack of capital is deadly."

The government does subsidize producers, at absurdly small levels compared with subsidies in the United States. Corn growers get about $30 an acre. Small programs exist to provide technical help and fertilizer to small producers, but most farmers I met hadn't even heard of them.

Mexico should be helping its corn farmers increase their productivity or move into new crops—especially since few new jobs have been created that could absorb these farmers. Mexicans fleeing the countryside are flocking to Houston and swelling Mexico's cities, already congested with the poor and unemployed. If Washington wants to reduce Mexico's immigration to the United States, ending subsidies for agribusiness would be far more effective than beefing up the border patrol.

19

Binational Citizens

Mexican Migrants Are Challenging Old Ideas about Assimilation

JONATHAN FOX

This past spring, more than three million immigrants—most of them originally from Mexico—marched through the streets of dozens of U.S. cities to support a comprehensive reform that would legalize the status of undocumented immigrants. The size and number of the rallies caught almost everyone by surprise, including many in immigrant communities. Never before had Mexican migrants demanded such a visible role in a national policy discussion.

The ultimate impact of these marches on immigration reform remains uncertain. But the huge wave of Mexican civic engagement revealed a previously silent but steady and potentially profound transformation of the American political terrain: the emergence of Mexican migrants as civic and political actors.

More than 10 million Mexicans now live and work in the United States, including roughly one in eight adults worldwide who were born in Mexico. While the growing population is widely recognized, the presence of Mexican *society* in the United States has not been fully appreciated. The conventional view, pressed with particular energy by conservative nationalists, is that Mexican immigrants are highly insular. These critics point to lower rates of naturalization, English-language acquisition, and social mobility compared to other national-origin groups, as well as persistent pride by Mexican immigrants in their language and ethnicity. The large concentrations of Spanish-speaking immigrants in major cities are, the critics conclude, inherently unassimilable.

While many different factors may account for these low rates of naturalization and social mobility, it is well known that many Mexican migrants retain deep ties to their country of origin. Many work together with their *paisanos* to promote "philanthropy from below," funding hundreds of community-development initiatives in their hometowns. And more than 40,000 signed up to exercise their newly won right to cast absentee ballots in Mexico's 2006 presidential election.

But as the recent demonstrations suggest, things may now be changing as Mexican migrants create new ways of becoming American, with membership

SOURCE: Jonathan Fox, "Binational Citizens: Mexican Migrants Are Challenging Old Ideas about Assimilation,"*Boston Review* (September/October 2006), pp. 26–27.

organizations playing a central role. Most migrant organizations—often with members from the same state in Mexico—started out focused exclusively on aid to their hometowns, but many have now developed programs for families in their new communities in the United States. They have thus become important arenas for migrants to hone the skills that allow them to enter U.S. civic life, as well as city and state politics. Migrants who participate in these associations often claim a kind of *civic binationality*—simultaneous membership in Mexican and U.S. society—with their initial engagement with hometowns abroad spurring their active engagement with adopted hometowns in the United States. These binational migrants have a great deal to teach us about new opportunities for encouraging immigrant integration into the United States today.

The most prominent group of membership organizations for migrants are the hometown associations (HTAs); but there are also worker organizations and religious congregations. The Mexican consulates have registered well over 600 HTAs, with some estimates exceeding 2,000. Typically the core membership is around two dozen families, though some have hundreds more. HTAs are primarily concentrated in metropolitan areas, and especially in Los Angeles and Chicago. Many HTA members are well established: the leaders are often economically stable and have legal status or citizenship. HTAs have in turn federated into associations that bring people from one state in Mexico together in one state in the United States, as in the flagship case of the numerous Zacatecas Federations.

A *Sacramento Bee* reporter recently estimated that Mexican HTAs have an active membership of between 250,000 and 500,000; and 14 percent of relatively recent Mexican migrants surveyed in 2005 said that they belonged to some kind of hometown association. HTAs have a long history, with the first Zacatecan club in California dating back to 1962, but their numbers and membership have boomed in the past 15 years. Within the United States, the massive regularization of undocumented workers that followed the 1986 immigration reform facilitated both economic improvement and increased cross-border freedom of movement for millions of migrants. On the Mexican side, the government deployed the convening power of its extensive consular apparatus, bringing together people from the same community of origin and offering three-to-one community-development matching funds to encourage collective social investments of remittances. Though these efforts began as a response to pressures from organized Zacatecan migrants, they also served as a powerful inducement for other migrants to come together in formal organizations for the first time. In addition, in 1996 the Mexican state changed the tone of its relationship with the diaspora by formally permitting dual nationality for the first time. While many clubs emerged from below, many of the state-level federations were formed through engagement with the Mexican state.

But beyond the boom in sheer numbers, many Mexican migrant organizations are changing their political focus, shedding their prior disengagement from U.S. society and politics. Mexican HTAs did relatively little in the 1994 campaign against California's notorious anti-immigrant Proposition 187. A decade later, when the state-level immigrant-rights advocacy campaign rallied for drivers' licenses for the undocumented, HTA members were very actively involved, working the phone banks at the headquarters of Los Angeles's formidable trade-union movement. The leadership of the Southern California Council of Presidents of Mexican Federations has now joined the fray of state politics. Some Mexican federations have also joined the migrant-led National Alliance of Latin American and Caribbean Communities, especially in the Midwest. The AFL-CIO and the immigrant-led National Network of Day Laborers recently announced a collaborative agreement. These kinds of alliances would have been hard to imagine a decade ago.

Mainstream Latino politicians and public-interest groups in the United States are also making new efforts to reach out to Mexican HTAs. It is worth quoting the *Sacramento Bee* account in detail:

> "Our goal has been to help our communities in Mexico. Now it is time to help our communities here," said Salvador Garcia, the owner of a Los Angeles–area demolition company who serves as president of the Consejo [of Mexican HTA federations] as well as the hometown Federacio' de Jalisco.
>
> Recently, Garcia was chairman of a meeting—conducted in Spanish—at Sebastian Dominguez's auto body shop with leaders representing collective hometown associations from eight Mexican states. They talked about raising money to pay for college scholarships for immigrant children and for soccer fields in Spanish-speaking communities in Los Angeles. A few weeks ago, Mexican politicians from the states of Michoaca'n and Oaxaca stopped by the auto body shop for the hometown associations' support in encouraging Los Angeles residents to vote in the 2006 Mexican presidential election. On this night, the featured visitor was Ann Marie Tallman, national president and general counsel for the Mexican American Legal Defense and Education Fund. Tallman proposed a partnership, offering the Consejo presidents use of office space at the legal defense and education fund's Los Angeles headquarters, business leadership classes and media training. . . "We really need to reconnect with our roots," Tallman said later. "They (hometown associations) are the eyes and ears of the community. This is a bona fide movement. . . Shame on us for not noticing before."

Traditional Latino organizations and Mexican migrant organizations often overlap in their issues and sometimes even membership, though they often have very different organizational structures, access to resources, and views on whether to pursue a binational or primarily U.S.-focused agenda. While traditional Latino organizations tend to be focused on civil-rights issues in the United States and questions of equal access to health care and education, migrant organizations tend to be focused on binational issues and on specific concerns of access to services that specifically affect immigrants. While U.S. Latino leaders are strongly committed to promoting immigrant incorporation, some have been skeptical about whether migrants' binational perspectives foster civic integration into U.S. society.

Nonetheless, the gap between these agendas is narrowing as Mexican migrant organizations become increasingly U.S.-focused and Latino organizations increasingly embrace concerns of the growing number of U.S. Latinos who are migrants. Thus, the July 2006 national conference of the National Council of La Raza involved an unprecedented degree of outreach to immigrants, including widespread interest in citizenship promotion and Spanish-language workshops.

Mexican migrants have also become increasingly active in traditional U.S. social organizations, such as affiliates of the Industrial Areas Foundation. In addition, both Catholic and evangelical Protestant churches have seen much of their growth come from Latin American migrants. Some religious social organizations, such as New York City's Asociación Tepeyac, see their role as building the social and political engagement of migrants to give them a voice in U.S. society while they continue to engage with their country of origin. These communities appropriate symbols and patterns of worship from migrants' hometowns while addressing issues that migrants face in the United States.

Worker organizations have also been important vehicles of social integration. Despite their lack of prior experience with democratic unions, Mexican migrant workers express a similar level of interest in unions as others in the United States. Many migrants work in non-unionized industries—especially agriculture, residential construction, and services—and the emergence of worker support centers across the United States has proved particularly important. For immigrant farm workers, who are often geographically and socially isolated, outreach to U.S. public opinion has often involved consumer boycotts, usually involving alliances with religious communities and university students—as in the case of the Coalition of Immokalee Workers' recent successful campaign against Taco Bell. (The spring 2006 marches constituted the largest mass mobilization of workers of any kind in the history of the United States.)

Finally, Spanish-language media also play a decisive role both in sharing information among migrants and creating pathways to engagement in U.S. society. Three major national television networks now broadcast in Spanish, along with dozens of local stations and cable channels, more than 300 radio stations, and over 700 newspapers. These media help address issues that matter particularly to migrants from Mexico and elsewhere in Latin America in a way that neither English-language nor home-country media do (although migrants do use both of these extensively as well). The spring 2006 immigrant-rights protests showed the capacity of Spanish-language media to help mobilize millions of people. In many cities, radio hosts—many engaging with civic issues for the first time—played a central role in generating mass interest among migrants in participating in these protests. In other cases, these media also provide information on voting, health campaigns, and issues in the educational system, among many other matters of concern to migrants. Some public media, such as Radio Bilingue, were specifically created to serve as an information source for migrants to share and address their concerns, and even mainstream Spanish-language media leaders tend to see this as part of their mission.

Despite extensive gains in civic engagement, Mexican migrants' electoral participation remains very low compared to their overall numbers. The large number of undocumented migrants—perhaps half of all Mexican migrants—is part of the reason for this. Even among those who are permanent residents and eligible for citizenship, however, the naturalization rate remains far below that of immigrants from other countries. This appears to be changing, though. Between 1995 and 2001 the estimated percentage of legal residents of Mexican origin who became citizens doubled from 17 to 34. For those who do become citizens, the voter-turnout rate tends to follow broader U.S. patterns, in which lower levels of formal education and income are associated with lower turnout. Yet studies that compared naturalized-citizen turnout in California with that of other states found that the state's politicized environment of the 1990s encouraged significantly higher voting rates. It will be important to observe to what extent the recent mobilization will lead to an increase in the interest of Mexican legal permanent residents in becoming full citizens with voting rights.

Mexican migrants have an even lower degree of formal engagement in Mexican elections. In 2005, the Mexican Congress for the first time allowed Mexicans abroad to register to vote in Mexico by absentee ballot. Just over one percent of those eligible registered for the 2006 presidential elections (though in comparative terms, this is normal for first-time diasporic voting). The low registration rate—though still not well-understood—undoubtedly reflects the numerous procedural challenges involved in the complicated registration process. Voters had to register by registered mail more than six months before the elections, and long-distance voting rights were conditioned on a ban on Mexican political-party or campaign activity abroad, which meant that migrants had to depend almost exclusively on U.S. Spanish-language media to become informed voters.

Since the 1990s, the Mexican government has been seeking to increase its ties to migrants abroad—for example, it formed the Council of Mexicans Abroad, a group of Mexican migrants charged with advising the Mexican government on policy related to migrant communities. Although the results of this process in terms of actual influence on policy decisions are mixed, the council has built bridges between local migrant leaders and the Mexican government. The council's membership, which is largely elected, also reflects a high degree of civic binationality, insofar as many of these leaders combine deep roots in U.S. civic, social, and business organizations with strong ties to migrant organizations and to Mexico.

The overall panorama of Mexican migrant civic participation is a hopeful one. The peaceful immigrant protests in dozens of U.S. cities in the spring of 2006 reflected an extraordinary level of civic discipline, largely modeled by the key mobilizing institutions—churches, the media, community organizations, and unions. Yet participation went far beyond these organizations and their members and drew in large numbers of normally unaffiliated migrants and their supporters. This suggests an even greater breadth of civic commitment beyond formal participation in existing organizations.

As the number of Mexicans in the United States grows, they are increasingly engagaed in U.S. civic life—and they are reshaping it. Moreover, they are developing their own forms of civic association that represent their own needs and interests—just as past waves of immigrants to the United States did before restrictive policies closed the door in the early 1920s. So what's new with Mexican migrants? After all, the percentage of foreign-born in the United States today is comparable to the peak of the last major wave. It is certainly new that migrants from one single country represent such a large share of the foreign-born population. It is also new that so many immigrants come from across the border rather than across an ocean, which encourages persistent home-country ties. What may be newest, however, is the challenge they pose to traditional conceptions of assimilation and nationalism. While conservative critics like Samuel Huntington assume that these trends will lead to insularity and pose a Quebec-style threat to the U.S. social fabric, Mexican migrants themselves are demonstrating their capacity to forge practices of civic binationality, campaigning for the right to be heard in both Mexico and the United States.

REFLECTION QUESTIONS FOR CHAPTER 5

1. Social scientists say that the elements of globalization—capitalism, trade, and technological revolutions—are breaking down old obstacles and mindsets. With that in mind, is the notion of national sovereignty an outmoded concept?

2. How do national leaders come to accept the idea expressed by Singer that their role is to reject narrow national goals that conflict with goals for all of humanity?

3. Using the essay by Stiglitz as a guide, what are the benefits of globalization? What countries have benefited the most from globalization and why?

4. Explain Rosenberg's contention that cheap food hurts poor people. How do the U.S. and Mexican governments contribute to this paradox?

5. What does Fox mean by *civic binationality?* How does he think Mexican immigrants are reshaping civic life in the United States?

6

Cultural Globalization

Tribes have a culture. Societies have a culture. The question: Is there a global culture? The concept, traditionally, has meant the knowledge that a people share. This knowledge includes all manner of shared meanings such as religion, myths, technology, political ideology, language, music, art, fashion, and consumption patterns. Using this conventional view of culture, there are no global equivalencies to global myths, legends, and symbols that unite the world's people. On the other hand, English is at least a second language throughout much of the world, Western movies and television shows are shown almost everywhere, and products such as McDonald's, Coca-Cola, Disney, and Nike proliferate worldwide. There are global icons from the worlds of entertainment (for example, Julia Roberts, Harrison Ford, Britney Spears, U2, Rolling Stones) and sports (for example, Muhammad Ali, Tiger Woods, Michael Jordan). In short, while there are real differences across national boundaries, some elements of culture transcend those boundaries.

While local or national culture is tied to place and time, "global culture is free of these constraints: as such it is 'disconnected,' 'disembedded' and 'deterritorialized,' existing outside the usual reference to geographical territory."[1] There is a continuous and rapid flow of images, ideas, information, tastes, and products across geographical, political, and linguistic borders.

Two major issues arise from this form of global integration. First, the world appears to be becoming culturally homogenized along a westernized model. Is this evidence of progressive cosmopolitanism or oppressive imperialism?[2] Second, will a global culture diminish or even demolish the uniqueness, diversity, and richness of local cultures? In other words, does globalization lead to cultural homogenization or does it result in heterogenization as various groups resist the global onslaught on their way of life?

The first selection in this chapter, by political scientist Manfred B. Steger, addresses the question of whether the global culture results in sameness or difference. Then labor historian Leon Fink gives us a case study of the effects of globalization on four representative transplanted Guatemalans living and working in Morganton, North Carolina.

The final essay addresses a different question: What is the effect of the media on the globalization process? Katharine Ainger says that we tend to think of the transnational corporations of industrial capitalism as the engines of globalization. "But the globalizing conquistadores of the twenty-first century are the media giants of cultural capitalism—Disney, AOL Time Warner, Sony, Bertelsmann, News Corporation, Viacom, Vivendi Universal."

ENDNOTES

1. John Beynon and David Dunkerley (eds.), *Globalization: The Reader* (New York: Routledge, 2000), p. 13.
2. Jan Aart Scholte, *Globalization: A Critical Introduction* (New York: Palgrave, 2000), p. 23.

20

Global Culture

Sameness or Difference?

MANFRED B. STEGER

Does globalization make people around the world more alike or more different? This is the question most frequently raised in discussions on the subject of cultural globalization. A group of commentators we might call "pessimistic hyperglobalizers" argue in favour of the former. They suggest that we are not moving towards a cultural rainbow that reflects the diversity of the world's existing cultures. Rather, we are witnessing the rise of an increasingly homogenized popular culture underwritten by a Western "culture industry" based in New York, Hollywood, London, and Milan. As evidence for their interpretation, these commentators point to Amazonian Indians wearing Nike training shoes, denizens of the Southern Sahara purchasing Texaco baseball caps, and Palestinian youths proudly displaying their Chicago Bulls sweatshirts in downtown Ramallah. Referring to the diffusion of Anglo-American values and consumer goods as the "Americanization of the world," the proponents of this cultural homogenization thesis argue that Western norms and lifestyles are overwhelming more vulnerable cultures. Although there have been serious attempts by some countries to resist these forces of "cultural imperialism"—for example, a ban on satellite dishes in Iran, and the French imposition of tariffs and quotas on imported film and television—the spread of American popular culture seems to be unstoppable.

But these manifestations of sameness are also evident inside the dominant countries of the global North. American sociologist George Ritzer coined the term "McDonaldization" to describe the wide-ranging sociocultural processes by which the principles of the fast-food restaurant are coming to dominate more and more sectors of American society as well as the rest of the world. On the surface, these principles appear to be rational in their attempts to offer efficient and predictable ways of serving people's needs. However, looking behind the façade of repetitive TV commercials that claim to "love to see you smile," we can identify a number of serious problems. For one, the generally low nutritional value of fast-food meals—and particularly their high fat content—has been implicated in the rise of serious health problems such as heart disease, diabetes, cancer, and juvenile obesity. Moreover, the impersonal, routine operations of "rational" fast-service

SOURCE: "Global Culture: Sameness or Difference?" by Manfred B. Steger from *Globalization—A Very Short Introduction*. By permission of Oxford University Press.

establishments actually undermine expressions of forms of cultural diversity. In the long run, the McDonaldization of the world amounts to the imposition of uniform standards that eclipse human creativity and dehumanize social relations.

Perhaps the most thoughtful analyst in this group of pessimistic hyperglobalizers is American political theorist Benjamin Barber. In his popular book on the subject, he warns his readers against the cultural imperialism of what he calls "McWorld"—a soulless consumer capitalism that is rapidly transforming the world's diverse populations into a blandly uniform market. For Barber, McWorld is a product of a superficial American popular culture assembled in the 1950s and 1960s, driven by expansionist commercial interests. Music, video, theatre, books, and theme parks are all constructed as American image exports that create common tastes around common logos, advertising slogans, stars, songs, brand names, jingles, and trademarks.

Barber's insightful account of cultural globalization also contains the important recognition that the colonizing tendencies of McWorld provoke cultural and political resistance in the form of "Jihad"—the parochial impulse to reject and repel the homogenizing forces of the West wherever they can be found. . . . Jihad draws on the furies of religious fundamentalism and ethnonationalism which constitute the dark side of cultural particularism. Fuelled by opposing universal aspirations, Jihad and McWorld are locked in a bitter cultural struggle for popular allegiance. Barber asserts that both forces ultimately work against a participatory form of democracy, for they are equally prone to undermine civil liberties and thus thwart the possibility of a global democratic future.

Optimistic hyperglobalizers agree with their pessimistic colleagues that cultural globalization generates more sameness, but they consider this outcome to be a good thing. For example, American social theorist Francis Fukuyama explicitly welcomes the global spread of Anglo-American values and lifestyles, equating the Americanization of the world with the expansion of democracy and free markets. But optimistic hyperglobalizers do not just come in the form of American chauvinists who apply the old theme of manifest destiny to the global arena. Some representatives of this camp consider themselves staunch cosmopolitans who celebrate the Internet as the harbinger of a homogenized "techno-culture." Others are free-market enthusiasts who embrace the values of global consumer capitalism.

It is one thing to acknowledge the existence of powerful homogenizing tendencies in the world, but it is quite another to assert that the cultural diversity existing on our planet is destined to vanish. In fact, several influential commentators offer a contrary assessment that links globalization to new forms of cultural expression. Sociologist Roland Robertson, for example, contends that global cultural flows often reinvigorate local cultural niches. Hence, rather than being totally obliterated by the Western consumerist forces of sameness, local difference and particularity still play an important role in creating unique cultural constellations. Arguing that cultural globalization always takes place in local contexts, Robertson rejects the cultural homogenization thesis and speaks instead of "glocalization"—a complex interaction of the global and local characterized by cultural borrowing. The resulting expressions of cultural "hybridity" cannot be reduced to clear-cut manifestations of "sameness" or "difference." [S]uch processes of hybridization have become most visible in fashion, music, dance, film, food, and language.

In my view, the respective arguments of hyperglobalizers and sceptics are not necessarily incompatible. The contemporary experience of living and acting across cultural borders means both the loss of traditional meanings and the creation of new symbolic expressions. Reconstructed feelings of belonging coexist in uneasy tension with a sense of placelessness. Cultural globalization has contributed to a remarkable shift in people's consciousness. In fact, it appears that the

The American Way of Life

Number of types of packaged bread available at a Safeway in Lake Ridge, Virginia	104
Number of those breads containing no hydrogenated fat or diglycerides	0
Amount of money spent by the fast-food industry on television advertising per year	$3 billion
Amount of money spent promoting the National Cancer Institute's "Five A Day" programme, which encourages the consumption of fruits and vegetables to prevent cancer and other diseases	$1 million
Number of "coffee drinks" available at Starbucks, whose stores accommodate a stream of over 5 million customers per week, most of whom hurry in and out	26
Number of "coffee drinks" in the 1950s coffee houses of Greenwich Village, New York City	2
Number of new models of cars available to suburban residents in 2001	197
Number of convenient alternatives to the car available to most such residents	0
Number of U.S. daily newspapers in 2000	1,483
Number of companies that control the majority of those newspapers	6
Number of leisure hours the average American has per week	35
Number of hours the average American spends watching television per week	28

SOURCES: Eric Schlosser, *Fast Food Nation* (Houghton Miffin, 2001), p. 47; www.naa.org/info/facts00/11.htm; *Consumer Reports Buying Guide 2001* (Consumers Union, 2001), pp. 147–163: Laurie Garrett, *Betrayal of Trust* (Hyperion, 2000), p. 353; www.roper.com/news/content/news169.htm; *The World Almanac and Book of Facts 2001* (World Almanac Books, 2001), p. 315; www.starbucks.com.

old structures of modernity are slowly giving way to a new "postmodern" frame-work characterized by a less stable sense of identity and knowledge.

Given the complexity of global cultural flows, one would actually expect to see uneven and contradictory effects. In certain contexts, these flows might change traditional manifestations of national identity in the direction of a popular culture characterized by sameness; in others they might foster new expressions of cultural particularism; in still others they might encourage forms of cultural hybridity. Those commentators who summarily denounce the homogenizing effects of Americanization must not forget that hardly any society in the world today possesses an "authentic," self-contained culture. Those who despair at the flourishing of cultural hybridity ought to listen to exciting Indian rock songs, admire the intricacy of Hawaiian pidgin, or enjoy the culinary delights of Cuban-Chinese cuisine. Finally, those who applaud the spread of consumerist capitalism need to pay attention to its negative consequences, such as the dramatic decline of communal sentiments as well as the commodification of society and nature.

21

The Place of Community in Globalization

LEON FINK

It was sometime in early spring 1997 when I first heard about a labor conflict in Morganton, N.C., normally a quiet industrial center of 16,000 people perched at the edge of the Great Smoky Mountains. I would soon learn that this was no isolated incident or temporary labor relations breakdown, but a decade-long war of position between determined and well-organized workers on the one hand and a classically recalcitrant employer on the other. Beginning with an overnight walkout in 1991, erupting into a mass work stoppage and multiple arrests in May 1993, recharging with a four-day strike and successful union election campaign in 1995, continuing with a weeklong walkout and hunger strike in 1996, and a subsequent six-year face-off with a company that absolutely refused to sign a collective-bargaining contract, the workers at Case Farms poultry plant etched a profile of uncommon courage by demanding a voice and degree of respect at the workplace.

The events in Morganton were remarkable enough in themselves. For more than 20 years, as a labor historian at the University of North Carolina at Chapel Hill, I had watched "organized labor" virtually disappear from the map in North Carolina, losing battle after battle in campaigns waged by textile, furniture, and meat-processing workers. The state was locked in a seesaw battle with its neighbor South Carolina for the dubious honor of being the least unionized state in the country, a country that, as a whole was experiencing a severe slippage of union representation within the private-sector work force. What is more, the food-processing industry—and especially makers of meat products—had earned a reputation as the most determined of union foes.

But there were two additional reasons to be intrigued by the Morganton story. The 500-person Case Farms plant, estimated in 1995 as 80-percent Spanish-speaking—of whom 80 to 90 percent were Guatemalan and the rest Mexican—highlights a demographic transformation of the labor force in the United States, nowhere more dramatic than in what one commentator has called the "nuevo New South" and, within the region, nowhere more so than in North Carolina, where the Hispanic population grew by a whopping 394 percent between 1990 and 2000. Even more remarkably, the fact that the Guatemalans were nearly all Highland Maya—people who traced their bloodline

SOURCE: Leon Fink, "The Place of Community in Globalization," from *The Chronicle of Higher Education* (July 11, 2003), pp. B12–B13. Reprinted with permission from the author.

and their languages back to the ancient "corn people"—suggested a dramatic confrontation. How was it that, in a state with not a single organized chicken-processing factory, a group of Central American refugees bucked the tide of history?

Four quick impressions. The first image is of Gaspar Francisco, a Q'anjob'al-speaking Case Farms worker from a tiny aldea outside San Miguel Acatán, Guatemala. In discussion, it is quickly obvious that, for Gaspar, Morganton (and, indeed, the whole world outside San Miguel Acatán) is but a means to a paycheck, a way to make ends meet for his family back home. Gaspar has lived in the United States since 1985, and he was among the first 25 Guatemalans to settle in Morganton in 1990. Despite a decade of residency, however, Gaspar has precious few friends in Morganton; indeed, he knows little about the people or place in which he currently lives. Of the English language, he has learned *casi nada*. The mammoth share of what he earns, he sends back monthly in $500 or $1,000 increments to Guatemala. He rarely indulges in pastimes other than rest. Each year, with the company's permission, he spends up to one month in his native village. Only occasionally frequenting the Roman Catholic Church, Gaspar watches Univision and dreams of one day going home with enough money to buy land (his family has always rented) near San Miguel. A man of simple tastes, Gaspar would like nothing more than to return to his corn patch to continue, if slightly less desperately, the life bequeathed to him by his ancestors.

The second image is of a sidewalk conversation, outside a private mail-service office in San Miguel Acatán during the summer of 1998. Marcos Miguel González runs the office, and, at the end of the working day, he warmly greets the North American travelers who have been deposited on his street after a long bus ride from the regional capital of Huehuetenango. The big news from Huehue is that the volcano Pacaya erupted the previous day, spilling gray ash over the streets of the capital. The story is all over the country's newspapers and national television channel. Yet Marcos knows nothing of the story. None of the metropolitan dailies has yet arrived in San Miguel (in retrospect, I am not sure how often they ever appear there), and his own satellite TV does not pick up the signal from the capital. While Marcos pumps the North American visitors for news about his own country, he has exciting news for us as well. He knows that Reggie Miller helped the Indiana Pacers defeat the Chicago Bulls, 107–105, in an NBA Eastern Conference semifinal game the previous evening.

The third image summons up Justo Herman Castro Lux and his wife, Tránsita Gutiérrez Solís Castro, K'iche'-speaking immigrants from the Guatemalan department of Totonicapán and two of the most devoted Pan-Mayanists in Morganton. Appointed to a community-development project in the summer of 1998, Justo and his Chilean-born "co-missioner," Francisco Risso, a Catholic Worker activist, co-founded a local Consejo Maya, a Maya Council, that worked at several levels: offering English-language classes, an alco-holics' self-help group, support for the Case Farms union and for immigrant rights, a class in leadership training, and a Maya cultural group featuring poetry declamations and the music of the marimba. That local effort was linked to a national Pan-Maya movement, made concrete in the First Maya Congress in North America, December 3–4, 1999, in Lake Worth, Fla., where Justo emerged as a member of the Congress's governing council. To celebrate their commitment to tradition, Justo and Tránsita have chosen names for their four children—Tojil, Canil, Ixchel, and Chilam—from the pantheon of Maya literature. For her part, Tránsita always wears her native traje, her traditional clothes, even while at work in a local convenience store.

The final image is an exchange with three daughters of Francisco "Pancho" Jose (the Q'anjob'al leader of Morganton's first Guatemalan labor protest) in the library of Morganton's St. Charles Catholic Church. Of the three, Ana, now 26, is the only one who has even a faint memory of her life in Guatemala. Yet Ana, Petrona, 23, and Isabel, 17, all identify very much as Guatemalans and Q'anjob'al, as well as Hispanics, though not as Latinos ("that's the Mexicans"). Isabel, the only one of the three with U.S. citizenship, avidly watches *Who Wants to Be a Millionaire* and is proud when she can guess the right answer. She is considering joining ROTC and embarking on a career in the U.S. Army, thinking it could be a path toward at once professional training as a nurse-midwife and a down payment on a house. Ultimately, she would like to live in Hawaii, for there, she dreams, are Americans who "look a lot like us!" Ana, for her part, takes pride in her native traje; unfortunately, however, she has "never learned" to put it on correctly, so simply throws it around her as a "jumper" when she goes to church. At the end of our interview, when they have answered all my questions, Petrona interjects, "Tell us about Guatemala! What is it like?"

Such stories only begin to suggest the complexity of identity among the contemporary Maya, whether in their native Guatemala or transplanted to the United States. Within the space of the four examples, we meet in Gaspar a classic "bird of passage," a temporary migrant worker who identifies overwhelmingly with his home country and is culturally all but unaffected by his economically forced emigration. Meanwhile, Pancho's daughters betray both the power and ambiguity of assimilation—on the one hand, the pull of an American-centered consumer melting pot; on the other, the unmistakable sense of difference, bred at once by loyalty to one's roots and imposed by racial and ethnic discrimination in the United States.

The story of Marcos—the only nonimmigrant of the bunch—suggests something rather different. By tuning into a globalized communications network (as well as making a living, in part, off remittance payments sent from the north), he points to a new "transnational" reality affecting Guatemalans—to a greater or lesser degree, in both host and sender countries. If globalization has not quite taken command in San Miguel, it certainly has made a difference. Yet, rather than simple absorption into the logic of the global marketplace, the case of Justo and Tránsita suggests an alternative form of transnational identity. They use the computer and Internet access precisely to *not* be appropriated into a Westernized nation.

The story of the Maya in Morganton effectively joins one of the newest buzz-words of social science, "globalization," with one of the oldest, "community." Juxtaposing the two concepts, in fact, permits us to cover a great deal of interpretive ground. The entry into the poultry plants of North Carolina of Guatemalan war refugees and, later, those we might well consider "economic refugees" reflects the increasing fluidity of both world investments and labor markets. Just as much of the core manufacturing of American-owned corporations has shifted to offshore and, especially, Third World sites, so an increasing percentage of low-wage labor within the United States is being performed by new-immigrant workers.

In one of the more schematic discussions of globalization, Peter L. Berger, in the essay "Four Faces of Global Culture," distinguishes four variations on global culture: "Davos" (or business) culture, "Faculty Club International" (or human-rights) culture, "McWorld" (or commercial pop) culture, and Evangelical Protestantism. Yet, the Guatemalan migrant poultry workers surely take their place among a fifth variation of what the anthropologist Arjun Appadurai calls the "global cultural economy." Moreover, though clearly harnessed to an employer-defined policy of neoliberal access to cheap labor, the new-immigrant labor pools constitute more than a mere site of exploitation.

The Maya of Morganton are also quintessential representatives of a community. In recent times, community has regularly appeared to America scholars as something "they" (the primitive, pre-modern, or non-Western) have, but that "we" have lost or are struggling to renew. Preoccupation with the decline of community has been at the root of much modern social-science investigation, and it continues today. Perhaps the most prominent recent version of the community-breakdown thesis is the political scientist Robert D. Putnam's *Bowling Alone: The Collapse and Revival of American Community,* which argues that civic engagement in the United States has dangerously declined over the past 30 years.

The appearance of the Maya in the heartland of solitary (or at least unorganized) workers in and of itself constitutes a kind of interpretive provocation. In ways that recall the experience of their 19th-century immigrant predecessors, the émigré Maya workers "use" community at once to defend themselves against employer exploitation and to advance the interests of family and friends across international borders. The new-immigrant carriers of a rural, communal culture may indeed offer instructive lessons to a more metropolitan labor movement in the United States.

Even as contemporary sociologists inveigh against the decline of a broad-based "civic culture" in today's consumer society, more and more of the hard work of the country is actually in the hands of people with a quite sturdy family and community structure—if little else. Rather than simply totting up the gains and losses of "modernization" for such traditional virtues as "community," therefore, the very presence of diverse social logics operating *within* contemporary American culture demands a more flexible interpretive instrument.

My research into the genesis of labor protest and union organization at Case Farms also points to subtle patterns of community mobilization that draw on connections between the Maya homeland and the diaspora of emigration. The union, for one, succeeded to the extent that it inserted itself (however unconsciously) to recreate a kind of *comité cívico,* the local decision-making bodies that have taken on increasing significance in Guatemala's indigenous centers since the mid-1980s. Trusting in the union leaders—and the union's strike votes—as they might honor a *comité's* verdict on a water or road-building decision for their pueblo, the workers (or at least a solid majority of them over several years) assumed an obligation to act together for the common good.

It is a stretch, but not an unwarranted one, to link the union discipline among Guatemalan workers in Morganton to the sense of togetherness and practice of community self-help—extending over centuries in Guatemala—manifest in the community *comité*, the *cofradía* (religious brotherhoods dedicated to the local Catholic saint), and the cargo system of public duties associated with earlier civil-religious hierarchies. Perhaps most telling, in that regard, was the fact that the lead organizers of the union in Morganton in 1995 had, only months before, established a hometown association aimed at sending the remains of those villagers who had died in the United States back to their native soil for proper burial.

The combination of group ties among new immigrants and the necessity of relying on those ties in an alien environment creates an opening for worker mobilization among the displaced population that ought to interest labor activists and their academic observers alike. That is not to deny the obstacles that confront today's bottom-rung labor force. To date, neither their internal ties nor their outer support have proved sufficient to gain Morganton's poultry workers a union contract. And, indeed, after six years of trying—including two court-mandated but totally frustrating bouts of direct negotiations with company lawyers—the Laborers' International Union of North America officially abandoned its commitment to the Case Farms workers in the summer of 2001. In the absence of a collective-bargaining agreement—which no court or labor-relations board can impose under American law—the workers again are left entirely to the resources of their own informal community ties.

It is almost a truism that both modern-day power and money are stacked against low-wage workers like those at Case Farms as never before. Moreover, since September 11, 2001, the tightening of U.S. borders and the collapse of the job market add immeasurably to their burden. Nor is it to suggest that members of the Mayan diaspora are not themselves undergoing profound cultural displacement and acting in ways different from their reverted ancestors. All of the above are true. The very richness of the mix, however, creates one of the more dramatic trials of the human spirit provoked by the faceless forces of globalization.

22

Empires of the Senseless

KATHARINE AINGER

By the end of the first day of the historic street protests against the World Trade Organization—meeting to decide the future of the global economy as the millennium turned—almost every newspaper-dispensing box in Seattle had been graffitied with a single word: *Lies.*

Leafing through the newspaper reports the next morning, it wasn't hard to see why.

The editorial pages brimmed with the apoplectic outpourings of the planet's leading opinion-formers. Thomas Friedman of the *New York Times* called the anti-globalization protesters, "a Noah's ark of flat-earth advocates, protectionist trade unions, and yuppies looking for their 1960s fix." "Andrew Marr, writing in liberal British broadsheet *The Observer,* described their demands as "the Communist Manifesto rewritten by Christopher Robin." The *Wall Street Journal* joined in to jeer the "global village idiots . . . bringing their bibs and bottles to [Washington] this week" when the movement targeted the World Bank in April 2000.

The media have not been "pro-globalization" so much as an integral part of the process. For most journalists neoliberalism is not an economic ideology whose fundamental assumptions can be challenged, but simply "reality." Though they are occasionally willing to cover isolated problems of market economics and corporate rule, they greet systemic critiques of the global power structure with derision and incomprehension. For as a wise person once said: "The first to discover water was unlikely to have been a fish."

When CEOs, heads of state and other luminaries met in Davos, Switzerland, this January for the "summit of business summits" the presence of the corporate media moguls came as no surprise. But also nibbling hors d'oeuvres with the architects of globalization were a group of specially chosen "Media Leaders"— about 200 editors, producers and commentators from around the world, who not only took part in the meetings but also attended special closed sessions. Among them: Bill Emmot of *The Economist,* Will Hutton of *The Observer,* Thomas Friedman of the *New York Times,* Alan Yentob of the BBC.

Meanwhile volunteer journalists with the Davos Independent Media Center were camped in the snow outside providing alternative reporting on the anti-globalizers'

SOURCE: Katharine Ainger, "Empires of the Senseless: The Media Don't Just Promote."
From *New Internationalist Magazine*, April 2001, Issue 333. Reprinted by kind permission
of the *New Internationalist.* Copyright *New Internationalist* (www.newint.org).

counter-summit. Many were harassed by armed Swiss police, others had their equipment confiscated. Clearly, if you're the wrong kind of media reporting on the wrong kind of meeting, you're not a journalist. You're a criminal.

THERE'S NO BUSINESS LIKE SHOWBUSINESS

When we think of the rapacious corporations of industrial capitalism, we tend to picture big oil companies opening up wells and driving indigenous peoples from their lands; giant fruit multinationals controlling vast, pesticide-filled plantations; steelworks, mines, roads across pristine wilderness.

But the globalizing conquistadores of the 21st century are the media giants of cultural capitalism—Disney, AOL Time Warner, Sony, Bertelsmann, News Corporation, Viacom, Vivendi Universal. According to Subcomandante Marcos, spokesperson of the indigenous Zapatista rebels of Mexico, the global media "present a virtual world, created in the image of what the globalization process requires."[1]

In a recent CNN discussion Gerry Levin, Chief Executive of AOL Time Warner, announced that global media would become *the* dominant industry of this century, more powerful than governments. U.S. citizens now spend more money on entertainment than on clothing or healthcare—and the pattern is being mirrored around the developed world. Michael J. Wolf, Davos schmoozer and adviser to the media moguls, says: "Entertainment—not autos, not steel, not financial services—is fast becoming the driving wheel of the new world economy." Forget the military-industrial complex—this is the media-entertainment complex.

The planet is encircled by an ever-expanding web of wires and cables and the paths of orbiting satellites, while new wealth is being made from words, ideas, knowledge, songs, stories, data, culture. The media corporations, too, are an extractive industry. As Jeremy Rifkin says, they are "mining local cultural resources in every part of the world and repackaging them as cultural commodities and entertainment."[2]

But information and culture are not just tradable commodities to be bought and sold in the global marketplace. Freedom of information is fundamental to democracy. Culture is the sum of the stories we tell about ourselves, stories that inform who we are and how we describe the world.

The true meaning of globalization is not so much about Indian yogis checking their share prices via their high-speed modems, or whether nomadic herders in the Gobi desert "should" watch *Baywatch*. Rather, it is about the undebated imposition of the *organizing logic*, the "anti-culture," of the marketplace into every corner of our lives, onto every culture on earth.

The Latin root of the world "culture" means cultivating the land; "broadcast" was once an agricultural term meaning "to scatter seed." Just as agribusinesses corner the seed business by breaking the natural reproductive cycle to create genetically modified Terminator seeds—which have to be bought rather than resown year-on-year—in every country media corporations help to break our relationships to our communities, educators, collective cultures, experiences. They turn us into isolated consumers—and then sell our stories back to us.

THE MEDIA-ENTERTAINMENT COMPLEX

When U.S. media critic Ben Bagdikian started tracking media ownership in 1982, there were 50 firms dominating the market. Now there are fewer than ten.... While each country's media scene still varies enormously, the creation of a single global commercial media model is the ultimate aim of the World Trade Organization, the International Monetary Fund and the World Bank.

And globalization means that regional media—from Mexico to India to Aotearoa/New Zealand—are following the same pattern of merging, converging, trying to compete with—and getting swallowed by—the big boys. Indian public-service channel Doordarshan has already sold Australian media mogul Kerry Packer a chunk of primetime TV. As a result of privatization over the past decade, 99 per cent of Hungary's 3,000 media outlets are largely controlled by Western business.[3]

While, self-evidently, tyrannical state control over the media is a profound assault on democracy and human rights in countries around the world, when a handful of corporations dominates the world's information this is called having a "free press." In the lingua franca of globaloney, freedom means the freedom to do business and because, by and large, they manage to entertain us, and because this is not a monolithic system of control, the profound lack of democracy at its heart goes unexamined. As Uruguayan novelist Eduardo Galeano says: "Never have so many been held incommunicado by so few." He describes this as "the dictatorship of the single word and the single image, much more devastating than that of the single party."[4] We are creating a world in which a small and shrinking commercial monopoly gets to tell all the stories while the rest of us get to watch and listen.

If we are witnessing the creation of a single global empire under one market logic, the world-spanning communications networks are its Roman roads—channels both of the ruling ideology and traded goods.

The fibre-optic cables threading into your homes bring you telephones but also Internet connections and cable TV. Liberalization of telecommunication in 1997 delivered up control of these Roman roads in nearly every country to global companies—who in this digital age are increasingly the result of inter-breeding between telecoms, media, and computer software transnationals. But as Noam Chomsky points out: "Concentration of communications in any hands (particularly foreign hands) raises some rather serious questions about meaningful democracy."[5] And if you control both the content and the "pipe" you can always price the competition out of the market. In an interview with the *Village Voice*, independent media advocate Anthony Riddle recalls the comment of a Russian man discussing control of the Soviet media: "All the wires ran through a switch on one man's desk. He could pull the switch at any time."

"WHAT'S UP, SCOOB?"

Today you can get news on your phone, listen to your newspaper correspondent on the Internet and watch television through a computer—and all from the same company. So AOL Time Warner with its range of "in-house brands" could give you news via CNN and *Time* magazine; its customer tracking could remember that Madonna album you downloaded last month and post a "reminder" for a concert onto your personal online calendar; while your medical records and bank balance are held by its partner companies. And you thought they just showed *Scooby Doo* reruns.

The databases that hold millions of detailed consumer profiles are a major corporate asset. According to *Fortune* magazine (another AOL Time Warner title): "Once captive, [customers] eyeballs will then be resold to advertisers and commerce partners."

In this entertainment economy, the definition of "media company" is increasingly hazy—some have speculated that AOL Time Warner could buy Wal-Mart next; internationally, Vivendi runs Universal Studio theme parks as well as water companies privatized by the World Bank; Disney has entered the real-estate market as privatized towns with corporate governments catch on in America. Rupert Murdoch of News Corporation loses more sleep over Microsoft than over his traditional media competitors, while Nike sees itself in greater competition with Disney than Reebok.[6]

This is saturation capitalism, where almost almost every aspect of our lives is "mediated" by commerce.

NEW WORLD INFORMATION ORDER

The deepest irony of all this is that, as the economy globalizes, we actually find out less and less about one another from the media. Increasing commercial pressure and cost-cutting means that coverage of international news in the West has dropped by an average of 50 per cent in the last ten years. On a single British channel, ITV, it has dropped by 80 per cent since the onset of satellite competition.

In contrast, viewers in the South are subject to a constant stream of cheap Western television formats that undercut local production and creativity.

And where imagery leads, trade follows—among the first foreign industries to enter India after the liberalization of the economy were transnational media and advertising corporations targeting wealthy elites.

Concerns over the torrential flow of information from North to South are hardly new. In 1980 the United Nations Education, Scientific and Cultural Organization (UNESCO) published *Many Voices, One World*, a study calling for a "new world information order" and the creation of an independent international news agency operating out of the South. In a move that laid bare the raw power politics of global communications, [the] U.S. and Britain—under Reagan and Thatcher—called this an attack on "free speech" and pulled out of UNESCO completely. The proposal was quietly shelved.

Today, the exuberant technicolour extravaganza that is Bollywood actually churns out seven times more films than the U.S. But despite Bollywood's output, it does not have the economic might of Hollywood, which makes 85 per cent of films watched anywhere in the world. Partly as a result of this imbalance, film and publishing industries in most developing countries are actually in decline. Ninety-five per cent of Latin American films are produced in the U.S., and Africa—which makes 42 films a year—imports an even higher percentage.

THE END OF GEOGRAPHY

Media consultant Michael J. Wolf describes how in the "underscreened, undermalled, still-waiting-for-cable world—in China, India, the Islamic nations . . . new ventures are starting on an almost hourly basis . . . So the entertainment economy is settling in new territories around the world."

Sure enough, a little way past the avenue of Arabian palm trees, and just across from the Hard Rock Cafe (motto: "In Rock We Trust"), a 500-acre "Media City" is rising out of the desert in Dubai. In this multimedia free-trade zone, international corporations are setting up regional offices, 100 per cent tax-free. In India, meanwhile, the local personnel of call centers servicing Western consumers are being trained to speak with American accents. Callers need never know they are talking to someone in the suburbs of Delhi rather than Dallas.

This globalized world under a single market economy is incorporating every culture and every place within its ever-expanding frontiers. This is not the "new world information order" UNESCO dreamed of, but a universalizing of the old world information order as it delinks from its own geographical and cultural roots.

In this context, regulating the media in order to nourish a diversity of viewpoints and cultures—goals of public-service broadcasting—is becoming an anachronism. Instead, as Gerry Levin of AOL Time Warner points out: "We're going to need to have corporations redefined as instruments of public service." Media companies are currently lobbying the World Trade Organization for a trade agreement that would spell the death of media regulation and public-service broadcasting—including the BBC. This is currently under negotiation, but according to leaked WTO documents, "Ensuring pluralism and a media system based on free and democratic principles" has already been ruled out as a legitimate government objective.[7]

Meanwhile the U.S. trade department has tucked in its napkin, held aloft its knife and fork, and drawn up a menu to take to the WTO. Australia's limits to foreign ownership of media could be the starter dish, followed by Europe's subsidized film industry. Dessert might be Sweden's ban on advertising to children and Canada's requirement that a proportion of its media be made locally.

The industry argument is that media pluralism, special needs and cultural diversity will be served by digital technology and the new multichannels. The market will serve all the needs of society—all those who can pay, that is.

THE MEDIA GUERRILLAS

Fortunately, culture and media are among the most contentious of issues at the WTO. If the arguments escalate, this could create an opening to voice our defiance against the worlds both Orwell and Huxley described—for we can rely neither on states nor markets to serve our communication needs.

Already in different places around the world people are organizing: to tell their own stories, expose lies and obfuscation, break up the monopolies. For we are more than just target markets for advertisers and eyeballs for the cheapest television they can get us, exhausted after a hard day's work, to watch.

Media activism is on the rise, and is not just confined to culture jammers in the North. When Filipino President Estrada's cronies bought up newspapers to stem criticism of his rule, citizens helped to topple him by posting evidence of his corruption anonymously on the Guerrilla Information Network website.

And as corporations globalize, resistance goes transnational. Media activists are creating international networks, from Indymedia [. . .] to the Media-Channel, which networks media-watch groups and communications NGOs from the U.S. to Africa. Just as activists protesting corporate power have targeted agribusinesses over GM, more and more people are not just creating alternative channels but targeting the mainstream media itself. Brazilian activists demonstrated outside the headquarters of their country's media conglomerates against their poor coverage of social issues. In India the National Alliance of People's Movements called for accurate reporting on how globalization harms the poor everywhere—North as well as the South; while in the U.S. protesters have demanded that the Federal Communications Committee "free the airwaves" for community broadcasting. French group MediaLibre piled a wall of television sets in front of France's Ministry of Culture and Communication in the name of public-interest media free from corporate and government influence; in Montevideo activists burned cardboard televisions as the agent of "consumer culture." Media reform has also become a political campaign issue in Australia, Aotearoa/New Zealand and the U.S. anti-trust laws such as those used to order the break-up of Microsoft are political tools that must be defended and strengthened with popular support.

Just as in the 1960s political consciousness began to be raised about the degradation of the natural environment, there is a nascent movement working to protect the diversity of our cultural and information environment. The challenges of globalization seem to have given even the much-chastened UNESCO a reason to sit up and take notice. It states that: "Over hundreds of millions of years, nature developed an astonishing variety of life forms which are tightly interwoven; the survival of all are necessary to ensure the continued existence of natural ecosystems. Similarly, 'cultural ecosystems,' made up of a rich and complex mosaic of cultures, more or less powerful, need diversity to preserve and pass on their valuable heritage to future generations."[8]

We need to reclaim our stories, reweave the web of cultural diversity and create channels of information in order to understand the globalization process.

We need to do this in order collectively to imagine and articulate alternative futures for ourselves.

ENDNOTES

1. Subcomandante Marcos, *Our Word is Our Weapon*, New York: Seven Stories, 2001.
2. Jeremy Rifkin, *The Age of Access*, New York: Penguin Putnam, 2000.
3. Communications Law in Transition Newsletter, Programme in Comparative Media Law and Policy, University of Oxford, 6 December 2000.
4. Eduardo Galeano, *Upside Down*, New York: Metropolitan Books, 2001.
5. Noam Chomsky, "The Passsion for Free Markets: Exporting American Values through the New World Trade Organization," *Z Magazine*, http://www.zmag.org.
6. Naomi Klein, *No Logo*, London: Flamingo, 2000
7. Murray Dobbin, "Trading away the Public Interest," *Financial Post*, Canada, 26 June 2000.
8. UNESCO, "Culture, trade and globalisation," http://www.unesco.org/culture/industries/trade.

REFLECTION QUESTIONS FOR CHAPTER 6

1. What evidence do you find in your community of the culture of other societies? Are they exactly as found in their countries of origin or have they adapted to the local culture?

2. Some have argued that rather than a global culture there is or will be a "clash of civilizations" as religious fundamentalists or nationalists resist (sometimes with violence) any attempt to change their cultures. What evidence do you see from world events to support or reject that argument?

3. Is the direction of the globalization of culture one way—that is, the westernization of the globe? Are there products from other countries that have changed our habits? Are there non-U.S. cultural icons from entertainment and sports who have captivated Americans? What about the influence of other countries on fashion, architecture, work arrangements, and the like?

7

The Restructuring of Social Arrangements

Gender, Families, and Relationships

By now it is a truism that some of the most devastating effects of globalization fall on women. We denounce female exploitation in third-world economies, where women and girls are forced to work in jobs with high turnover, low wages, and hazardous working conditions. Indeed, the standard view of the "global assembly line" comes packaged with gendered images. Other features of globalization are well known for their harmful effects on women, including sex trafficking and female enslavement. Yet these realities tell us little about the impact of gender on the global economy.

Few macro structural discussions of globalization put gender at the center of the analysis. Although they "stir" women into the macro picture, they fail to examine how "producers, consumers, and bystanders of globalization are all caught up in processes that are gendered themselves."[1]

The readings in this chapter challenge and broaden mainstream views of the emerging global order by exploring several dimensions of the relationship between globalization and gender. They uncover hidden social and economic global processes that rest on expectations of masculine and feminine in locations around the world. Yet in the process of moving across national boundaries through global systems of production, immigration, and cultural images, gender inequalities are reconstructed and take new shape. At the same time, global movements of gender transform the social institutions with which they come in contact.[2] Looking at global connections and interdependencies through the lens of gender reveals that inequalities between women and men serve as building blocks of the global order and they are simultaneously restructured by global processes.

In the first article, Barbara Ehrenreich and Arlie Russell Hochschild provide an overview of women's migration from the third world to do "women's work" for families around the world. The authors unmask the transfer of women's traditional work—child care, homemaking, and sex—from poor countries to rich ones. While the globalization of women's traditional work supports the professional careers and lifestyles of women in affluent countries, it leaves immigrant women of color disenfranchised and separated from their own families. The global transfer of "care work" throws new light on the entire process of globalization and its contradictory effects on women, men, and children in different parts of the world.

In the next article, Rhacel Parreñas takes the analysis and critique of the global migration of care work a step further. Her study of Filipina migrant mothers and the children they leave behind draws on the stories of family members who struggle to remain "family" even as they are separated by time and space. Their own words are compelling and they offer new insights about transnational families and the costs extracted from them by the global economy.

Saskia Sassen gives us a big-picture explanation of gender dynamics in the global economy. Her essay departs from the usual globalization themes of hyper mobile capital and rising information economies. Instead, Sassen shifts her emphasis to global processes that operate in *specific* workplaces and work processes. Global cities and survival circuits are two concrete instances where women are crucial economic actors. This reading is abstract, but it is well worth the effort. It gives us a new understanding of how the global system relies on women workers while also creating new demands for their labor.

While the first three readings point to women's paid and unpaid work as a key component of globalization, the final reading makes gender in the global arena visible for *men*. R. W. Connell asks the question, "What does masculinity have to do with globalization?" He answers the question by untangling key strands in "the world gender order" and unearthing multiple masculinities, all of which are intertwined with global forces.

ENDNOTES

1. Freeman, Carla. "Is Local: Global as Feminine : Masculine Rethinking the Gender of Globalization," *SIGNS: Journal of Women in Culture and Society*, Vol. 26, No. 4 (Summer 2001):1007–1038.
2. Baca Zinn, Maxine, Pierrette Hondagneu-Sotelo, and Michael A. Messner, eds. *Gender Through the Prism of Difference*, 3d Edition (New York: Oxford University Press, 2005).

23

Global Woman

Nannies, Maids, and Sex Workers in the New Economy

BARBARA EHRENREICH
ARLIE RUSSELL HOCHSCHILD

"Whose baby are you?" Josephine Perera, a nanny from Sri Lanka, asks Isadora, her pudgy two-year-old charge in Athens, Greece.

Thoughtful for a moment, the child glances toward the closed door of the next room, in which her mother is working, as if to say, "That's my mother in there."

"No, you're *my* baby," Josephine teases, tickling Isadora lightly. Then, to settle the issue, Isadora answers, "Together!" She has two mommies—her mother and Josephine. And surely a child loved by many adults is richly blessed.

In some ways, Josephine's story—which unfolds in an extraordinary documentary film, *When Mother Comes Home for Christmas*, directed by Nilita Vachani—describes an unparalleled success. Josephine has ventured around the world, achieving a degree of independence her mother could not have imagined, and amply supporting her three children with no help from her ex-husband, their father. Each month she mails a remittance check from Athens to Hatton, Sri Lanka, to pay the children's living expenses and school fees. On her Christmas visit home, she bears gifts of pots, pans, and dishes. While she makes payments on a new bus that Suresh, her oldest son, now drives for a living, she is also saving for a modest dowry for her daughter, Norma. She dreams of buying a new house in which the whole family can live. In the meantime, her work as a nanny enables Isadora's parents to devote themselves to their careers and avocations.

But Josephine's story is also one of wrenching global inequality. While Isadora enjoys the attention of three adults, Josephine's three children in Sri Lanka have been far less lucky. According to Vachani, Josephine's youngest child, Suminda, was two—Isadora's age—when his mother first left home to work in Saudi Arabia. Her middle child, Norma, was nine; her oldest son, Suresh, thirteen. From Saudi Arabia, Josephine found her way first to Kuwait, then to Greece. Except for one two-month trip home, she has lived apart from her children for ten years. She writes them weekly letters, seeking news of relatives, asking about school, and complaining that Norma doesn't write back.

SOURCE: Introduction by Barbara Ehrenreich and Arlie Russell Hochschild from *Global Woman: Nannies, Maids, and Sex Workers in the New Economy,* edited by Barbara Ehrenreich and Arlie Russell Hochschild. © 2002 by Barbara Ehrenreich and Arlie Russell Hochschild. Reprinted by permission of Henry Holt and Company, LLC.

Although Josephine left the children under her sister's supervision, the two youngest have shown signs of real distress. Norma has attempted suicide three times. Suminda, who was twelve when the film was made, boards in a grim, Dickensian orphanage that forbids talk during meals and showers. He visits his aunt on holidays. Although the oldest, Suresh, seems to be on good terms with his mother, Norma is tearful and sullen, and Suminda does poorly in school, picks quarrels, and otherwise seems withdrawn from the world. Still, at the end of the film, we see Josephine once again leave her three children in Sri Lanka to return to Isadora in Athens. For Josephine can either live with her children in desperate poverty or make money by living apart from them. Unlike her affluent First World employers, she cannot both live with her family and support it.

Thanks to the process we loosely call "globalization," women are on the move as never before in history. In images familiar to the West from television commercials for credit cards, cell phones, and airlines, female executives jet about the world, phoning home from luxury hotels and reuniting with eager children in airports. But we hear much less about a far more prodigious flow of female labor and energy: the increasing migration of millions of women from poor countries to rich ones, where they serve as nannies, maids, and sometimes sex workers. In the absence of help from male partners, many women have succeeded in tough "male world" careers only by turning over the care of their children, elderly parents, and homes to women from the Third World. This is the female underside of globalization, whereby millions of Josephines from poor countries in the south migrate to do the "women's work" of the north—work that affluent women are no longer able or willing to do. These migrant workers often leave their own children in the care of grandmothers, sisters, and sisters-in-law. Sometimes a young daughter is drawn out of school to care for her younger siblings.

This pattern of female migration reflects what could be called a worldwide gender revolution. In both rich and poor countries, fewer families can rely solely on a male breadwinner. In the United States, the earning power of most men has declined since 1970, and many women have gone out to "make up the difference." By one recent estimate, women were the sole, primary, or coequal earners in more than half of American families.[1] So the question arises: Who will take care of the children, the sick, the elderly? Who will make dinner and clean house?

While the European or American woman commutes to work an average twenty-eight minutes a day, many nannies from the Philippines, Sri Lanka, and India cross the globe to get to their jobs. Some female migrants from the Third World do find something like "liberation," or at least the chance to become independent breadwinners and to improve their children's material lives. Other, less fortunate migrant women end up in the control of criminal employers—their passports stolen, their mobility blocked, forced to work without pay in brothels or to provide sex along with cleaning and child-care services in affluent homes. But even in more typical cases, where benign employers pay wages on time, Third World migrant women achieve their success only by assuming the cast-off domestic roles of middle- and high-income women in the First World—roles that have been previously rejected, of course, by men. And their "commute" entails a cost we have yet to fully comprehend.

The migration of women from the Third World to do "women's work" in affluent countries has so far received little media attention—for reasons that are easy enough to guess. First, many, though by no means all, of the new female migrant workers are women of color, and therefore subject to the racial "discounting" routinely experienced by, say, Algerians in France, Mexicans in the United States, and Asians in the United Kingdom. Add to racism the private "indoor" nature of so much of the new migrants' work. Unlike factory workers, who congregate in large numbers, or taxi drivers, who are visible on the street, nannies and maids are often hidden away, one or two at a time, behind closed doors in private homes. Because of the illegal nature of their work, most sex workers are even further concealed from public view.

At least in the case of nannies and maids, another factor contributes to the invisibility of migrant women and their work—one that, for their affluent employers, touches closer to home. The Western culture of individualism, which finds extreme expression in the United States, militates against acknowledging help or human interdependency of nearly any kind. Thus, in the time-pressed upper middle class, servants are no longer displayed as status symbols, decked out in white caps and aprons, but often remain in the background, or disappear when company comes. Furthermore, affluent careerwomen increasingly earn their status not through leisure, as they might have a century ago, but by apparently "doing it all"—producing a full-time career, thriving children, a contented spouse, and a well-managed home. In order to preserve this illusion, domestic workers and nannies make the house hotel-room perfect, feed and bathe the children, cook and clean up—and then magically fade from sight.

The lifestyles of the First World are made possible by a global transfer of the services associated with a wife's traditional role—child care, homemaking, and sex—from poor countries to rich ones. To generalize and perhaps oversimplify: [I]n an earlier phase of imperialism, northern countries extracted natural resources and agricultural products—rubber, metals, and sugar, for example—from lands they conquered and colonized. Today, while still relying on Third World countries for agricultural and industrial labor, the wealthy countries also seek to extract something harder to measure and quantify, something that can look very much like love. Nannies like Josephine bring the distant families that employ them real maternal affection, no doubt enhanced by the heartbreaking absence of their own children in the poor countries they leave behind. Similarly, women who migrate from country to country to work as maids bring not only their muscle power but an attentiveness to detail and to the human relationships in the household that might otherwise have been invested in their own families. Sex workers offer the simulation of sexual and romantic love, or at least transient sexual companionship. It is as if the wealthy parts of the world are running short on precious emotional and sexual resources and have had to turn to poorer regions for fresh supplies.

There are plenty of historical precedents for this globalization of traditional female services. In the ancient Middle East, the women of populations defeated in war were routinely enslaved and hauled off to serve as household workers and concubines for the victors. Among the Africans brought to North America as slaves in the sixteenth through nineteenth centuries, about a third were women and children, and many of those women were pressed to be concubines, domestic servants, or both. Nineteenth-century Irishwomen—along with many rural Englishwomen—migrated to English towns and cities to work as domestics in the homes of the growing upper middle class. Services thought to be innately feminine—child care, housework, and sex—often win little recognition or pay. But they have always been sufficiently in demand to transport over long distances if necessary. What is new today is the sheer number of female migrants and the very long distances they travel. Immigration statistics show huge numbers of women in motion, typically from poor countries to rich. Although the gross statistics give little clue as to the jobs women eventually take, there are reasons to infer that much of their work is "caring work," performed either in private homes or in institutional settings such as hospitals, hospices, child-care centers, and nursing homes.

The statistics are, in many ways, frustrating. We have information on legal migrants but not on illegal migrants, who, experts tell us, travel in equal if not greater numbers. Furthermore, many Third World countries lack data for past years, which makes it hard to trace trends over time; or they use varying methods of gathering information, which makes it hard to compare one country with another. Nevertheless, the trend is clear enough for some scholars, including Stephen Castles, Mark Miller, and Janet Momsen, to speak of a "feminization of migration."[2] From 1950 to 1970, for example, men predominated in labor migration to northern Europe from Turkey, Greece, and North Africa. Since then, women have been replacing men. In 1946, women were fewer than 3 percent of the Algerians and Moroccans living in France; by 1990, they were more than 40 percent.[3] Overall, half of the world's 120 million legal and illegal migrants are now believed to be women.

Patterns of international migration vary from region to region, but women migrants from a surprising number of sending countries actually outnumber men, sometimes by a wide margin. For example, in the 1990s, women make up over half of Filipino migrants to all countries and 84 percent of Sri Lankan migrants to the Middle East.[4] Indeed, by 1993 statistics, Sri Lankan women such as Josephine vastly outnumbered Sri Lankan men as migrant workers who'd left for Saudi Arabia, Kuwait, Lebanon, Oman, Bahrain, Jordan, and Qatar, as well as to all countries of the Far East, Africa, and Asia.[5] About half of the migrants leaving Mexico, India, Korea, Malaysia, Cyprus, and Swaziland to work elsewhere are also women. Throughout the 1990s women outnumbered men among migrants to the United States, Canada, Sweden, the United Kingdom, Argentina, and Israel.[6]

Most women, like men, migrate from the south to the north and from poor countries to rich ones. Typically, migrants go to the nearest comparatively rich country, preferably one whose language they speak or whose religion and culture they share. There are also local migratory flows: from northern to southern Thailand, for instance, or from East Germany to West. But of the regional or cross-regional flows, four stand out. One goes from Southeast Asia to the oil-rich Middle and Far East—from Bangladesh, Indonesia, the Philippines, and Sri

Lanka to Bahrain, Oman, Kuwait, Saudi Arabia, Hong Kong, Malaysia, and Singapore. Another stream of migration goes from the former Soviet bloc to western Europe—from Russia, Romania, Bulgaria, and Albania to Scandinavia, Germany, France, Spain, Portugal, and England. A third goes from south to north in the Americas, including the stream from Mexico to the United States, which scholars say is the longest-running labor migration in the world. A fourth stream moves from Africa to various parts of Europe. France receives many female migrants from Morocco, Tunisia, and Algeria. Italy receives female workers from Ethiopia, Eritrea, and Cape Verde.

Female migrants overwhelmingly take up work as maids or domestics. As women have become an ever greater proportion of migrant workers, receiving countries reflect a dramatic influx of foreign-born domestics. In the United States, African-American women, who accounted for 60 percent of domestics in the 1940s, have been largely replaced by Latinas, many of them recent migrants from Mexico and Central America. In England, Asian migrant women have displaced the Irish and Portuguese domestics of the past. In French cities, North African women have replaced rural French girls. In western Germany, Turks and women from the former East Germany have replaced rural native-born women. Foreign females from countries outside the European Union made up only 6 percent of all domestic workers in 1984. By 1987, the percentage had jumped to 52, with most coming from the Philippines, Sri Lanka, Thailand, Argentina, Colombia, Brazil, El Salvador, and Peru.[7]

The governments of some sending countries actively encourage women to migrate in search of domestic jobs, reasoning that migrant women are more likely than their male counterparts to send their hard-earned wages to their families rather than spending the money on themselves. In general, women send home anywhere from half to nearly all of what they earn. These remittances have a significant impact on the lives of children, parents, siblings, and wider networks of kin—as well as on cash-strapped Third World governments. Thus, before Josephine left for Athens, a program sponsored by the Sri Lankan government taught her how to use a microwave oven, a vacuum cleaner, and an electric mixer. As she awaited her flight, a song piped into the airport departure lounge extolled the opportunity to earn money abroad. The songwriter was in the pay of the Sri Lanka Bureau of Foreign Employment, an office devised to encourage women to migrate. The lyrics say:

> After much hardship, such difficult times
>
> How lucky I am to work in a foreign land.
>
> As the gold gathers so do many greedy flies.
>
> But our good government protects us from them.
>
> After much hardship, such difficult times,
>
> How lucky I am to work in a foreign land.
>
> I promise to return home with treasures for everyone.

Why this transfer of women's traditional services from poor to rich parts of the world? The reasons are, in a crude way, easy to guess. Women in Western countries have increasingly taken on paid work, and hence need other—paid domestics and caretakers for children and elderly people—to replace them.[8] For their part, women in poor countries have an obvious incentive to migrate: relative and absolute poverty. The "care deficit" that has emerged in the wealthier countries as women enter the workforce *pulls* migrants from the Third World and postcommunist nations; poverty *pushes* them.

In broad outline, this explanation holds true. Throughout western Europe, Taiwan, and Japan, but above all in the United States, England, and Sweden, women's employment has increased dramatically since the 1970s. In the United States, for example, the proportion of women in paid work rose from 15 percent of mothers of children six and under in 1950 to 65 percent today. Women now make up 46 percent of the U.S. labor force. Three-quarters of mothers of children eighteen and under and nearly two-thirds of mothers of children age one and younger now work for pay. Furthermore, according to a recent International Labor Organization study, working Americans averaged longer hours at work in the late 1990s than they did in the 1970s. By some measures, the number of hours spent at work have increased more for women than for men, and especially for women in managerial and professional jobs.

Meanwhile, over the last thirty years, as the rich countries have grown much richer, the poor countries have become—in both absolute and relative terms— poorer. Global inequalities in wages are particularly striking. In Hong Kong, for instance, the wages of a Filipina domestic are about fifteen times the amount she could make as a schoolteacher back in the Philippines. In addition, poor countries turning to the IMF or World Bank for loans are often forced to undertake measures of so-called structural adjustment, with disastrous results for the poor and especially for poor women and children. To qualify for loans, governments are usually required to devalue their currencies, which turns the hard currencies of rich countries into gold and the soft currencies of poor countries into straw. Structural adjustment programs also call for cuts in support for "noncompetitive industries," and for the reduction of public services such as health care and food subsidies for the poor. Citizens of poor countries, women as well as men, thus have a strong incentive to seek work in more fortunate parts of the world.

But it would be a mistake to attribute the globalization of women's work to a simple synergy of needs among women—one group, in the affluent countries, needing help and the other, in poor countries, needing jobs. For one thing, this formulation fails to account for the marked failure of First World governments to meet the needs created by its women's entry into the workforce. The downsized American—and to a lesser degree, western European—welfare state has become a "deadbeat dad." Unlike the rest of the industrialized world, the United States does not offer public child care for working mothers, nor does it ensure paid family and medical leave. Moreover, a series of state tax revolts in the 1980s reduced the number of hours public libraries were open and slashed school-enrichment and after-school programs. Europe did not experience anything comparable. Still, tens of millions of western European women are in the workforce who were not before—and there has been no proportionate expansion in public services.

Secondly, any view of the globalization of domestic work as simply an arrangement among women completely omits the role of men. Numerous studies, including some of our own, have shown that as American women took on paid employment, the men in their families did little to increase their contribution to the work of the home. For example, only one out of every five men among the working couples whom Hochschild interviewed for *The Second Shift* in the 1980s shared the work at home, and later studies suggest that while working mothers are doing somewhat less housework than their counterparts twenty years ago, most men are doing only a little more.[9] With divorce, men frequently abdicate their child-care responsibilities to their ex-wives. In most cultures of the First World outside the United States, powerful traditions even more firmly discourage husbands from doing "women's work." So, strictly speaking, the presence of immigrant nannies does not enable affluent women to enter the workforce; it enables affluent *men* to continue avoiding the second shift.

The men in wealthier countries are also, of course, directly responsible for the demand for immigrant sex workers—as well as for the sexual abuse of many migrant women who work as domestics. Why, we wondered, is there a particular demand for "imported" sexual partners? Part of the answer may lie in the fact that new immigrants often take up the least desirable work, and, thanks to the AIDS epidemic, prostitution has become a job that ever fewer women deliberately choose. But perhaps some of this demand . . . grows out of the erotic lure of the "exotic." Immigrant women may seem desirable sexual partners for the same reason that First World employers believe them to be especially gifted as caregivers: [T]hey are thought to embody the traditional feminine qualities of nurturance, docility, and eagerness to please. Some men feel nostalgic for these qualities, which they associate with a bygone way of life. Even as many wage-earning Western women assimilate to the competitive culture of "male" work and ask respect for making it in a man's world, some men seek in the "exotic Orient" or "hot-blooded tropics" a woman from the imagined past.

Of course, not all sex workers migrate voluntarily. An alarming number of women and girls are trafficked by smugglers and sold into bondage. Because trafficking is illegal and secret, the numbers are hard to know with any certainty. Kevin Bales estimates that in Thailand alone, a country of 60 million, half a million to a million women are prostitutes, and one out of every twenty of these is enslaved.[10] . . . [M]any of these women are daughters whom northern hill-tribe families have sold to brothels in the cities of the south. Believing the promises of jobs and money, some begin the voyage willingly, only to discover days later that the "arrangers" are traffickers who steal their passports, define them as debtors, and enslave them as prostitutes. Other women and girls are kidnapped, or sold by their impoverished families, and then trafficked to brothels. Even worse fates befall women from neighboring Laos and Burma, who flee crushing poverty and repression at home only to fall into the hands of Thai slave traders.

If the factors that pull migrant women workers to affluent countries are not as simple as they at first appear, neither are the factors that push them. Certainly relative poverty plays a major role, but, interestingly, migrant women often do not come from the poorest classes of their societies.[11] In fact, they are typically more affluent and better educated than male migrants. Many female migrants from the Philippines and Mexico, for example, have high school or college diplomas and have held middle-class—albeit low-paid—jobs back home. One study of Mexican migrants suggests that the trend is toward increasingly better-educated female migrants. Thirty years ago, most Mexican-born maids in the United States had been poorly educated maids in Mexico. Now a majority have high school degrees and have held clerical, retail, or professional jobs before leaving for the United States.[12] Such women are likely to be enterprising and adventurous enough to resist the social pressures to stay home and accept their lot in life.

Noneconomic factors—or at least factors that are not immediately and directly economic—also influence a woman's decision to emigrate. By migrating, a woman may escape the expectation that she care for elderly family members, relinquish her paycheck to a husband or father, or defer to an abusive husband. Migration may also be a practical response to a failed marriage and the need to provide for children without male help. In the Philippines, . . . Rhacel Salazar Parreñas tells us, migration is sometimes called a "Philippine divorce." And there are forces at work that may be making the men of poor countries less desirable as husbands. Male unemployment runs high in the countries that supply female domestics to the First World. Unable to make a living, these men often grow demoralized and cease contributing to their families in other ways. Many female migrants, [. . .] tell of unemployed husbands who drink or gamble their remittances away. Notes one study of Sri Lankan women working as maids in the Persian Gulf: "It is not unusual . . . for the women to find upon their return that their Gulf wages by and large have been squandered on alcohol, gambling and other dubious undertakings while they were away."[13]

To an extent then, the globalization of child care and housework brings the ambitious and independent women of the world together: the career-oriented upper-middle-class woman of an affluent nation and the striving woman from a crumbling Third World or postcommunist economy. Only it does not bring them together in the way that second-wave feminists in affluent countries once liked to imagine—as sisters and allies struggling to achieve common goals. Instead, they come together as mistress and maid, employer and employee, across a great divide of privilege and opportunity.

This trend toward global redivision of women's traditional work throws new light on the entire process of globalization. Conventionally, it is the poorer countries that are thought to be dependent on the richer ones—a dependency symbolized by the huge debt they owe to global financial institutions. What we explore in this book, however, is a dependency that works in the other direction, and it is a dependency of a particularly intimate kind. Increasingly often, as affluent and middle-class families in the First World come to depend on migrants from poorer regions to provide child care, homemaking, and sexual services, a global relationship arises that in some ways mirrors the traditional relationship between the sexes. The First World takes on a role like that of the old-fashioned male in the family—pampered, entitled, unable to cook, clean, or find his socks. Poor countries take on a role like that of the traditional woman within the family—patient, nurturing, and self-denying. A division of labor feminists critiqued when it was "local" has now, metaphorically speaking, gone global.

To press this metaphor a bit further, the resulting relationship is by no means a "marriage," in the sense of being openly acknowledged. In fact, it is striking how invisible the globalization of women's work remains, how little it is noted or discussed in the First World. Trend spotters have had almost nothing to say about the fact that increasing numbers of affluent First World children and elderly persons are tended by immigrant care workers or live in homes cleaned by immigrant maids. Even the political groups we might expect to be concerned about this trend—antiglobalization and feminist activists—often seem to have noticed only the most extravagant abuses, such as trafficking and female enslavement. So if a metaphorically gendered relationship has developed between rich and poor countries, it is less like a marriage and more like a secret affair.

But it is a "secret affair" conducted in plain view of the children. Little Isadora and the other children of the First World raised by "two mommies" may be learning more than their ABCs from a loving surrogate parent. In their own living rooms, they are learning a vast and tragic global politics.[14] Children see. But they also learn how to disregard what they see. They learn how adults make the visible invisible. That is their "early childhood education."

ENDNOTES

1. See Ellen Galinsky and Dana Friedman, *Women: The New Providers*, Whirlpool Foundation Study, Part 1 (New York: Families and Work Institute, 1995), p. 37.

2. Special thanks to Roberta Espinoza, [. . .] In addition to material directly cited, this introduction draws from the following works: Kathleen M. Adams and Sara Dickey, eds., *Home and Hegemony: Domestic Service and Identity Politics in South and Southeast Asia* (Ann Arbor: University of Michigan Press, 2000); Floya Anthias and Gabriella Lazaridis, eds., *Gender and Migration in Southern Europe: Women on the Move* (Oxford and New York: Berg, 2000); Stephen Castles and Mark J. Miller, *The Age of Migration: International Population Movements in the Modern World* (New York and London: The Guilford Press, 1998); Noeleen Heyzer, Geertje Lycklama à Nijehold, and Nedra Weerakoon, eds., *The Trade in Domestic Workers: Causes, Mechanisms, and Consequences of International Migration* (London: Zed Books, 1994); Eleanore Kofman, Annie Phizacklea, Parvati Raghuram, and Rosemary Sales, *Gender and International Migration in Europe: Employment, Welfare, and Politics* (New York and London: Routledge, 2000); Douglas S. Massey, Joaquin Arango, Graeme Hugo, Ali Kouaouci, Adela Pellegrino, and J. Edward Taylor, *Worlds in Motion: Understanding International Migration at the End of the Millennium* (Oxford: Clarendon Press, 1999); Janet Henshall Momsen, ed., *Gender, Migration, and Domestic Service* (London: Routledge, 1999); Katie Willis and Brenda Yeoh, eds., *Gender and Immigration* (London: Edward Elgar Publishers, 2000).

3. Illegal migrants are said to make up anywhere from 60 percent (as in Sri Lanka) to 87 percent (as in Indonesia) of all migrants. In Singapore in 1994, 95 percent of

Filipino overseas contract workers lacked work permits from the Philippine government. The official figures based on legal migration therefore severely underestimate the number of migrants. See Momsen, 1999, p. 7.

4. Momsen, 1999, p. 9.

5. Sri Lanka Bureau of Foreign Employment, 1994, as cited in G. Gunatilleke, "The Economic, Demographic, Sociocultural and Political Setting for Emigration from Sri Lanka." *International Migration*, vol. 23 (3/4), 1995, pp. 667–98.

6. Anthias and Lazaridis, 2000; Heyzer, Nijehold, and Weerakoon, 1994, pp. 4–27; Momsen, 1999, p. 21; "Wistat: Women's Indicators and Statistics Database," version 3, CD-ROM (United Nations, Department for Economic and Social Information and Policy Analysis, Statistical Division, 1994).

7. Geovanna Campani, "Labor Markets and Family Networks: Filipino Women in Italy," in Hedwig Rudolph and Mirjana Morokvasic, eds., *Bridging States and Markets: International Migration in the Early 1990s* (Berlin: Edition Sigma, 1993), p. 206.

8. This "new" source of the Western demand for nannies, maids, child-care, and elder-care workers does not, of course, account for the more status-oriented demand in the Persian Gulf states, where most affluent women don't work outside the home.

9. For information on male work at home during the 1990s, see Arlie Russell Hochschild and Anne Machung, *The Second Shift: Working Parents and the Revolution at Home* (New York: Avon, 1997), p. 277.

10. Kevin Bales, *Disposable People: New Slavery in the Global Economy* (Berkeley: University of California Press, 1999), p. 43.

11. Andrea Tyree and Katharine M. Donato, "A Demographic Overview of the International Migration of Women," in *International Migration: The Female Experience*, eds. Rita Simon and Caroline Bretell (Totowa, N.J.: Rowman & Allanheld, 1986), p. 29. Indeed, many immigrant maids and nannies are more educated than the people they work for. See Pei-Chia Lan's paper in this volume.

12. Momsen, 1999, pp. 10, 73.

13. Grete Brochmann, *Middle East Avenue: Female Migration from Sri Lanka to the Gulf* (Boulder, Colo.: Westview Press, 1993), pp. 179, 215.

14. On this point, thanks to Raka Ray, Sociology Department at the University of California, Berkeley.

24

The Care Crisis in the Philippines

Children and Transnational Families in the New Global Economy

RHACEL SALAZAR PARREÑAS

A growing crisis of care troubles the world's most developed nations. Even as demand for care has increased, its supply has dwindled. The result is a care deficit, to which women from the Philippines have responded in force. Roughly two-thirds of Filipino migrant workers are women, and their exodus, usually to fill domestic jobs, has generated tremendous social change in the Philippines. When female migrants are mothers, they leave behind their own children, usually in the care of other women. Many Filipino children now grow up in divided households, where geographic separation places children under serious emotional strain. And yet it is impossible to overlook the significance of migrant labor to the Philippine economy. Some 34 to 54 percent of the Filipino population is sustained by remittances from migrant workers.

Women in the Philippines, just like their counterparts in postindustrial nations, suffer from a "stalled revolution." Local gender ideology remains a few steps behind the economic reality, which has produced numerous female-headed, transnational households. Consequently, a far greater degree of anxiety attends the quality of family life for the dependents of migrant mothers than for those of migrant fathers. The dominant gender ideology, after all, holds that a woman's rightful place is in the home, and the households of migrant mothers present a challenge to this view. In response, government officials and journalists denounce migrating mothers, claiming that they have caused the Filipino family to deteriorate, children to be abandoned, and a crisis of care to take root in the Philippines. To end this crisis, critics admonish, these mothers must return. Indeed, in May 1995, Philippine president Fidel Ramos called for initiatives to keep migrant mothers at home. He declared, "We are not against overseas employment of Filipino women. We are against overseas employment at the cost of family solidarity." Migration, Ramos strongly implied, is morally acceptable only when it is undertaken by single, childless women.

SOURCE: Rhacel Salazar Parreñas, "The Care Crisis in the Philippines: Children and Transnational Families in the New Global Economy," pp. 39–54 39–54 in *Global Woman: Nannies, Maids, and Sex Workers in the New Economy*, Barbara Ehrenreich and Arlie Russell Hochschild (eds.) (New York: Metropolitan, 2003). Reprinted with permission from the author. Notes have been deleted from this reading.

The Philippine media reinforce this position by consistently publishing sensationalist reports on the suffering of children in transnational families. These reports tend to vilify migrant mothers, suggesting that their children face more profound problems than do those of migrant fathers; and despite the fact that most of the children in question are left with relatives, journalists tend to refer to them as having been "abandoned." One article reports, "A child's sense of loss appears to be greater when it is the mother who leaves to work abroad." Others link the emigration of mothers to the inadequate child care and unstable family life that eventually lead such children to "drugs, gambling, and drinking." Writes one columnist, "Incest and rapes within blood relatives are alarmingly on the rise not only within Metro Manila but also in the provinces. There are some indications that the absence of mothers who have become OCWs [overseas contract workers] has something to do with the situation." The same columnist elsewhere expresses the popular view that the children of migrants become a burden on the larger society: "Guidance counselors and social welfare agencies can show grim statistics on how many children have turned into liabilities to our society because of absentee parents."

From January to July 2000, I conducted sixty-nine in-depth interviews with young adults who grew up in transnational households in the Philippines. Almost none of these children have yet reunited with their migrant parents. I interviewed thirty children with migrant mothers, twenty-six with migrant fathers, and thirteen with two migrant parents. The children I spoke to certainly had endured emotional hardships; but contrary to the media's dark presentation, they did not all experience their mothers' migration as abandonment. The hardships in their lives were frequently diminished when they received support from extended families and communities, when they enjoyed open communication with their migrant parents, and when they clearly understood the limited financial options that led their parents to migrate in the first place.

To call for the return of migrant mothers is to ignore the fact that the Philippines has grown increasingly dependent on their remittances. To acknowledge this reality could lead the Philippines toward a more egalitarian gender ideology. Casting blame on migrant mothers, however, serves only to divert the society's attention away from these children's needs, finally aggravating their difficulties by stigmatizing their family's choices.

The Philippine media has certainly sensationalized the issue of child welfare in migrating families, but that should not obscure the fact that the Philippines faces a genuine care crisis. Care is now the country's primary export. Remittances —mostly from migrant domestic workers—constitute the economy's largest source of foreign currency, totaling almost $7 billion in 1999. With limited choices in the Philippines, women migrate to help sustain their families financially, but the price is very high. Both mothers and children suffer from family separation, even under the best of circumstances.

Migrant mothers who work as nannies often face the painful prospect of caring for other people's children while being unable to tend to their own. One such mother in Rome, Rosemarie Samaniego, describes this predicament:

When the girl that I take care of calls her mother "Mama," my heart jumps all the time because my children also call me "Mama." I feel the gap caused by our physical separation especially in the morning, when I pack [her] lunch, because that's what I used to do for my children. . . . I used to do that very same thing for them. I begin thinking that at this hour I should be taking care of my very own children and not someone else's, someone who is not related to me in any way, shape, or form. . . . The work that I do here is done for my family, but the problem is they are not close to me but are far away in the Philippines. Sometimes, you feel the separation and you start to cry. Some days, I just start crying while I am sweeping the floor because I am thinking about my children in the Philippines. Sometimes, when I receive a letter from my children telling me that they are sick, I look up out the window and ask the Lord to look after them and make sure they get better even without me around to care after them. [*Starts crying.*] If I had wings, I would fly home to my children. Just for a moment, to see my children and take care of their needs, help them, then fly back over here to continue my work.

The children of migrant workers also suffer an incalculable loss when a parent disappears overseas. As Ellen Seneriches, a twenty-one-year-old daughter of a domestic worker in New York, says:

There are times when you want to talk to her, but she is not there. That is really hard, very difficult. . . . There are times when I want to call her, speak to her, cry to her, and I cannot. It is difficult. The only thing that I can do is write to her. And I cannot cry through the e-mails and sometimes I just want to cry on her shoulder.

Children like Ellen, who was only ten years old when her mother left for New York, often repress their longings to reunite with their mothers. Knowing that their families have few financial options, they are left with no choice but to put their emotional needs aside. Often, they do so knowing that their mothers' care and attention have been diverted to other children. When I asked her how she felt about her mother's wards in New York, Ellen responded:

Very jealous. I am very, very jealous. There was even a time when she told the children she was caring for that they are very lucky that she was taking care of them, while her children back in the Philippines don't even have a mom to take care of them. It's pathetic, but it's true. We were left alone by ourselves and we had to be responsible at a very young age without a mother. Can you imagine?

Children like Ellen do experience emotional stress when they grow up in transnational households. But it is worth emphasizing that many migrant mothers attempt to sustain ties with their children, and their children often recognize and appreciate these efforts. Although her mother, undocumented in the United States, has not returned once to the Philippines in twelve years, Ellen does not doubt that she has struggled to remain close to her children despite the distance. In fact, although Ellen lives only three hours away from her father, she feels closer to and communicates more frequently with her mother. Says Ellen:

> I realize that my mother loves us very much. Even if she is far away, she would send us her love. She would make us feel like she really loved us. She would do this by always being there. She would just assure us that whenever we have problems to just call her and tell her. [Pauses.] And so I know that it has been more difficult for her than other mothers. She has had to do extra work because she is so far away from us.

Like Ellen's mother, who managed to "be there" despite a vast distance, other migrant mothers do not necessarily "abandon" their traditional duty of nurturing their families. Rather, they provide emotional care and guidance from afar. Ellen even credits her mother for her success in school. Now a second-year medical school student, Ellen graduated at the top of her class in both high school and college. She says that the constant, open communication she shares with her mother provided the key to her success. She reflects:

> We communicate as often as we can, like twice or thrice a week through e-mails. Then she would call us every week. And it is very expensive, I know. . . . My mother and I have a very open relationship. We are like best friends. She would give me advice whenever I had problems. . . . She understands everything I do. She understands why I would act this or that way. She knows me really well. And she is also transparent to me. She always knows when I have problems, and like-wise I know when she does. I am closer to her than to my father.

Ellen is clearly not the abandoned child or social liability the Philippine media describe. She not only benefits from sufficient parental support—from both her geographically distant mother and her nearby father—but also exceeds the bar of excellence in schooling. Her story indicates that children of migrant parents can overcome the emotional strains of transnational family life, and that they can enjoy sufficient family support, even from their geographically distant parent.

Of course, her good fortune is not universal. But it does raise questions about how children withstand such geographical strains; whether and how they maintain solid ties with their distant parents; and what circumstances lead some children to feel that those ties have weakened or given out. The Philippine media tend to equate the absence of a child's biological mother with abandonment, which leads to the assumption that all such children, lacking familial support, will become social liabilities. But I found that positive surrogate parental figures and open communication with the migrant parent, along with acknowledgment of the migrant parent's contribution to the collective mobility of the family, allay many of the emotional insecurities that arise from transnational household arrangements. Children who lack these resources have greater difficulty adjusting.

Extensive research bears out this observation. The Scalabrini Migration Center, a nongovernmental organization for migration research in the Philippines, surveyed 709 elementary-school-age Filipino children in 2000, comparing the experiences of those with a father absent, a mother absent, both parents absent, and both parents present. While the researchers observed that parental absence does prompt feelings of abandonment and loneliness among children, they concluded that "it does not necessarily become an occasion for laziness and unruliness." Rather, if the extended family supports the child and makes him or her aware of the material benefits migration brings, the child may actually be spurred toward greater self-reliance and ambition, despite continued longings for family unity.

Jeek Pereno's life has been defined by those longings. At twenty-five, he is a merchandiser for a large department store in the Philippines. His mother more than adequately provided for her children, managing with her meager wages first as a domestic worker and then as a nurse's aide, to send them $200 a month and even to purchase a house in a fairly exclusive neighborhood in the city center. But Jeek still feels abandoned and insecure in his mother's affection; he believes that growing up without his parents robbed him of the discipline he needed. Like other children of migrant workers, Jeek does not feel that his faraway mother's financial support has been enough. Instead, he wishes she had offered him more guidance, concern, and emotional care.

Jeek was eight years old when his parents relocated to New York and left him, along with his three brothers, in the care of their aunt. Eight years later, Jeek's father passed away, and two of his brothers (the oldest and youngest) joined their mother in New York. Visa complications have prevented Jeek and his other brother from following—but their mother has not once returned to visit them in the Philippines. When I expressed surprise at this, Jeek solemnly replied: "Never. It will cost too much, she said."

Years of separation breed unfamiliarity among family members, and Jeek does not have the emotional security of knowing that his mother has genuinely tried to lessen that estrangement. For Jeek, only a visit could shore up this security after seventeen years of separation. His mother's weekly phone calls do not suffice. And because he experiences his mother's absence as indifference, he does not feel comfortable communicating with her openly about his unmet needs. The result is repression, which in turn aggravates the resentment he feels. Jeek told me:

> I talk to my mother once in a while. But what happens, whenever she asks how I am doing, I just say okay. It's not like I am really going to tell her that I have problems here. . . . It's not like she can do anything about my problems if I told her about them. Financial problems, yes she can help. But not the other problems, like emotional problems.... She will try to give advice, but I am not very interested to talk to her about things like that. . . . Of course, you are still young, you don't really know what is going to happen in the future. Before you realize that your parents left you, you can't do anything about it anymore. You are not in a position to tell them not to leave you. They should have not left us. (*Sobs.*)

I asked Jeek if his mother knew he felt this way. "No," he said, "she doesn't know." Asked if he received emotional support from anyone, Jeek replied, "As much as possible, if I can handle it, I try not to get emotional support from anyone. I just keep everything inside me."

Jeek feels that his mother not only abandoned him but failed to leave him with an adequate surrogate. His aunt had a family and children of her own. Jeek recalled, "While I do know that my aunt loves me and she took care of us to the best of her ability, I am not convinced that it was enough. . . . Because we were not disciplined enough. She let us do whatever we wanted to do." Jeek feels that his education suffered from this lack of discipline, and he greatly regrets not having concentrated on his studies. Having completed only a two-year vocational program in electronics, he doubts his competency to pursue a college degree. At twenty-five, he feels stuck, with only the limited option of turning from one low-paying job to another.

Children who, unlike Jeek, received good surrogate parenting managed to concentrate on their studies and in the end to fare much better. Rudy Montoya, a nineteen-year-old whose mother has done domestic work in Hong Kong for more than twelve years, credits his mother's brother for helping him succeed in high school:

> My uncle is the most influential person in my life. Well, he is in Saudi Arabia now. . . . He would tell me that my mother loves me and not to resent her, and that whatever happens, I should write her. He would encourage me and he would tell me to trust the Lord. And then, I remember in high school, he would push me to study. I learned a lot from him in high school. Showing his love for me, he would help me with my school work. . . . The time that I spent with my uncle was short, but he is the person who helped me grow up to be a better person.

Unlike Jeek's aunt, Rudy's uncle did not have a family of his own. He was able to devote more time to Rudy, instilling discipline in his young charge as well as reassuring him that his mother, who is the sole income provider for her family, did not abandon him. Although his mother has returned to visit him only twice—once when he was in the fourth grade and again two years later—Rudy, who is now a college student, sees his mother as a "good provider" who has made tremendous sacrifices for his sake. This knowledge offers him emotional security, as well as a strong feeling of gratitude. When I asked him about the importance of education, he replied, "I haven't given anything back to my mother for the sacrifices that she has made for me. The least I could do for her is graduate, so that I can find a good job, so that eventually I will be able to help her out, too."

Many children resolve the emotional insecurity of being left by their parents the way that Rudy has: by viewing migration as a sacrifice to be repaid by adult children. Children who believe that their migrant mothers are struggling for the sake of the family's collective mobility, rather than leaving to live the "good life," are less likely to feel abandoned and more likely to accept their mothers' efforts to sustain close relationships from a distance. One such child is Theresa Bascara, an eighteen-year-old college student whose mother has worked as a domestic in Hong Kong since 1984. As she puts it, "[My inspiration is] my mother, because she is the one suffering over there. So the least I can give back to her is doing well in school."

For Ellen Seneriches, the image of her suffering mother compels her to reciprocate. She explained:

> Especially after my mother left, I became more motivated to study harder. I did because my mother was sacrificing a lot and I had to compensate for how hard it is to be away from your children and then crying a lot at night, not knowing what we are doing. She would tell us in voice tapes. She would send us voice tapes every month, twice a month, and we would hear her cry in these tapes.

Having witnessed her mother's suffering even from a distance, Ellen can acknowledge the sacrifices her mother has made and the hardships she has endured in order to be a "good provider" for her family. This knowledge assuaged the resentment Ellen frequently felt when her mother first migrated.

Many of the children I interviewed harbored images of their mothers as martyrs, and they often found comfort in their mothers' grief over not being able to nurture them directly. The expectation among such children that they will continue to receive a significant part of their nurturing from their mothers, despite the distance, points to the conservative gender ideology most of them maintain. But whether or not they see their mothers as martyrs, children of migrant women feel best cared for when their mothers make consistent efforts to show parental concern from a distance. As Jeek's and Ellen's stories indicate, open communication with the migrant parent soothes feelings of abandonment; those who enjoy such open channels fare much better than those who lack them. Not only does communication ease children's emotional difficulties; it also fosters a sense of family unity, and it promotes the view that migration is a survival strategy that requires sacrifices from both children and parents for the good of the family.

For daughters of migrant mothers, such sacrifices commonly take the form of assuming some of their absent mothers' responsibilities, including the care of younger siblings. As Ellen told me:

> It was a strategy, and all of us had to sacrifice for it. . . . We all had to adjust, every day of our lives. . . . Imagine waking up without a mother calling you for breakfast. Then there would be no one to prepare the clothes for my brothers. We are all going to school. . . . I had to wake up earlier. I had to prepare their clothes. I had to wake them up and help them prepare for school. Then I also had to help them with their homework at night. I had to tutor them.

Asked if she resented this extra work, Ellen replied, "No. I saw it as training, a training that helped me become a leader. It makes you more of a leader doing that every day. I guess that is an advantage to me, and to my siblings as well."

Ellen's effort to assist in the household's daily maintenance was another way she reciprocated for her mother's emotional and financial support. Viewing her added work as a positive life lesson, Ellen believes that these responsibilities enabled her to develop leadership skills. Notably, her high school selected her as its first ever female commander for its government-mandated military training corps.

Unlike Jeek, Ellen is secure in her mother's love. She feels that her mother has struggled to "be there"; Jeek feels that his has not. Hence, Ellen has managed to successfully adjust to her household arrangement, while Jeek has not. The continual open communication between Ellen and her mother has had ramifications for their entire family: in return for her mother's sacrifices, Ellen assumed the role of second mother to her younger siblings, visiting them every weekend during her college years in order to spend quality time with them.

In general, eldest daughters of migrant mothers assume substantial familial responsibilities, often becoming substitute mothers for their siblings. Similarly, eldest sons stand in for migrant fathers. Armando Martinez, a twenty-nine-year-old entrepreneur whose father worked in Dubai for six months while he was in high school, related his experiences:

> I became a father during those six months. It was like, ugghhh, I made the rules. . . . I was able to see that it was hard if your family is not complete, you feel that there is something missing. . . . It's because the major decisions, sometimes, I was not old enough for them. I was only a teenager, and I was not that strong in my convictions when it came to making decisions. It was like work that I should not have been responsible for. I still wanted to play. So it was an added burden on my side.

Even when there is a parent left behind, children of migrant workers tend to assume added familial responsibilities, and these responsibilities vary along gender lines. Nonetheless, the weight tends to fall most heavily on children of migrant mothers, who are often left to struggle with the lack of male responsibility for care work in the Philippines. While a great number of children with migrant fathers receive full-time care from stay-at-home mothers, those with migrant mothers do not receive the same amount of care. Their fathers are likely to hold full-time jobs, and they rarely have the time to assume the role of primary caregiver. Of thirty children of migrant mothers I interviewed, only four had stay-at-home fathers. Most fathers passed the caregiving responsibilities on to other relatives, many of whom, like Jeek's aunt, already had families of their own to care for and regarded the children of migrant relatives as an extra burden. Families of migrant fathers are less likely to rely on the care work of extended kin. Among my interviewees, thirteen of twenty-six children with migrant fathers lived with and were cared for primarily by their stay-at-home mothers.

Children of migrant mothers, unlike those of migrant fathers, have the added burden of accepting nontraditional gender roles in their families. The Scalabrini Migration Center reports that these children "tend to be more angry, confused, apathetic, and more afraid than other children." They are caught within an "ideological stall" in the societal acceptance of female-headed transnational households. Because her family does not fit the traditional nuclear household model, Theresa Bascara sees her family as "broken," even though she describes her relationship to her mother as "very close." She says, "A family, I can say, is only whole if your father is the one working and your mother is only staying at home. It's okay if your mother works too, but somewhere close to you."

Some children in transnational families adjust to their household arrangements with greater success than others do. Those who feel that their mothers strive to nurture them as well as to be good providers are more likely to be accepting. The support of extended kin, or perhaps a sense of public accountability for their welfare, also helps children combat feelings of abandonment. Likewise, a more gender-egalitarian value system enables children to appreciate their mothers as good providers, which in turn allows them to see their mothers' migrations as demonstrations of love.

Even if they are well-adjusted, however, children in transnational families still suffer the loss of family intimacy. They are often forced to compensate by accepting commodities, rather than affection, as the most tangible reassurance of their parents' love. By putting family intimacy on hold, children can only wait for the opportunity to spend quality time with their migrant parents. Even when that time comes, it can be painful. As Theresa related:

> When my mother is home, I just sit next to her. I stare at her face, to see the changes in her face, to see how she aged during the years that she was away from us. But when she is about to go back to Hong Kong, it's like my heart is going to burst. I would just cry and cry. I really can't explain the feeling. Sometimes, when my mother is home, preparing to leave for Hong Kong, I would just start crying, because I already start missing her. I ask myself, how many more years will it be until we see each other again?
>
> . . . . Telephone calls. That's not enough. You can't hug her, kiss her, feel her, everything. You can't feel her presence. It's just words that you have. What I want is to have my mother close to me, to see her grow older, and when she is sick, you are the one taking care of her and when you are sick, she is the one taking care of you.

Not surprisingly when asked if they would leave their own children to take jobs as migrant workers, almost all of my respondents answered, "Never." When I asked why not, most said that they would never want their children to go through what they had gone through, or to be denied what they were denied, in their childhoods. Armando Martinez best summed up what children in transnational families lose when he said:

> You just cannot buy the times when your family is together. Isn't that right? Time together is something that money can neither buy nor replace. . . .
> The first time your baby speaks, you are not there. Other people would

experience that joy. And when your child graduates with honors, you are also not there . . . . Is that right? When your child wins a basketball game, no one will be there to ask him how his game went, how many points he made. Is that right? Your family loses, don't you think?

Children of transnational families repeatedly stress that they lack the pleasure and comfort of daily interaction with their parents. Nonetheless, these children do not necessarily become "delinquent," nor are their families necessarily broken, in the manner the Philippine media depicts. Armando mirrored the opinion of most of the children in my study when he defended transnational families: "Even if [parents] are far away, they are still there. I get that from basketball, specifically zone defense." [He laughed.] "If someone is not there, you just have to adjust. It's like a slight hindrance that you just have to adjust to. Then when they come back, you get a chance to recover. It's like that."

Recognizing that the family is an adaptive unit that responds to external forces, many children make do, even if doing so requires tremendous sacrifices. They give up intimacy and familiarity with their parents. Often, they attempt to make up for their migrant parents' hardships by maintaining close bonds across great distances, even though most of them feel that such bonds could never possibly draw their distant parent close enough. But their efforts are frequently sustained by the belief that such emotional sacrifices are not without meaning—that they are ultimately for the greater good of their families and their future. Jason Halili's mother provided care for elderly persons in Los Angeles for fifteen years. Jason, now twenty-one, reasons, "If she did not leave, I would not be here right now. So it was the hardest route to take, but at the same time, the best route to take."

Transnational families were not always equated with "broken homes" in the Philippine public discourse. Nor did labor migration emerge as a perceived threat to family life before the late 1980s, when the number of migrant women significantly increased. This suggests that changes to the gendered division of family labor may have as much as anything else to do with the Philippine care crisis.

The Philippine public simply assumes that the proliferation of female-headed transnational households will wreak havoc on the lives of children. The Scalabrini Migration Center explains that children of migrant mothers suffer more than those of migrant fathers because child rearing is "a role women are more adept at, are better prepared for, and pay more attention to." The center's study, like the Philippine media, recommends that mothers be kept from migrating. The researchers suggest that "economic programs should be targeted particularly toward the absorption of the female labor force, to facilitate the possibility for mothers to remain in the family." Yet the return migration of mothers is neither a plausible nor a desirable solution. Rather, it implicitly accepts gender inequities in the family, even as it ignores the economic pressures generated by globalization.

As national discourse on the care crisis in the Philippines vilifies migrant women, it also downplays the contributions these women make to the country's economy. Such hand-wringing merely offers the public an opportunity to discipline women morally and to resist reconstituting family life in a manner that reflects the country's increasing dependence on women's foreign remittances.

This pattern is not exclusive to the Philippines. As Arjun Appadurai observes, globalization has commonly led to "ideas about gender and modernity that create large female work forces at the same time that cross-national ideologies of 'culture,' 'authenticity,' and national honor put increasing pressure on various communities to morally discipline working women."

The moral disciplining of women, however, hurts those who most need protection. It pathologizes the children of migrants, and it downplays the emotional difficulties that mothers like Rosemarie Samaniego face. Moreover, it ignores the struggles of migrant mothers who attempt to nurture their children from a distance. Vilifying migrant women as bad mothers promotes the view that the return to the nuclear family is the only viable solution to the emotional difficulties of children in transnational families. In so doing, it directs attention away from the special needs of children in transnational families—for instance, the need for community projects that would improve communication among far-flung family members, or for special school programs, the like of which did not exist at my field research site. It's also a strategy that sidelines the agency and adaptability of the children themselves.

To say that children are perfectly capable of adjusting to nontraditional households is not to say that they don't suffer hardships. But the overwhelming public support for keeping migrant mothers at home does have a negative impact on these children's adjustment. Implicit in such views is a rejection of the division of labor in families with migrant mothers, and the message such children receive is that their household arrangements are simply wrong. Moreover, calling for the return migration of women does not necessarily solve the problems plaguing families in the Philippines. Domestic violence and male infidelity, for instance—two social problems the government has never adequately addressed—would still threaten the well-being of children.

Without a doubt, the children of migrant Filipina domestic workers suffer from the extraction of care from the global south to the global north. The plight of these children is a timely and necessary concern for nongovernmental, governmental, and academic groups in the Philippines. Blaming migrant mothers, however, has not helped, and has even hurt, those whose relationships suffer most from the movement of care in the global economy. Advocates for children in transnational families should focus their attention not on calling for a return to the nuclear family but on trying to meet the special needs transnational families possess. One of those needs is for a reconstituted gender ideology in the Philippines; another is for the elimination of legislation that penalizes migrant families in the nations where they work.

If we want to secure quality care for the children of transnational families, gender egalitarian views of child rearing are essential. Such views can be fostered by recognizing the economic contributions women make to their families and by redefining motherhood to include providing for one's family. Gender should be recognized as a fluid social category, and masculinity should be redefined, as the larger society questions the biologically based assumption that only women have an aptitude to provide care. Government officials and the media could then stop vilifying migrant women, redirecting their attention, instead, to men. They could question the lack of male accountability for care work, and they could demand that men, including migrant fathers, take more responsibility for the emotional welfare of their children.

The host societies of migrant Filipina domestic workers should also be held more accountable for their welfare and for that of their families. These women's work allows First World women to enter the paid labor force. As one Dutch employer states, "There are people who would look after children, but other things are more fun. Carers from other countries, if we can use their surplus carers, that's a solution."

Yet, as we've seen, one cannot simply assume that the care leaving disadvantaged nations is surplus care. What is a solution for rich nations creates a problem in poor nations. Mothers like Rosemarie Samaniego and children like Ellen Seneriches and Jeek Pereno bear the brunt of this problem, while the receiving countries and the employing families benefit.

Most receiving countries have yet to recognize the contributions of their migrant care workers. They have consistently ignored these workers' rights and limited their full incorporation into society. The wages of migrant workers are so low that they cannot afford to bring their own families to join them, or to regularly visit their children in the Philippines; relegated to the status of guest workers, they are restricted to the low-wage employment sector, and with very few exceptions, the migration of their spouses and children is also restricted. These arrangements work to the benefit of employers, since migrant care workers can give the best possible care for their employers' families when they are free of care-giving responsibilities to their own families. But there is a dire need to lobby for more inclusive policies, and for employers to develop a sense of accountability for their workers' children. After all, migrant workers significantly help their employers to reduce *their* families' care deficit.

25

Global Cities and Survival Circuits

SASKIA SASSEN

When today's media, policy, and economic analysts define globalization, they emphasize hypermobility, international communication, and the neutralization of distance and place. This account of globalization is by far the dominant one. Central to it are the global information economy, instant communication, and electronic markets—all realms within which place no longer makes a difference, and where the only type of worker who matters is the highly educated professional. Globalization thus conceived privileges global transmission over the material infrastructure that makes it possible; information over the workers who produce it, whether these be specialists or secretaries; and the new transnational corporate culture over the other jobs upon which it rests, including many of those held by immigrants. In brief, the dominant narrative of globalization concerns itself with the upper circuits of global capital, not the lower ones, and with the hypermobility of capital rather than with capital that is bound to place.

The migration of maids, nannies, nurses, sex workers, and contract brides has little to do with globalization by these lights. Migrant women are just individuals making a go of it, after all, and the migration of workers from poor countries to wealthier ones long predates the current phase of economic globalization. And yet it seems reasonable to assume that there are significant links between globalization and women's migration, whether voluntary or forced, for jobs that used to be part of the First World woman's domestic role. Might the dynamics of globalization alter the course or even reinscribe the history of the migration and exploitation of Third World laborers? There are two distinct issues here. One is whether globalization has enabled formerly national or regional processes to go global. The other is whether globalization has produced a new kind of migration, with new conditions and dynamics of its own.

GLOBAL CITIES AND SURVIVAL CIRCUITS

When today's women migrate from south to north for work as nannies, domestics, or sex workers, they participate in two sets of dynamic configurations. One of these

SOURCE: Saskia Sassen, "Global Cities and Survival Circuits," pp. 254–274 in *Global Woman: Nannies, Maids, and Sex Workers in the New Economy*, Barbara Ehrenreich and Arlie Russell Hochschild (eds.) (New York: Metropolitan, 2002). Reprinted with permission from the author.

is the global city. The other consists of survival circuits that have emerged in response to the deepening misery of the global south.[1]

Global cities concentrate some of the global economy's key functions and resources. There, activities implicated in the management and coordination of the global economy have expanded, producing a sharp growth in the demand for highly paid professionals. Both this sector's firms and the lifestyles of its professional workers in turn generate a demand for low-paid service workers. In this way, global cities have become places where large numbers of low-paid women and immigrants get incorporated into strategic economic sectors. Some are incorporated directly as low-wage clerical and service workers, such as janitors and repairmen. For others, the process is less direct, operating instead through the consumption practices of high-income professionals, who employ maids and nannies and who patronize expensive restaurants and shops staffed by low-wage workers. Traditionally, employment in growth sectors has been a source of workers' empowerment; this new pattern undermines that linkage, producing a class of workers who are isolated, dispersed, and effectively invisible.

Meanwhile, as Third World economies on the periphery of the global system struggle against debt and poverty, they increasingly build survival circuits on the backs of women—whether these be trafficked low-wage workers and prostitutes or migrant workers sending remittances back home. Through their work and remittances, these women contribute to the revenue of deeply indebted countries. "Entrepreneurs" who have seen other opportunities vanish as global firms entered their countries see profit-making potential in the trafficking of women; so, too, do longtime criminals who have seized the opportunity to operate their illegal trade globally. These survival circuits are often complex; multiple locations and sets of actors constitute increasingly far-reaching chains of traders and "workers."

Through their work in both global cities and survival circuits, women, so often discounted as valueless economic actors, are crucial to building new economies and expanding existing ones. Globalization serves a double purpose here, helping to forge links between sending and receiving countries, and enabling local and regional practices to assume a global scale. On the one hand, the dynamics that converge in the global city produce a strong demand for low-wage workers, while the dynamics that mobilize women into survival circuits produce an expanding supply of migrants who can be pushed—or sold—into such jobs. On the other hand, the very technological infrastructure and transnationalism that characterize global industries also enable other types of actors to expand onto the global stage, whether these be money launderers or people traffickers.[2] It seems, then, that in order to understand the extraction from the Third World of services that used to define women's domestic role in the First, we must depart from the mainstream view of globalization.

TOWARD AN ALTERNATIVE NARRATIVE
ABOUT GLOBALIZATION

The spatial dispersal of economic activities and the neutralization of place constitute half of the globalization story. The other half involves the territorial centralization of top-level management, control operations, and the most advanced specialized services. Markets, whether national or global, and companies, many of which have gone global, require central locations where their most complex tasks are accomplished. Furthermore, the information industry rests on a vast physical infrastructure, which includes strategic nodes where facilities are densely concentrated. Even the most advanced sectors of the information industry employ many different types of workplaces and workers.

If we expand our analysis of globalization to include this production process, we can see that secretaries belong to the global economy, as do the people who clean professionals' offices and homes. An economic configuration very different from the one suggested by the concept of "information economy" emerges— and it is one that includes material conditions, production sites, and activities bounded by place.

The mainstream account of globalization tends to take for granted the existence of a global economic system, viewing it as a function of the power of transnational corporations and communications. But if the new information technologies and transnational corporations can be operated, coordinated, and controlled globally, it's because that capacity has been produced. By focusing on its production, we shift our emphasis to the *practices* that constitute economic globalization: the work of producing and reproducing the organization and management of a global production system and a global marketplace for finance.

This focus on practices draws the categories of place and work process into the analysis of economic globalization. In so broadening our analysis, we do not deny the importance of hypermobility and power. Rather, we acknowledge that many of the resources necessary for global economic activities are not hypermobile and are, on the contrary, deeply embedded in place, including such sites as global cities and export processing zones. Global processes are structured by local constraints, including the work culture, political culture, and composition of the workforce within a particular nation state.[3]

If we recapture the geography behind globalization, we might also recapture its workers, communities, and work cultures (not just the corporate ones). By focusing on the global city, for instance, we can study how global processes become localized in specific arrangements, from the high-income gentrified urban neighborhoods of the transnational professional class to the work lives of the foreign nannies and maids in those same neighborhoods.

WOMEN IN THE GLOBAL CITY

Globalization has greatly increased the demand in global cities for low-wage workers to fill jobs that offer few advancement possibilities. The same cities have seen an explosion of wealth and power, as high-income jobs and high-priced urban space have noticeably expanded. How, then, can workers be hired at low wages and with few benefits even when there is high demand and the jobs belong to high-growth sectors? The answer, it seems, has involved tapping into a growing new labor supply—women and immigrants—and in so doing, breaking the historical nexus that would have empowered workers under these conditions. The fact that these workers tend to be women and immigrants also lends cultural legitimacy to their nonempowerment. In global cities, then, a majority of today's resident workers are women, and many of these are women of color, both native and immigrant.

At the same time, global cities have seen a gathering trend toward the informalization of an expanding range of activities, as low-profit employers attempt to escape the costs and constraints of the formal economy's regulatory apparatus. They do so by locating commercial or manufacturing operations in areas zoned exclusively for residential use, for example, or in buildings that violate fire and health standards; they also do so by assigning individual workers industrial home-work. This allows them to remain in these cities. At its best, informalization re-introduces the community and the household as important economic spaces in global cities. It is in many ways a low-cost (and often feminized) equivalent to deregulation at the top of the system. As with deregulation (for example, financial deregulation), informalization introduces flexibility, reduces the "burdens" of regulation, and lowers costs, in this case of labor. In the cities of the global north—including New York, London, Paris, and Berlin—informalization serves to downgrade a variety of activities for which there is often a growing local demand. Immigrant women, in the end, bear some of the costs.

As the demand for high-level professional workers has skyrocketed, more and more women have found work in corporate professional jobs.[4] These jobs place heavy demands on women's time, requiring long work hours and intense engagement. Single professionals and two-career households therefore tend to prefer urban to suburban residence. The result is an expansion of high-income residential areas in global cities and a return of family life to urban centers. Urban professionals want it all, including dogs and children, whether or not they have the time to care for them. The usual modes of handling household tasks often prove inadequate. We can call this type of household a "professional household without a 'wife,'" regardless of whether its adult couple consists of a man and a woman, two men, or two women. A growing share of its domestic tasks are relocated to the market: [T]hey are bought directly as goods and services or in-directly through hired labor. As a consequence, we see the return of the so-called serving classes in all of the world's global cities, and these classes are largely made up of immigrant and migrant women.

This dynamic produces a sort of double movement: a shift to the labor market of functions that used to be part of household work, but also a shift of what used to be labor market functions in standardized workplaces to the household and, in the case of informalization, to the immigrant community.[5] This reconfig-uration of economic spaces has had different impacts on women and men, on male-typed and female-typed work cultures, and on male- and female-centered forms of power and empowerment.

For women, such transformations contain the potential, however limited, for autonomy and empowerment. Might informalization, for example, reconfigure certain economic relationships between men and women? With informalization, the neighborhood and the household reemerge as sites for economic activity, creating "opportunities" for low-income women and thereby reordering some of the hierarchies in which women find themselves. This becomes particularly clear in the case of immigrant women, who often come from countries with traditionally male-centered cultures.

A substantial number of studies now show that regular wage work and improved access to other public realms has an impact on gender relations in the lives of immigrant women. Women gain greater personal autonomy and independence, while men lose ground. More control over budgeting and other domestic decisions devolves to women, and they have greater leverage in requesting help from men in domestic chores. Access to public services and other public resources also allows women to incorporate themselves into the mainstream society; in fact, women often mediate this process for their households. Some women likely benefit more than others from these circumstances, and with more research we could establish the impact of class, education, and income. But even aside from relative empowerment in the household, paid work holds out another significant possibility for women: their greater participation in the public sphere and their emergence as public actors.

Immigrant women tend to be active in two arenas: institutions for public and private assistance, and the immigrant or ethnic community. The more women are involved with the migration process, the more likely it is that migrants will settle in their new residences and participate in their communities. And when immigrant women assume active public and social roles, they further reinforce their status in the household and the settlement process.[6] Positioned differently from men in relation to the economy and state, women tend to be more involved in community building and community activism. They are the ones who will likely handle their families' legal vulnerabilities as they seek public and social services. These trends suggest that women may emerge as more forceful and visible actors in the labor market as well.

And so two distinct dynamics converge in the lives of immigrant women in global cities. On the one hand, these women make up an invisible and disempowered class of workers in the service of the global economy's strategic sectors. Their invisibility keeps immigrant women from emerging as the strong proletariat that followed earlier forms of economic organization, when workers' positions in leading sectors had the effect of empowering them. On the other hand, the access to wages and salaries, however low; the growing feminization of the job supply; and the growing feminization of business opportunities thanks to informalization, all alter the gender hierarchies in which these women find themselves.

NEW EMPLOYMENT REGIMES IN CITIES

Most analysts of postindustrial society and advanced economies report a massive growth in the need for highly educated workers but little demand for the type of labor that a majority of immigrants, perhaps especially immigrant women, have tended to supply over the last two or three decades. But detailed empirical studies of major cities in highly developed countries contradict this conventional view of the postindustrial economy. Instead, they show an ongoing demand for immigrant workers and a significant supply of old and new low-wage jobs that require little education.[7]

Three processes of change in economic and spatial organization help explain the ongoing, indeed growing, demand for immigrant workers, especially immigrant women. One is the consolidation of advanced services and corporate headquarters in the urban economic core, especially in global cities. While the corporate headquarters-and-services complex may not account for the majority of jobs in these cities, it establishes a new regime of economic activity, which in turn produces the spatial and social transformations evident in major cities. Another relevant process is the downgrading of the manufacturing sector, as some manufacturing industries become incorporated into the postindustrial economy. Downgrading is a response to competition from cheap imports, and to the modest profit potential of manufacturing compared to telecommunications, finance, and other corporate services.[8] The third process is informalization, a notable example of which is the rise of the sweatshop. Firms often take recourse to informalized arrangements when they have an effective local demand for their goods and services but they cannot compete with cheap imports, or cannot compete for space and other business needs with the new high-profit firms of the advanced corporate service economy.

In brief, that major cities have seen changes in their job supplies can be chalked up both to the emergence of new sectors and to the reorganization of work in sectors new and old. The shift from a manufacturing to a service-dominated economy, particularly evident in cities, destabilizes older relationships between jobs and economic sectors. Today, much more than twenty years ago, we see an expansion of low-wage jobs associated with growing sectors rather than with declining ones. At the same time, a vast array of activities that once took place under standardized work arrangements have become increasingly informalized, as some manufacturing relocates from unionized factories to sweatshops and private homes. If we distinguish the characteristics of jobs from those of the sectors in which they are located, we can see that highly dynamic, technologically advanced growth sectors may well contain low-wage, dead-end jobs. Similarly, backward sectors like downgraded manufacturing can reflect the major growth trends in a highly developed economy.

It seems, then, that we need to rethink two assumptions: that the postindustrial economy primarily requires highly educated workers, and that informalization and downgrading are just Third World imports or anachronistic holdovers. Service-dominated urban economies do indeed create low-wage jobs with minimal education requirements, few advancement opportunities, and low pay for demanding work. For workers raised in an ideological context that emphasizes success, wealth, and career, these are not attractive positions; hence the growing demand for immigrant workers. But given the provenance of the jobs these immigrant workers take, we must resist assuming that they are located in the backward sectors of the economy.

THE OTHER WORKERS IN THE ADVANCED
CORPORATE ECONOMY

Low-wage workers accomplish a sizable portion of the day-to-day work in global cities' leading sectors. After all, advanced professionals require clerical, cleaning, and repair workers for their state-of-the-art offices, and they require truckers to bring them their software and their toilet paper. In my research on New York and other cities, I have found that between 30 and 50 percent of workers in the leading sectors are actually low-wage workers.[9]

The similarly state-of-the-art lifestyles of professionals in these sectors have created a whole new demand for household workers, particularly maids and nannies, as well as for service workers to cater to their high-income consumption habits.[10] Expensive restaurants, luxury housing, luxury hotels, gourmet shops, boutiques, French hand laundries, and special cleaning services, for example, are more labor-intensive than their lower-priced equivalents. To an extent not seen in a very long time, we are witnessing the reemergence of a "serving class" in contemporary high-income households and neighborhoods. The image of the immigrant woman serving the white middle-class professional woman has replaced that of the black female servant working for the white master in centuries past. The result is a sharp tendency toward social polarization in today's global cities.

We are beginning to see how the global labor markets at the top and at the bottom of the economic system are formed. The bottom is mostly staffed through the efforts of individual workers, though an expanding network of organizations has begun to get involved. (So have illegal traffickers, as we'll see later.) Kelly Services, a *Fortune* 500 global staffing company that operates in twenty-five countries, recently added a home-care division that is geared toward people who need assistance with daily living but that also offers services that in the past would have been taken care of by the mother or wife figure in a household. A growing range of smaller global staffing organizations offer day care, including dropping off and picking up schoolchildren, as well as completion of in-house tasks from child care to cleaning and cooking.[11] One international agency for nannies and au pairs (EF Au Pair Corporate Program) advertises directly to corporations, urging them to include the service in their offers to potential hires.

Meanwhile, at the top of the system, several global *Fortune* 500 staffing companies help firms fill high-level professional and technical jobs. In 2001, the largest of these was the Swiss multinational Adecco, with offices in fifty-eight countries; in 2000 it provided firms worldwide with 3 million workers. Manpower, with offices in fifty-nine different countries, provided 2 million workers. Kelly Services provided 750,000 employees in 2000.

The top and the bottom of the occupational distribution are becoming internationalized and so are their labor suppliers. Although midlevel occupations are increasingly staffed through temporary employment agencies, these companies have not internationalized their efforts. Occupations at the top and at the bottom are, in very different but parallel ways, sensitive. Firms need reliable and hopefully talented professionals, and they need them specialized but standardized so that they can use them globally. Professionals seek the same qualities in the workers they employ in their homes. The fact that staffing organizations have moved into providing domestic services signals both that a global labor market has emerged in this area and that there is an effort afoot to standardize the services maids, nannies, and home-care nurses deliver.

PRODUCING A GLOBAL SUPPLY OF THE NEW CARETAKERS: THE FEMINIZATION OF SURVIVAL

The immigrant women described in the first half of this chapter enter the migration process in many different ways. Some migrate in order to reunite their families; others migrate alone. Many of their initial movements have little to do with globalization. Here I am concerned with a different kind of migration experience, and it is one that is deeply linked to economic globalization: migrations organized by third parties, typically governments or illegal traffickers. Women who enter the migration stream this way often (though not always) end up in different sorts of jobs than those described above. What they share with the women described earlier in this chapter is that they, too, take over tasks previously associated with housewives.

The last decade has seen a growing presence of women in a variety of cross-border circuits. These circuits are enormously diverse, but they share one feature: they produce revenue on the backs of the truly disadvantaged. One such circuit consists in the illegal trafficking in people for the sex industry and for various types of labor. Another circuit has developed around cross-border migrations, both documented and not, which have become an important source of hard currency for the migrants' home governments. Broader structural conditions are largely responsible for forming and strengthening circuits like these. Three major actors emerge from those conditions, however: women in search of work, illegal traffickers, and the governments of the home countries.

These circuits make up, as it were, countergeographies of globalization. They are deeply imbricated with some of globalization's major constitutive dynamics: the formation of global markets, the intensifying of transnational and translocal networks, and the development of communication technologies that easily escape conventional surveillance. The global economic system's institutional support for cross-border markets and money flows has contributed greatly to the formation and strengthening of these circuits.[12] The countergeographies are dynamic and mobile; to some extent, they belong to the shadow economy, but they also make use of the regular economy's institutional infrastructure.[13]

Such alternative circuits for survival, profit, and hard currency have grown at least partly in response to the effects of economic globalization on developing countries. Unemployment is on the rise in much of the developing world; small and medium-sized enterprises oriented to the national, rather than the export, market have closed; and government debt, already large, is in many cases rising. The economies frequently grouped under the label "developing" are often struggling, stagnant, or even shrinking. These conditions have pressed additional responsibilities onto women, as men have lost job opportunities and governments have cut back on social services.[14] In other words, it has become increasingly important to find alternative ways of making a living, producing profits, and generating government revenues, as developing countries have faced the following concurrent trends: diminishing job prospects for men, a falloff in traditional business opportunities as foreign firms and export industries displace previous economic mainstays, and a concomitant decrease in government revenues, due both to the new conditions of globalization and to the burden of servicing debts.[15]

The major dynamics linked to economic globalization have significantly affected developing economies, including the so-called middle-income countries of the global south. These countries have had not only to accommodate new conditions but to implement a bundle of new policies, including structural adjustment programs, which require that countries open up to foreign firms and eliminate state subsidies. Almost inevitably, these economies fall into crisis; they then implement the International Monetary Fund's programmatic solutions. It is now clear that in most of the countries involved, including Mexico, South Korea, Ghana, and Thailand, these solutions have cost certain sectors of the economy and population enormously, and they have not fundamentally reduced government debt.

Certainly, these economic problems have affected the lives of women from developing countries. Prostitution and migrant labor are increasingly popular ways to make a living; illegal trafficking in women and children for the sex industry, and in all kinds of people as laborers, is an increasingly popular way to make a profit; and remittances, as well as the organized export of workers, have become increasingly popular ways for governments to bring in revenue. Women are by far the majority group in prostitution and in trafficking for the sex industry, and they are becoming a majority group in migration for labor.

Such circuits, realized more and more frequently on the backs of women, can be considered a (partial) feminization of survival. Not only are households, indeed whole communities, increasingly dependent on women for their survival, but so too are governments, along with enterprises that function on the margins of the legal economy. As the term *circuits* indicates, there is a degree of institutionalization in these dynamics; that is to say, they are not simply aggregates of individual actions.

GOVERNMENT DEBT: SHIFTING RESOURCES FROM WOMEN TO FOREIGN BANKS

Debt and debt-servicing problems have been endemic in the developing world since the 1980s. They are also, I believe, crucial to producing the new countergeographies of globalization. But debt's impact on women, and on the feminization of survival, has more to do with particular features of debt than with debt *tout court*.

A considerable amount of research indicates that debt has a detrimental effect on government programs for women and children, notably education and health care. Further, austerity and adjustment programs, which are usually implemented in order to redress government debt, produce unemployment, which also adversely affects women[16] by adding to the pressure on them to ensure household survival. In order to do so, many women have turned to subsistence food production, informal work, emigration, and prostitution.[17]

Most of the countries that fell into debt in the 1980s have found themselves unable to climb out of it. In the 1990s, a whole new set of countries joined the first group in this morass. The IMF and the World Bank responded with their structural adjustment program and structural adjustment loans, respectively. The latter tied loans to economic policy reform rather than to particular projects. The idea was to make these states more "competitive," which typically meant inducing sharp cuts in various social programs.

Rather than becoming "competitive," the countries subjected to structural adjustment have remained deeply indebted, with about fifty of them now categorized as "highly indebted poor countries." Moreover, a growing number of middle-income countries are also caught in this debt trap. Argentina became the most dramatic example when it defaulted on $140 billion in debt in December 2001—the largest ever sovereign default. Given the structure and servicing of these debts, as well as their weight in debtor countries' economies, it is not likely that many of these countries will ever be able to pay off their debts in full. Structural adjustment programs seem to have made this even less likely; the economic reforms these programs demanded have added to unemployment and the bankruptcy of many small, nationally oriented firms.

It has been widely recognized that the south has already paid its debt several times over. According to some estimates, from 1982 to 1998, indebted countries paid four times their original debts, and at the same time their debt increased four times.[18] Nonetheless, these countries continue to pay a significant share of their total revenue to service their debt. Thirty-three of the officially named forty-one highly indebted poor countries paid $3 in debt service to the north for every $1 they received in development assistance. Many of these countries pay more than 50 percent of their government revenues toward debt service, or 20 to 25 percent of their export earnings.

The ratios of debt to GNP in many of the highly indebted poor countries exceed sustainable limits; many are far more extreme than the level considered unmanageable during the Latin American debt crisis of the 1980s. Such ratios are especially high in Africa, where they stand at 123 percent, compared with 42 percent in Latin America and 23 percent in Asia.[19] Such figures suggest that most of these countries will not get out of their indebtedness through structural adjustment programs. Indeed, it would seem that in many cases the latter have had the effect of intensifying debt dependence. Furthermore, together with various other factors, structural adjustment programs have contributed to an increase in unemployment and in poverty.

ALTERNATIVE SURVIVAL CIRCUITS

It is in this context—marked by unemployment, poverty, bankruptcies of large numbers of firms, and shrinking state resources to meet social needs—that alternative circuits of survival emerge, and it is to these conditions that such circuits are articulated. Here I want to focus on the growing salience of the trafficking of women as a profit-making option and on the growing importance of the emigrants' remittances to the bottom lines of the sending states.

Trafficking, or the forced recruitment and transportation of people for work, is a violation of human, civil, and political rights. Much legislative effort has gone into addressing trafficking: international treaties and charters, U.N. resolutions, and various bodies and commissions have all attempted to put a stop to this practice.[20] Nongovernmental organizations have also formed around this issue.[21]

Trafficking in women for the sex industry is highly profitable for those running the trade. The United Nations estimates that 4 million people were trafficked in 1998, producing a profit of $7 billion for criminal groups.[22] These funds include remittances from prostitutes' earnings as well as payments to organizers and facilitators. In Poland, police estimate that for each woman delivered, the trafficker receives about $700. Ukrainian and Russian women, highly prized in the sex market, earn traffickers $500 to $1,000 per woman delivered. These women can be expected to service fifteen clients a day on average, and each can be expected to make about $215,000 per month for the criminal gang that trafficked her.[23]

It is estimated that in recent years, several million women and girls have been trafficked from and within Asia and the former Soviet Union, both of which are major trafficking areas. The growing frequency of trafficking in these two regions can be linked to increases in poverty, which may lead some parents to sell their daughters to brokers. In the former Soviet republics and Eastern Europe, unemployment has helped promote the growth of criminal gangs, some of which traffic women. Unemployment rates hit 70 percent among women in Armenia, Russia, Bulgaria, and Croatia after the implementation of market policies; in Ukraine, the rate was 80 percent. Some research indicates that need is the major motivation for entry into prostitution.[24]

The sex industry is not the only trafficking circuit: [M]igrant workers of both sexes can also be profitably trafficked across borders. According to a U.N. report, criminal organizations in the 1990s generated an estimated $3.5 billion per year in profits from trafficking migrants. Organized crime has only recently entered this business; in the past, trafficking was mostly the province of petty criminals. Some recent reports indicate that organized-crime groups are creating strategic intercontinental alliances through networks of coethnics in various countries; this facilitates transport, local distribution, provision of false documents, and the like. These international networks also allow traffickers to circulate women and other migrants among third countries; they may move women from Burma, Laos, Vietnam, and China to Thailand, while moving Thai women to Japan and the United States.[25] The Global Survival Network reported on these practices after it conducted a two-year investigation, establishing a dummy company in order to enter the illegal trade.[26]

Once trafficked women reach their destination countries, some features of immigration policy and its enforcement may well make them even more vulnerable. Such women usually have little recourse to the law. If they are undocumented, which they are likely to be, they will not be treated as victims of abuse but as violators of entry, residence, and work laws. As countries of the global north attempt to address undocumented immigration and trafficking by clamping down on entry at their borders, more women are likely to turn to traffickers to help them get across. These traffickers may turn out to belong to criminal organizations linked to the sex industry.

Moreover, many countries forbid foreign women to work as prostitutes, and this provides criminal gangs with even more power over the women they traffic. It also eliminates one survival option for foreign women who may have limited access to jobs. Some countries, notably the Netherlands and Switzerland, are far more tolerant of foreign women working as prostitutes than as regular laborers. According to International Organization for Migration data, in the European Union, a majority of prostitutes are migrant women: 75 percent in Germany and 80 percent in the Italian city of Milan.

Some women know that they are being trafficked for prostitution, but for many the conditions of their recruitment and the extent of the abuse and bondage they will suffer only become evident after they arrive in the receiving country.[27] Their confinement is often extreme—akin to slavery—and so is their abuse, including rape, other forms of sexual violence, and physical punishment. Their meager wages are often withheld. They are frequently forbidden to protect themselves against AIDS, and they are routinely denied medical care. If they seek help from the police, they may be taken into detention for violating immigration laws; if they have been provided with false documents, there will be criminal charges.

With the sharp growth of tourism over the last decade, the entertainment sector has also grown, becoming increasingly important in countries that have adopted tourism as a strategy for development.[28] In many places, the sex trade is part of the entertainment industry, and the two have grown in tandem. Indeed, the sex trade itself has become a development strategy in some areas where unemployment and poverty are widespread, and where governments are desperate for revenue and hard currency. When local manufacturing and agriculture no longer provide jobs, profits, or government revenue, a once marginal economic wellspring becomes a far more important one. The IMF and the World Bank sometimes recommend tourism as a solution to the troubles of poor countries, but when they provide loans for its development or expansion, they may well inadvertently contribute to the expansion of the entertainment industry and, indirectly, of the sex trade. Because it is linked to development strategies in this way, the trafficking of women may continue to expand in these countries.

Indeed, the global sex industry is likely to expand in any case, given the involvement of organized crime in the sex trade, the formation of cross-border ethnic networks, and the growing transnationalization of tourism. These factors may well lead to a sex trade that reaches out to more and more "markets." It's a worrisome possibility, especially as growing numbers of women face few if any employment options. Prostitution becomes—in certain kinds of economies— crucial to expanding the entertainment industry, and thereby to tourism as a development strategy that will in turn lead to increased government revenue. These links are structural; the significance of the sex industry to any given economy rises in the absence of other sources of jobs, profits, and revenues.

Women, and migrants generally, are crucial to another development strategy as well: the remittances migrant workers send home are a major source of hard-currency reserves for the migrant's home country. While remittances may seem minor compared to the financial markets' massive daily flow of capital, they are often very significant for struggling economies. In 1998, the latest year for which we have data, the remittances migrants sent home topped $70 billion globally. To understand the significance of this figure, compare it to the GDP and foreign currency reserves in the affected countries, rather than to the global flow of capital. For instance, in the Philippines, a major sender of migrants generally and of women for the entertainment industry in particular, remittances were the third largest source of foreign currency over the last several years. In Bangladesh, which sends significant numbers of workers to the Middle East, Japan, and several European countries, remittances totaled about a third of foreign-currency transactions.

Exporting workers is one means by which governments cope with unemployment and foreign debt. The benefits of this strategy come through two channels, one of which is highly formalized and the other a simple byproduct of the migration process. South Korea and the Philippines both furnish good examples of formal labor-export programs. In the 1970s, South Korea developed extensive programs to promote the export of workers, initially to the Middle Eastern OPEC countries and then worldwide, as an integral part of its growing overseas construction industry. When South Korea's economy boomed, exporting workers became a less necessary and less attractive strategy. The Philippine government, by contrast, expanded and diversified its labor exports in order to deal with unemployment and to secure needed foreign-currency reserves through remittances.

The Philippines Overseas Employment Administration (POEA) has played an important role in the emigration of Filipina women to the United States, the Middle East, and Japan. Established by the Filipino government in 1982, POEA organized and supervised the export of nurses and maids to high-demand areas. Foreign debt and unemployment combined to make the export of labor an attractive option. Filipino workers overseas send home an average of almost $1 billion a year. For their parts, labor-importing countries had their own reasons to welcome the Filipino government's policy. The OPEC countries of the Middle East saw in the Filipina migrants an answer to their growing demand for domestic workers following the 1973 oil boom. Confronted with an acute shortage of nurses, a profession that demanded years of training yet garnered low wages and little prestige, the United States passed the Immigration Nursing Relief Act of 1989, which allowed for the importation of nurses.[29] And in booming 1980s Japan, which witnessed rising expendable incomes but marked labor shortages, the government passed legislation permitting the entry of "entertainment workers."[30]

The largest number of migrant Filipinas work overseas as maids, particularly in other Asian countries.[31] The second largest group, and the fastest growing, consists of entertainers, who migrate mostly to Japan. The rapid increase in the number of women migrating as entertainers can be traced to the more than five hundred "entertainment brokers" that now operate in the Philippines outside the state umbrella. These brokers provide women for the Japanese sex industry, which is basically controlled by organized gangs rather than through the government-sponsored program for the entry of entertainers. Recruited for singing and entertaining, these women are frequently forced into prostitution as well.[32]

The Filipino government, meanwhile, has also passed regulations that permit mail-order-bride agencies to recruit young Filipinas to marry foreign men. This trade rapidly picked up pace thanks to the government's organized support. The United States and Japan are two of the most common destinations for mail-order brides. Demand was especially high in Japan's agricultural communities in the 1980s, given that country's severe shortage of people in general and of young women in particular, as the demand for labor boomed in the large metropolitan areas. Municipal governments in Japanese towns made it a policy to accept Filipina brides.

A growing body of evidence indicates that mail-order brides frequently suffer physical abuse. In the United States, the Immigration and Naturalization Service has recently reported acute domestic violence against mail-order wives. Again, the law discourages these women from seeking recourse, as they are liable to be detained if they do so before they have been married for two years. In Japan, foreign mail-order wives are not granted full legal status, and considerable evidence indicates that many are subject to abuse not only by their husbands but by their husbands' extended families as well. The Philippine government approved most mail-order-bride brokers before 1989, but during Corazon Aquino's presidency, the stories of abuse by foreign husbands led the Philippine government to ban the mail-order-bride business. Nonetheless, such organizations are almost impossible to eliminate, and they continue to operate in violation of the law.

The Philippines may have the most developed programs for the export of its women, but it is not the only country to have explored similar strategies. After its 1997–1998 financial crisis, Thailand started a campaign to promote migration for work and to encourage overseas firms to recruit Thai workers. Sri Lanka's government has tried to export another 200,000 workers in addition to the 1 million it already has overseas; Sri Lankan women remitted $880 million in 1998, mostly from their earnings as maids in the Middle East and Far East. Bangladesh organized extensive labor-export programs to the OPEC countries of the Middle East in the 1970s. These programs have continued, becoming a significant source of foreign currency along with individual migrations to these and other countries, notably the United States and Great Britain. Bangladesh's workers remitted $1.4 billion in each of the last few years.[33]

CONCLUSION

Globalization is not only about the hypermobility of capital and the ascendance of information economies. It is also about specific types of places and work processes. In order to understand how economic globalization relates to the extraction of services from the Third World to fulfill what was once the First World woman's domestic role, we must look at globalization in a way that emphasizes some of these concrete conditions.

The growing immiserization of governments and economies in the global south is one such condition, insofar as it enables and even promotes the migration and trafficking of women as a strategy for survival. The same infrastructure designed to facilitate cross-border flows of capital, information, and trade also makes possible a range of unintended cross-border flows, as growing numbers of traffickers, smugglers, and even governments now make money off the backs of women. Through their work and remittances, women infuse cash into the economies of deeply indebted countries, and into the pockets of "entrepreneurs" who have seen other opportunities vanish. These survival circuits are often complex, involving multiple locations and sets of actors, which altogether constitute increasingly global chains of traders and "workers."

But globalization has also produced new labor demand dynamics that center on the global cities of the north. From these places, global economic processes are managed and coordinated by increasing numbers of highly paid professionals. Both the firms and the lifestyles of these professionals are maintained by low-paid service workers, who are in growing demand. Large numbers of low-wage women and immigrants thus find themselves incorporated into strategic economic sectors in global cities. This incorporation happens directly, as in the case of low-wage clerical and blue collar workers, such as janitors and repair workers. And it happens indirectly, through the consumption practices of high-income professionals, which generate a demand for maids and nannies as well as low-wage workers in expensive restaurants and shops. Low-wage workers are then incorporated into the leading sectors, but under conditions that render them invisible.

Both in global cities and in survival circuits, women emerge as crucial economic actors. It is partly through them that key components of new economies have been built. Globalization allows links to be forged between countries that send migrants and countries that receive them; it also enables local and regional practices to go global. The dynamics that come together in the global city produce a strong demand for migrant workers, while the dynamics that mobilize women into survival circuits produce an expanding supply of workers who can be pushed or sold into those types of jobs. The technical infrastructure and transnationalism that underlie the key globalized industries also allow other types of activities, including money laundering and trafficking, to assume a global scale.

ENDNOTES

1. For more detailed accounts of each of these configurations please see my "Towards a Feminist Analytics of Globalization," in Saskia Sassen, *Globalization and Its Discontents: Essays on the Mobility of People and Money* (New York: The New Press, 1998); and my article, "Women's Burden: Countergeographies of Globalization and the Feminization of Survival," *Journal of International Affairs*, vol. 53, no. 2 (spring 2000) pp. 503–24.

2. In my larger research project, I also focus on a range of liberating activities and practices that globalization enables; for example, some aspects of the human-rights and environmental movements, as well as of the antiglobalization network. In this sense, globalization enables the production of its own countergeographies, some of which are exploitative, others emancipatory.

3. By emphasizing that global processes are at least partly embedded in national territories, such a focus introduces new variables into current conceptions of economic globalization and the shrinking regulatory role of the state. That is to say, new transnational economic processes do not necessarily occur within the global/national spatial duality that many analysts of the global economy presuppose. That duality suggests two mutually exclusive spaces, one beginning where the other ends. National states play a role in the implementation of global economic systems, and this role can assume different forms, depending on the level of development, political culture, and mode of articulation with global processes. By reintroducing the state into our analysis of globalization, we open the way toward examining how this transformed state articulates the gender question. One way in which states have been reconfigured is through the political ascendance of ministries of finance and the decline of departments dealing with social concerns, including housing, health, and welfare.

4. Indeed, women in many of these settings are seen, rightly or wrongly, as better cultural brokers, and these skills matter to global firms. In the financial-services industry, women are considered crucial to interfacing with consumers, because they are believed to inspire more trust and thereby to make it easier for individual investors to put their money in what are often known to be highly speculative endeavors. See Melissa Fisher, "Wall Street Women's 'Herstories' in Late Financial Corporate Capitalism," in *Constructing Corporate America: History, Politics, Culture*, ed. Kenneth Lipartito and David B. Sicilia (New York: Oxford University Press, 2002).

5. I have developed this at length in *Globalization and Its Discontents*.

6. Pierrette Hondagneu-Sotelo, *Gendered Transitions: Mexican Experiences of Immigration* (Berkeley: University of California Press, 1994); Sarah Mahler, *American Dreaming: Immigrant Life on the Margins* (Princeton, N.J.: Princeton University Press, 1995).

7. Frank Munger, ed., *Laboring Under the Line* (New York: Russell Sage Foundation, 2002); Laurance Roulleau-Berger, ed., *Youth and Work in the Postindustrial City of North America and Europe* (Leiden and New York: Brill, 2002); Hector R. Cordero-Guzman, Robert C. Smith, and Ramon Grosfoguel, eds., *Migration, Transnationalization, and Race in a Changing New York* (Philadelphia: Temple University Press, 2001); see generally for data and sources, Saskia Sassen, *The Global City* (Princeton, N.J.: Princeton University Press, 2001), chapters 8 and 9.

8. At issue here is a type of manufacturing that requires an urban location because it is geared toward urban markets and belongs to a fairly developed network of contractors and subcontractors. We have called this type of manufacturing "urban manufacturing" to distinguish it from sectors that respond to very different constraints and advantages. It generally consists of design-linked manufacturing done on contract: jewelry making, woodwork and metalwork for architecture and real estate firms, the production of fashion, furniture, lamps, and so on. Many components of urban manufacturing are not downgraded, or at least not yet. One major policy implication is that city governments should support this type of manufacturing and cease to subsidize the kind that will leave the city sooner or later anyhow (see Matthew T. Mitchell, "Urban Manufacturing in New York City" [Master's thesis, Department of Urban Planning, Columbia University, 1996]). Women, especially immigrant women, are the key labor force in urban manufacturing.

9. For evidence and multiple sources, see Sassen, 2001, chapters 8 and 9.

10. The consumption needs of the growing low-income population in large cities are also increasingly often met through labor-intensive, rather than standardized and unionized, forms of producing goods and services: manufacturing and retail establishments that are small, rely on family labor, and often fall below minimum safety and health standards. Cheap, locally produced sweatshop garments and bedding, for example, can compete with low-cost Asian imports. A growing range of products and services, from low-cost furniture made in basements to "gypsy cabs" and family day care, are available to meet the demands of the growing low-income population in these cities. Inequality reshapes the consumption structure of cities in innumerable ways, and this in turn has feedback effects on the organization of work, in both the formal and the informal economy.

11. Very prominent in this market are the International Nanny and Au Pair Agency, headquartered in Britain; Nannies Incorporated, based in London and Paris; and the International Au Pair Association (IAPA) based in Canada.

12. I have argued this for the case of international labor migrations (e.g., Saskia Sassen, *Guests and Aliens* [New York: The New Press, 1999]). See also Max Castro, ed., *Free Markets, Open Societies, Closed Borders?* (Berkeley: University of California Press, 2000); and Frank Bonilla, Edwin Melendez, Rebecca Morales, and Maria de Los Angeles Torres, eds., *Borderless Borders* (Philadelphia: Temple University Press, 1998).

13. Once there is an institutional infrastructure for globalization, processes that have previously operated at the national level can scale up to the global level, even when they do not need to. This phenomenon contrasts with processes that are by their very nature global, such as the network of financial centers underlying the formation of a global capital market.

14. An older literature on women and debt links the implementation of a first generation of structural adjustment programs to the growth of government debt in the 1980s; this literature documents the disproportionate burden these programs put on women. It is a large literature in many different languages, and including a vast number of limited-circulation items produced by various activist and support organizations. For overviews, see Kathryn Ward, *Women Workers and Global Restructuring* (Ithaca, N.Y.: School of Industrial and Labor Relations Press, 1990); Kathryn Ward and Jean Pyle, "Gender, Industrialization and Development," in *Women in the Latin American Development Process: From Structural Subordination to Empowerment*, eds. Christine E. Bose and Edna Acosta-Belen (Philadelphia: Temple University Press, 1995), pp. 37–64; Christine E. Bose and Edna Acosta-Belen, eds., *Women in the Latin American Development Process* (Philadelphia: Temple University Press, 1995); Lourdes Beneria and Shelley Feldman, eds., *Unequal Burden: Economic Crises, Persistent Poverty, and Women's Work* (Boulder, Colo.: Westview Press, 1992); York Bradshaw, Rita Noonan, Laura Gash, and Claudia Buchmann, "Borrowing Against the Future: Children and Third World Indebtness," *Social Forces*, vol. 71, no. 3 (1993), pp. 629–656; Irene Tinker, ed., *Persistent Inequalities: Women and World Development* (New York: Oxford University Press, 1990); and Carolyn Moser, "The Impact of Recession and Structural Adjustment Policies at the Micro-Level: Low-Income Women and Their Households in Guayaquil, Ecuador," *Invisible Adjustment*, UN1CEF, vol. 2 (1989). Now there is also a new literature on structural adjustment's second generation. These studies are more directly linked to globalization; I will cite them later in this article.

15. In many of these countries, a large number of firms in traditional sectors oriented to the local or national market have closed, and export-oriented cash crops have increasingly often replaced survival agriculture and food production for local or national markets.

16. See Michel Chossudovsky, *The Globalisation of Poverty* (London: Zed/TWN. 1997); Guy Standing, "Global Feminization Through Flexible Labor: A Theme Revisited," *World Development*, vol. 27, no. 3 (1999), pp. 583–602; Aminur Rahman, "Microcredit Initiatives for Equitable and Sustainable Development: Who Pays?" *World Development*, vol. 27, no. 1 (1999), pp. 67–82; Diane Elson, *Male Bias in Development*, 2nd ed. (Manchester: Metropolitan, 1995). For an excellent overview of the literature on the impact of the debt on women, see Kathryn Ward, "Women and the Debt," paper presented at the Colloquium on Globalization and the Debt, Emory University, Atlanta: Metropolitan (1999). On file with author at kbward@siu.edu.

17. On these various issues, see Diana Alarcon-Gonzalez and Terry McKinley, "The Adverse Effects of Structural Adjustment on Working Women in Mexico," *Latin American Perspectives*, vol. 26, no. 3 (1999), pp. 103–17; Claudia Buchmann, "The Debt Crisis, Structural Adjustment and Women's Education," *International Journal of Comparative Studies*, vol. 37, nos. 1-2 (1996), pp. 5–30; Helen I. Safa, *The Myth of the Male Breadwinner: Women and Industrialization in the Caribbean* (Boulder, Colo.: Westview Press, 1995); Nilufer Cagatay and Sule Ozler, "Feminization of the Labor Force: The Effects of Long-term Development and Structural Adjustment," *World Development,* vol. 23. no. 11 (1995), pp. 1883–94; Erika Jones, "The Gendered Toll of Global Debt Crisis," *Sojourner*, vol. 25, no. 3, pp. 20–38; and several of the references cited in the preceding footnotes.

18. Eric Toussaint, "Poor Countries Pay More Under Debt Reduction Scheme" (July 1999), www.twnside.org.sg/souths/twn/title/1921-cn.htm. According to Susan George, the south has paid back the equivalent of six Marshall Plans to the north (Asoka Bandarage, *Women, Population, and Crisis* [London: Metropolitan: 7th ed, 1997]).

19. The IMF asks HIPCs to pay 20 to 25 percent of their export earnings toward debt service. In contrast, in 1953 the Allies canceled 80 percent of Germany's war debt and only insisted on 3 to 5 percent of export-earnings debt service. These general terms were also evident as Central Europe emerged from communism. For one of the best critical examinations of globalization, Richard C. Longworth, *Global Squeeze: The Coming Crisis for First World Nations* (Chicago: Contemporary Books, 1998).

20. See Janie Chuang, "Redirecting the Debate over Trafficking in Women: Definitions, Paradigms, and Contexts," *Harvard Human Rights Journal*, vol. 10 (winter 1998). Trafficking has become sufficiently recognized as an issue that it was addressed in the G8 meeting in Birmingham in May 1998, a first for the G8 (*Trafficking in Migrants*, International Office of Migration quarterly bulletin, Geneva: IOM, 1998). The heads of the eight major industrialized countries stressed the importance of cooperating against international organized crime and people trafficking. President Clinton issued a set of directives to his administration in order to strengthen efforts against trafficking in women and girls. This in turn generated a legislative initiative by Senator Paul Wellstone, which led to a Senate bill in 1999. (For a good critical analysis, see Dayan, "Policy Initiatives in the U.S. against the Illegal Trafficking of Women for the Sex Industry," Department of Sociology, University of Chicago, 1999, on file with the author).

21. The Coalition Against Trafficking in Women has centers and representatives in Australia, Bangladesh, Europe, Latin America, North America, Africa, and Asia Pacific. The Women's Rights Advocacy Program has established the Initiative Against Trafficking in Persons to combat the global trade in persons. Other organizations are referred to throughout this article.

22. See, generally, the Foundation Against Trafficking in Women (STV) and the Global Alliance Against Traffic in Women (GAATW). For regularly updated sources of information on trafficking, see http://www.hrlawgroup.org/site/programs/traffic.html. See also Sietske Altink, *Stolen Lives: Trading Women into Sex and Slavery* (New York: Harrington Park Press, 1995); Kamala Kempadoo and Jo Doezema, *Global Sex Workers: Rights, Resistance, and Redefinition* (London: Routledge, 1998); Susan Shannon, "The Global Sex Trade: Humans as the Ultimate Commodity," *Crime and Justice International* (May 1999), pp. 5–25; Lap-Chew Lin and Wijers Marian, *Trafficking in Women, Forced Labour and Slavery-Like Practices in Marriage, Domestic Labour and Prostitution* (Utrecht: Foundation Against Trafficking in Women [STV], and Bangkok: Global Alliance Against Traffic in Women [GAATW], 1997); Lin Lim, *The Sex Sector: The Economic and Social Bases of Prostitution in Southeast Asia* (Geneva: International Labor Office, 1998).

23. For more detailed information, see the STV-GAATW reports; IOM 1996; CIA, "International Trafficking in Women to the United States: A Contemporary Manifestation of Slavery and Organized Crime," prepared by Amy O'Neill Richard (Washington, D.C.: Center for the Study of Intelligence, 2000), www.cia.gov/csi/monograph/women/trafficking.pdf.

24. There is also a growing trade in children for the sex industry. This has long been the case in Thailand, but it is now present in several other Asian countries, eastern Europe, and Latin America.

25. There are various reports on trafficking routes. Malay brokers sell Malay women into prostitution in Australia. Women from Albania and Kosovo have been trafficked by gangs into prostitution in London. Teens from Paris and other European cities have been sold to Arab and African customers; see Susan Shannon, "The Global Sex Trade: Humans as the Ultimate Commodity," *Crime and Justice International* (May 1999), pp. 5–25. In the United States, the police broke up an international Asian ring that imported women from China, Thailand, Korea, Malaysia, and Vietnam; see William Booth, "Thirteen Charged in Gang Importing Prostitutes," *Washington Post*, August 21, 1999. The women were charged between $30,000 and $40,000 in contracts to be paid through their work in the sex trade or the garment industry. The women in the sex trade were shuttled around several states in the United States in order to bring continuing variety to the clients.

26. See Global Survival Network, "Crime and Servitude: An Expose of the Traffic in Women for Prostitution from the Newly Independent States," at www.globalsurvival.net/femaletrade.html, November 1997.

27. A fact sheet by the Coalition to Abolish Slavery and Trafficking reports that one survey of Asian sex workers found that rape often preceded their being sold into prostitution and that about one-third had been falsely led into prostitution.

28. Nancy A. Wonders and Raymond Michalowski, "Bodies, Borders, and Sex Tourism in a Globalized World: A Tale of Two Cities—Amsterdam and Havana," *Social Problems*, vol. 48, no. 4 (2001), pp. 545–71. See also Dennis Judd and Susan Fainstein, *The Tourist City* (New Haven: Yale University Press, 1999).

29. About 80 percent of the nurses imported under the new act were from the Philippines. See generally, Satomi Yamamoto, "The Incorporation of Women Workers into a Global City: A Case Study of Filipina Nurses in the Metropolitan New York Area," (2000). On file with the author at syamamot@uiuc.edu.

30. Japan passed a new immigration law—strictly speaking an amendment of an older law—that radically redrew the conditions for entry of foreign workers. It allowed professionals linked to the new service economy—specialists in Western-style finance, accounting, law, et cetera—but made the entry of what is termed "simple labor" illegal. The latter provision generated a rapid increase in the entry of undocumented workers for low-wage jobs. But the new law did make special provisions for the entry of "entertainers."

31. Brenda Yeoh, Shirlena Huang, and Joaquin Gonzalez, "Migrant Female Domestic Workers: Debating the Economic, Social and Political Impacts in Singapore," *International Migration Review*, vol. 33, no. 1 (1999), pp. 114–136; Christine Chin, "Walls of Silence and Late 20th-century Representations of Foreign Female Domestic Workers: The Case of Filipina and Indonesian Houseservants in Malaysia," *International Migration Review*, vol. 31, no. 1 (1997), pp. 353–85: Noeleen Heyzer, *The Trade in Domestic Workers* (London: Zed Books, 1994).

32. These women are recruited and transported both through formal legal channels and through informal or illegal ones. Either way, they have little power to resist. Even as they are paid below minimum wage, they produce significant profits for their

brokers and employers. There has been an enormous increase in so-called enter-tainment businesses in Japan.

33. Natacha David, "Migrants Made the Scapegoats of the Crisis," *ICFTU Online* (International Confederation of Free Trade Unions, 1999), www .hartford-hwp.com/archives/50/012.html.

26

Masculinities and Globalization

R. W. CONNELL

THE WORLD GENDER ORDER

Masculinities do not first exist and then come into contact with femininities; they are produced together, in the process that constitutes a gender order. Accordingly, to understand the masculinities on a world scale, we must first have a concept of the globalization of gender.

This is one of the most difficult points in current gender analysis because the very conception is counterintuitive. We are so accustomed to thinking of gender as the attribute of an individual, even as an unusually intimate attribute, that it requires a considerable wrench to think of gender on the vast scale of global society. Most relevant discussions, such as the literature on women and development, fudge the issue. They treat the entities that extend internationally (markets, corporations, intergovernmental programs, etc.) as ungendered in principle—but affecting unequally gendered recipients of aid in practice, because of bad policies. Such conceptions reproduce the familiar liberal-feminist view of the state as in principle gender-neutral, though empirically dominated by men.

But if we recognize that very large scale institutions such as the state are themselves gendered, in quite precise and specifiable ways (Connell 1990b), and if we recognize that international relations, international trade, and global markets are inherently an arena of gender formation and gender politics (Enloe 1990), then we can recognize the existence of a world gender order. The term can be defined as the structure of relationships that interconnect the gender regimes of institutions, and the gender orders of local society, on a world scale. That is, however, only a definition. The substantive questions remain: what is the shape of that structure, how tightly are its elements linked, how has it arisen historically, what is its trajectory into the future?

Current business and media talk about globalization pictures a homogenizing process sweeping across the world, driven by new technologies, producing vast unfettered global markets in which all participate on equal terms. This is a misleading image. As Hirst and Thompson (1996) show, the global economy is highly unequal and the current degree of homogenization is often overestimated. Multinational corporations based in the three major economic powers (the

SOURCE: R. W. Connell, *The Ethnographic Moment in Studies of Masculinity* (July 1998), pp. 3–23. Copyright © 1998 by Sage Publications, Inc. Reprinted by permission of Sage Publications, Inc.

United States, European Union, and Japan) are the major economic actors worldwide.

The structure bears the marks of its history. Modern global society was historically produced, as Wallerstein (1974) argued, by the economic and political expansion of European states from the fifteenth century on and by the creation of colonial empires. It is in this process that we find the roots of the modern world gender order. Imperialism was, from the start, a gendered process. Its first phase, colonial conquest and settlement, was carried out by gender-segregated forces, and it resulted in massive disruption of indigenous gender orders. In its second phase, the stabilization of colonial societies, new gender divisions of labor were produced in plantation economies and colonial cities, while gender ideologies were linked with racial hierarchies and the cultural defense of empire. The third phase, marked by political decolonization, economic neocolonialism, and the current growth of world markets and structures of financial control, has seen gender divisions of labor remade on a massive scale in the "global factory" (Fuentes and Ehrenreich 1983), as well as the spread of gendered violence alongside Western military technology.

The result of this history is a partially integrated, highly unequal and turbulent world society, in which gender relations are partly but unevenly linked on a global scale. The unevenness becomes clear when different substructures of gender (Connell 1987; Walby 1990) are examined separately.

The Division of Labor

A characteristic feature of colonial and neocolonial economies was the restructuring of local production systems to produce a male wage worker–female domestic worker couple (Mies 1986). This need not produce a "housewife" in the Western suburban sense, for instance, where the wage work involved migration to plantations or mines (Moodie 1994). But it has generally produced the identification of masculinity with the public realm and the money economy and of femininity with domesticity, which is a core feature of the modern European gender system (Holter 1997).

Power Relations

The colonial and postcolonial world has tended to break down purdah systems of patriarchy in the name of modernization, if not of women's emancipation (Kandiyoti 1994). At the same time, the creation of a westernized public realm has seen the growth of large-scale organizations in the form of the state and corporations, which in the great majority of cases are culturally masculinized and controlled by men. In *comprador* capitalism, however, the power of local elites depends on their relations with the metropolitan powers, so the hegemonic masculinities of neocolonial societies are uneasily poised between local and global cultures.

Emotional Relations

Both religious and cultural missionary activity has corroded indigenous homosexual and cross-gender practice, such as the native American *berdache* and the Chinese "passion of the cut sleeve" (Hinsch 1990). Recently developed Western models of romantic heterosexual love as the basis for marriage and of gay identity as the main alternative have now circulated globally—though as Altman (1996) observes, they do not simply displace indigenous models, but interact with them in extremely complex ways.

Symbolization

Mass media, especially electronic media, in most parts of the world follow North American and European models and relay a great deal of metropolitan content; gender imagery is an important part of what is circulated. A striking example is the reproduction of a North American imagery of femininity by Xuxa, the blonde television superstar in Brazil (Simpson 1993). In counterpoint, exotic gender imagery has been used in the marketing strategies of newly industrializing countries (e.g., airline advertising from Southeast Asia)—a tactic based on the longstanding combination of the exotic and the erotic in the colonial imagination (Jolly 1997).

Clearly, the world gender order is not simply an extension of a traditional European-American gender order. That gender order was changed by colonialism, and elements from other cultures now circulate globally. Yet in no sense do they mix on equal terms, to produce a United Colours of Benetton gender order. The culture and institutions of the North Atlantic countries are hegemonic within the emergent world system. This is crucial for understanding the kinds of masculinities produced within it.

THE REPOSITIONING OF MEN AND THE RECONSTITUTION OF MASCULINITIES

The positioning of men and the constitution of masculinities may be analyzed at any of the levels at which gender practice is configured: in relation to the body, in personal life, and in collective social practice. At each level, we need to consider how the processes of globalization influence configurations of gender.

Men's bodies are positioned in the gender order, and enter the gender process, through body reflexive practices in which bodies are both objects and agents (Connell 1995)—including sexuality, violence, and labor. The conditions of such practice include where one is and who is available, for interaction. So it is a fact of considerable importance for gender relations that the global social order distributes and redistributes bodies, through migration, and through political controls over movement and interaction.

The creation of empire was the original "elite migration," though in certain cases mass migration followed. Through settler colonialism, something close to the gender order of Western Europe was reassembled in North America and in Australia. Labor migration within the colonial systems was a means by which gender practices were spread, but also a means by which they were reconstructed, since labor migration was itself a gendered process—as we have seen in relation to the gender division of labor. Migration from the colonized world to the metropole became (except for Japan) a mass process in the decades after

World War II. There is also migration within the periphery, such as the creation of a very large immigrant labor force, mostly from other Muslim countries, in the oil-producing Gulf states.

These relocations of bodies create the possibility of hybridization in gender imagery, sexuality, and other forms of practice. The movement is not always toward synthesis, however, as the race ethnic hierarchies of colonialism have been recreated in new contexts, including the politics of the metropole. Ethnic and racial conflict has been growing in importance in recent years, and as Klein (1997) and Tillner (1997) argue, this is a fruitful context for the production of masculinities oriented toward domination and violence. Even without the context of violence, there can be an intimate interweaving of the formation of masculinity with the formation of ethnic identity, as seen in the study by Poynting, Noble, and Tabar (1997) of Lebanese youths in the Anglo-dominant culture of Australia.

At the level of personal life as well as in relation to bodies, the making of masculinities is shaped by global forces. In some cases, the link is indirect, such as the working-class Australian men caught in a situation of structural unemployment (Connell 1995), which arises from Australia's changing position in the global economy. In other cases, the link is obvious, such as the executives of multinational corporations and the financial sector servicing international trade. The requirements of a career in international business set up strong pressures on domestic life: [A]lmost all multinational executives are men, and the assumption in business magazines and advertising directed toward them is that they will have dependent wives running their homes and bringing up their children.

At the level of collective practice, masculinities are reconstituted by the remaking of gender meanings and the reshaping of the institutional contexts of practice. Let us consider each in turn.

The growth of global mass media, especially electronic media, is an obvious "vector" for the globalization of gender. Popular entertainment circulates stereotyped gender images, deliberately made attractive for marketing purposes. The example of Xuxa in Brazil has already been mentioned. International news media are also controlled or strongly influenced from the metropole and circulate Western definitions of authoritative masculinity, criminality, desirable femininity, and so on. But there are limits to the power of global mass communications. Some local centers of mass entertainment differ from the Hollywood model, such as the Indian popular film industry centered in Bombay. Further, media research emphasizes that audiences are highly selective in their reception of media messages, and we must allow for popular recognition of the fantasy in mass entertainment. Just as economic globalization can be exaggerated, the creation of a global culture is a more turbulent and uneven process than is often assumed (Featherstone 1995).

More important, I would argue, is a process that began long before electronic media existed, the export of institutions. Gendered institutions not only circulate definitions of masculinity (and femininity), as sex role theory notes. The functioning of gendered institutions, creating specific conditions for social practice, calls into existence specific patterns of practice. Thus, certain patterns of collective violence are embedded in the organization and culture of a Western-style army, which are different from the patterns of precolonial violence. Certain patterns of calculative egocentrism are embedded in the working of a stock market; certain patterns of rule following and domination are embedded in a bureaucracy.

Now, the colonial and postcolonial world saw the installation in the periphery, on a very large scale, of a range of institutions on the North Atlantic model: armies, states, bureaucracies, corporations, capital markets, labor markets, schools, law courts, transport systems. These are gendered institutions and their functioning has directly reconstituted masculinities in the periphery. This has not necessarily meant photocopies of European masculinities. Rather, pressures for change are set up that are inherent in the institutional form.

To the extent that particular institutions become dominant in world society, the patterns of masculinity embedded in them may become global standards. Masculine dress is an interesting indicator: almost every political leader in the world now wears the uniform of the Western business executive. The more common pattern, however, is not the complete displacement of local patterns but the articulation of the local gender order with the gender regime of global-model institutions. Case studies such as Hollway's (1994) account of bureaucracy in Tanzania illustrate the point; there; domestic patriarchy articulated with masculine authority in the state in ways that subverted the government's formal commitment to equal opportunity for women.

We should not expect the overall structure of gender relations on a world scale simply to mirror patterns known on the smaller scale. In the most vital of respects, there is continuity. The world gender order is unquestionably patriarchal, in the sense that it privileges men over women. There is a patriarchal dividend for men arising from unequal wages, unequal labor force participation, and a highly unequal structure of ownership, as well as cultural and sexual privileging. This has been extensively documented by feminist work on women's situation globally (e.g., Taylor 1985), though its implications for masculinity have mostly been ignored. The conditions thus exist for the production of a hegemonic masculinity on a world scale, that is to say, a dominant form of masculinity that embodies, organizes, and legitimates men's domination in the gender order as a whole.

The conditions of globalization, which involve the interaction of many local gender orders, certainly multiply the forms of masculinity in the global gender order. At the same time, the specific shape of globalization, concentrating economic and cultural power on an unprecedented scale, provides new resources for dominance by particular groups of men. This dominance may become institutionalized in a pattern of masculinity that becomes, to some degree, standardized across localities. I will call such patterns *globalizing masculinities*, and it is among them, rather than narrowly within the metropole, that we are likely to find candidates for hegemony in the world gender order.

GLOBALIZING MASCULINITIES

In this section, I will offer a sketch of major forms of globalizing masculinity in the three historical phases identified above in the discussion of globalization.

Masculinities of Conquest and Settlement

The creation of the imperial social order involved peculiar conditions for the gender practices of men. Colonial conquest itself was mainly carried out by segregated groups of men—soldiers, sailors, traders, administrators, and a good many who were all these by turn (such as the Rum Corps in early New South Wales, Australia). They were drawn from the more segregated occupations and milieu in the metropole, and it is likely that the men drawn into colonization tended to be the more rootless. Certainly the process of conquest could produce frontier masculinities that combined the occupational culture of these groups with an unusual level of violence and egocentric individualism. The vehement contemporary debate about the genocidal violence of the Spanish conquistadors—who in fifty years completely exterminated the population of Hispaniola—points to this pattern (Bitterli 1989).

The political history of empire is full of evidence of the tenuous control over the frontier exercised by the state—the Spanish monarchs unable to rein in the conquistadors, the governors in Sydney unable to hold back the squatters and in Capetown unable to hold back the Boers, gold rushes breaking boundaries everywhere, even an independent republic set up by escaped slaves in Brazil. The point probably applies to other forms of social control too, such as customary controls on men's sexuality. Extensive sexual exploitation of indigenous women was a common feature of conquest. In certain circumstances, frontier masculinities might be reproduced as a local cultural tradition long after the frontier had passed, such as the gauchos of southern South America, the cowboys of the western United States.

In other circumstances, however, the frontier of conquest and exploitation was replaced by a frontier of settlement. Sex ratios in the colonizing population changed, as women arrived and locally born generations succeeded. A shift back toward the family patterns of the metropole was likely. As Cain and Hopkins (1993) have shown for the British empire, the ruling group in the colonial worlds as a whole was an extension of the dominant class in the metropole, the landed gentry, and tended to reproduce its social customs and ideology. The creation of a settler masculinity might be the goal of state policy, as it seems to have been in late nineteenth-century New Zealand, as part of a general process of pacification and the creation of an agricultural social order (Phillips 1987). Or it might be undertaken through institutions created by settler groups, such as the elite schools in Natal studied by Morrell (1994).

The impact of colonialism on the construction of masculinity among the colonized is much less documented, but there is every reason to think it was severe. Conquest and settlement disrupted all the structures of indigenous society, whether or not this was intended by the colonizing powers (Bitterli 1989).

Indigenous gender orders were no exception. Their disruption could result from the pulverization of indigenous communities (as in the seizure of land in eastern North America and southeastern Australia), through gendered labor migration (as in gold mining with Black labor in South Africa; see Moodie 1994), to ideological attacks on local gender arrangements (as in the missionary assault on the berdache tradition in North America; see Williams 1986). The varied course of resistance to colonization is also likely to have affected the making of masculinities. This is clear in the region of Natal in South Africa, where sustained resistance to colonization by the Zulu kingdom was a key to the mobilization of ethnicnational masculine identities in the twentieth century (Morrell 1996).

Masculinities of Empire

The imperial social order created a hierarchy of masculinities, as it created a hierarchy of communities and races. The colonizers distinguished "more manly" from "less manly" groups among their subjects. In British India, for instance, Bengali men were supposed effeminate while Pathans and Sikhs were regarded as strong and warlike. Similar distinctions were made in South Africa between Hottentots and Zulus, in North America between Iroquois, Sioux, and Cheyenne on one side, and southern and southwestern tribes on the other.

At the same time, the emerging imagery of gender difference in European culture provided general symbols of superiority and inferiority. Within the imperial "poetics of war" (MacDonald 1994), the conqueror was virile, while the colonized were dirty, sexualized, and effeminate or childlike. In many colonial situations, indigenous men were called "boys" by the colonizers (e.g., in Zimbabwe; see Shire 1994). Sinha's (1995) interesting study of the language of political controversy in India in the 1880s and 1890s shows how the images of "manly Englishman" and "effeminate Bengali" were deployed to uphold colonial privilege and contain movements for change. In the late nineteenth century, racial barriers in colonial societies were hardening rather than weakening, and gender ideology tended to fuse with racism in forms that the twentieth century has never untangled.

The power relations of empire meant that indigenous gender orders were generally under pressure from the colonizers, rather than the other way around. But the colonizers too might change. The barriers of late colonial racism were not only to prevent pollution from below but also to forestall "going native," a well-recognized possibility—the starting point, for instance, of Kipling's famous novel *Kim* ([1901] 1987). The pressures, opportunities, and profits of empire might also work changes in gender arrangements among the colonizers, for instance, the division of labor in households with a large supply of indigenous workers as domestic servants (Bulbeck 1992). Empire might also affect the gender order of the metropole itself by changing gender ideologies, divisions of labor, and the nature of the metropolitan state. For instance, empire figured prominently as a source of masculine imagery in Britain, in the Boy Scouts, and in the cult of Lawrence of Arabia (Dawson 1991). Here we see examples of an important principle: the interplay of gender dynamics between different parts of the world order.

The world of empire created two very different settings for the modernization of masculinities. In the periphery, the forcible restructuring of economics and workforces tended to individualize, on one hand, and rationalize, on the other. A widespread result was masculinities in which the rational calculation of self-interest was the key to action, emphasizing the European gender contrast of rational man/irrational woman. The specific form might be local—for instance, the Japanese "salaryman," a type first recognized in the 1910s, was specific to the Japanese context of large, stable industrial conglomerates (Kinmonth 1981). But the result generally was masculinities defined around economic action, with both workers and entrepreneurs increasingly adapted to emerging market economies.

In the metropole, the accumulation of wealth made possible a specialization of leadership in the dominant classes, and struggles for hegemony in which masculinities organized around domination or violence were split from masculinities organized around expertise. The class compromises that allowed the development of the welfare state in Europe and North America were paralleled by gender compromises—gender reform movements (most notably the women's suffrage movement) contesting the legal privileges of men and forcing concessions from the state. In this context, agendas of reform in masculinity emerged: the temperance movement, compassionate marriage, homosexual rights movements, leading eventually to the pursuit of androgyny in "men's liberation" in the 1970s (Kimmel and Mosmiller 1992). Not all reconstructions of masculinity, however, emphasized tolerance or moved toward androgyny. The vehement masculinity politics of fascism, for instance, emphasized dominance and difference and glorified violence, a pattern still found in contemporary racist movements (Tillner 1997).

Masculinities of Postcolonialism and Neoliberalism

The process of decolonization disrupted the gender hierarchies of the colonial order and, where armed struggle was involved, might have involved a deliberate cultivation of masculine hardness and violence (as in South Africa; see Xaba 1997). Some activists and theorists of liberation struggles celebrated this, as a necessary response to colonial violence and emasculation; women in liberation struggles were perhaps less impressed. However one evaluates the process, one of the consequences of decolonization was another round of disruptions of community-based gender orders and another step in the reorientation of masculinities toward national and international contexts.

Nearly half a century after the main wave of decolonization, the old hierarchies persist in new shapes. With the collapse of Soviet communism, the decline of postcolonial socialism, and the ascendancy of the new right in Europe and North America, world politics is more and more organized around the needs of transnational capital and the creation of global markets.

The neoliberal agenda has little to say, explicitly, about gender: it speaks a gender-neutral lanuage of "markets," "individuals," and "choice." But the world in which neoliberalism is ascendant is still a gendered world, and neoliberalism has an implicit gender politics. The "individual" of neoliberal theory has in general the attributes and interests of a male entrepreneur, the attack on the welfare state generally weakens the position of women, while the increasingly unregulated power of transnational corporations places strategic power in the hands of particular groups of men. It is not surprising, then, that the installation of capitalism in Eastern Europe and the former Soviet Union has been accompanied by a reassertion of dominating masculinities and, in some situations, a sharp worsening in the social position of women.

We might propose, then, that the hegemonic form of masculinity in the current world gender order is the masculinity associated with those who control its dominant institutions: the business executives who operate in global markets, and the political executives who interact (and in many contexts, merge) with them. I will call this *transnational business masculinity*. This is not readily available for ethnographic study, but we can get some clues to its character from its reflections in management literature, business journalism, and corporate self-promotion, and from studies of local business elites (e.g., Donaldson 1997).

As a first approximation, I would suggest this is a masculinity marked by increasing egocentrism, very conditional loyalties (even to the corporation), and a declining sense of responsibility for others (except for purposes of image making). Gee, Hull and Lankshear (1996), studying recent management textbooks, note the peculiar construction of the executive in "fast capitalism" as a person with no permanent commitments, except (in effect) to the idea of accumulation itself. Transnational business masculinity is characterized by a limited technical rationality (management theory), which is increasingly separate from science.

Transnational business masculinity differs from traditional bourgeois masculinity by its increasingly libertarian sexuality, with a growing tendency to commodity relations with women. Hotels catering to businessmen in most parts of the world now routinely offer pornographic videos, and in some parts of the world, there is a well-developed prostitution industry catering for international businessmen. Transnational business masculinity does not require bodily force, since the patriarchal dividend on which it rests is accumulated by impersonal, institutional means. But corporations increasingly use the exemplary bodies of elite sportsmen as a marketing tool (note the phenomenal growth of corporate "sponsorship" of sport in the last generation) and indirectly as a means of legitimation for the whole gender order.

MASCULINITY POLITICS ON A WORLD SCALE

Recognizing global society as an arena of masculinity formation allows us to pose new questions about masculinity politics. What social dynamics in the global arena give rise to masculinity politics, and what shape does global masculinity politics take?

The gradual creation of a world gender order has meant many local instabilities of gender. Gender instability is a familiar theme of poststructuralist theory, but this school of thought takes as a universal condition a situation that is historically specific. Instabilities range from the disruption of men's local cultural dominance as women move into the public realm and higher education, through the disruption of sexual identities that produced "queer" politics in the metropole, to the shifts in the urban intelligentsia that produced "the new sensitive man" and other images of gender change.

One response to such instabilities, on the part of groups whose power is challenged but still dominant, is to reaffirm *local* gender orthodoxies and hierarchies. A masculine fundamentalism is, accordingly, a common response in gender politics at present. A soft version, searching for an essential masculinity among myths and symbols, is offered by the mythopoetic men's movement in the United States and by the religious revivalists of the Promise Keepers (Messner 1997). A much harder version is found, in that country, in the right-wing militia movement brought to world attention by the Oklahoma City bombing (Gibson 1994), and in contemporary Afghanistan, if we can trust Western media reports, in the militant misogyny of the Taliban. It is no coincidence that in the two latter cases, hardline masculine fundamentalism goes together with a marked anti-internationalism. The world system—rightly enough—is seen as the source of pollution and disruption.

Not that the emerging global order is a hotbed of gender progressivism. Indeed, the neoliberal agenda for the reform of national and international economics involves closing down historic possibilities for gender reform. I have noted how it subverts the gender compromise represented by the metropolitan welfare state. It has also undermined the progressive-liberal agendas of sex role reform represented by affirmative action programs, anti discrimination provisions, child care services, and the like. Right-wing parties and governments have been persistently cutting such programs, in the name of either individual liberties or global competitiveness. Through these means, the patriarchal dividend to men is defended or restored, without an *explicit* masculinity politics in the form of a mobilization of men.

Within the arenas of international relations, the international state, multinational corporations, and global markets, there is nevertheless a deployment of masculinities and a reasonably clear hegemony. The transnational business masculinity described above has had only one major competitor for hegemony in recent decades, the rigid, control-oriented masculinity of the military, and the military-style bureaucratic dictatorships of Stalinism. With the collapse of Stalinism and the end of the cold war, Big Brother (Orwell's famous parody of this form of masculinity) is a fading threat, and the more flexible, calculative, egocentric masculinity of the fast capitalist entrepreneur holds the world stage.

We must, however, recall two important conclusions of the ethnographic moment in masculinity research: that different forms of masculinity exist together and that hegemony is constantly subject to challenge. These are possibilities in the global arena too. Transnational business masculinity is not completely homogeneous; variations of it are embedded in different parts of the world system, which may not be completely compatible. We may distinguish a Confucian variant, based in East Asia, with a stronger commitment to hierarchy and social consensus, from a secularized Christian variant, based in North America, with more hedonism and individualism and greater tolerance for social conflict. In certain arenas, there is already conflict between the business and political leaderships embodying these forms of masculinity: initially over human rights versus Asian values, and more recently over the extent of trade and investment liberalization.

If these are contenders for hegemony, there is also the possibility of opposition to hegemony. The global circulation of "gay" identity (Altman 1996) is an important indication that nonhegemonic masculinities may operate in global arenas, and may even find a certain political articulation, in this case around human rights and AIDS prevention.

REFERENCES

Altman, Dennis. 1996. Rupture of continuity? The internationalisation of gay identities. *Social Text* 48 (3): 77–94.

Barrett, Frank J. 1996. The organizational construction of hegemonic masculinity: The case of the U.S. Navy. *Gender Work and Organization* 3 (3): 129–42.

BauSteineMaenner, ed. 1996. *Kritische Maennerforschung* [Critical research on men]. Berlin: Argument.

Bitterli, Urs. 1989. *Cultures in conflict: Encounters between European and non-European cultures*. 1492–1800, Stanford, CA: Stanford University Press.

Bolin, Anne. 1988. *In search of Eve: Transexual rites of passage*. Westport, CT: Bergin & Garvey.

Bulbeck, Chilla. 1992. *Australian women in Papua New Guinea: Colonial passages* 1920–1960. Cambridge, U.K.: Cambridge University Press.

Cain, P. J., and A. G. Hopkins. 1993. *British Imperialism: Innovation and expansion*, 1688–1914. New York: Longman.

Carrigan, Tim, Bob Connell, and John Lee. 1985. Toward a new sociology of masculinity. *Theory and Society* 14 (5): 551–604.

Chodorow, Nancy. 1994. *Femininities, masculinities, sexualities: Freud and beyond*. Lexington: University Press of Kentucky.

Cockburn, Cynthia. 1983. Brothers: *Male dominance and technological change*. London: Pluto.

Cohen, Jon. 1991. NOMAS: Challenging male supremacy. *Changing Men* (Winter/Spring): 45–46.

Connell, R. W. 1987. *Gender and power*. Cambridge, MA: Polity.

———. 1990a. An iron man: The body and some contradictions of hegemonic masculinity. In *Sport, men and the gender order: Critical feminist perspectives*, edited by Michael A. Messner and Donald F. Sabo, 83–95. Champaign. IL: Human Kinetics Books.

———. 1990b. The state, gender and sexual politics: Theory and appraisal. *Theory and Society* 19: 507–44.

_____. 1992. A very straight gay: Masculinity, homosexual experience and the dynamics of gender. *American Sociological Review* 57 (6): 735–51.

_____. 1995. *Masculinities.* Cambridge, MA: Polity.

_____. 1996. Teaching the boys: New research on masculinity, and gender strategies for schools. *Teachers College Record* 98 (2): 206–35.

Cornwall, Andrea, and Nancy Lindisfame, eds. 1994. *Dislocating masculinity: Comparative ethnographies.* London: Routledge.

Dawson, Graham. 1991. The blond Bedouin: Lawrence of Arabia, imperial adventure and the imagining of English-British masculinity. In *Manful assertions: Masculinities in Britain since 1800,* edited by Michael Roper and John Tosh, 113–44. London: Routledge.

Donaldson, Mike. 1991. *Time of our lives: Labour and love in the working class.* Sydney: Allen & Unwin.

_____. 1997. *Growing up very rich: The masculinity of the hegemonic.* Paper presented at the conference Masculinities: Renegotiating Genders, June, University of Wollongong, Australia.

Enloe, Cynthia. 1990. *Bananas, beaches and bases: Making feminist sense of international politics.* Berkeley: University of California Press.

Featherstone, Mike. 1995. *Undoing culture: Globalization, postmodernism and identity.* London: Sage.

Foley, Douglas E. 1990. *Learning capitalist culture: Deep in the heart of Tejas.* Philadelphia: University of Pennsylvania Press.

Fuentes, Annette, and Barbara Ehrenreich. 1983. *Women in the global factory.* Boston: South End.

Gee, James Paul, Glynda Hull, and Colin Lankshear. 1996. *The new work order: Behind the language of the new capitalism.* Sydney: Allen & Unwin.

Gender Equality Ombudsman. 1997. *The father's quota.* Information sheet on parental leave entitlements, Oslo.

Gibson, J. William. 1994. *Warrior dreams: Paramilitary culture in post-Vietnam America.* New York: Hill and Wang.

Hagemann-White, Carol, and Maria S. Rerrich, eds. 1988. *FrauenMaennerBilder* (Women, Imaging, Men). Bielefeld: AJZ-Verlag.

Hearn, Jeff. 1987. *The gender of oppression: Men, masculinity and the critique of Marxism.* Brighton, U.K.: Wheatsheaf.

Herdt, Gilbert H. 1981. *Guardians of the flutes: Idioms of masculinity.* New York: McGraw-Hill.

_____. ed. 1984. *Ritualized homosexuality in Melanesia.* Berkeley: University of California Press.

Heward, Christine. 1988. *Making a man of him: Parents and their sons' education at an English public school 1929–1950.* London: Routledge.

Hinsch, Bret. 1990. *Passions of the cut sleeve: The male homosexual tradition in China.* Berkeley: University of California Press.

Hirst, Paul, and Grahame Thompson. 1996. *Globalization in question: The international economy and the possibilities of governance.* Cambridge, MA: Polity.

Hollstein, Walter. 1992. *Machen Sie Platz, mein Herr! Teilen statt Herrschen* [Sharing instead of dominating]. Hamburg: Rowohlt.

Hollway, Wendy. 1994. Separation, integration and difference: Contradictions in a gender regime. In *Power/gender: Social relations in theory and practice*, edited by H. Lorraine Radtke and Henderikus Stam, 247–69. London: Sage.

Holter, Oystein Gullvag. 1997. *Gender, patriarchy and capitalism: A social forms analysis.* Ph.D. diss., University of Oslo, Faculty of Social Science.

Hondagneu-Sotelo, Pierrette, and Michael A. Messner. 1994. Gender displays and men's power. The "new man" and the Mexican immigrant man. In *Theorizing masculinities*, edited by Harry Brod and Michael Kaufman, 200–218. Twin Oaks, CA: Sage.

Ito Kimio. 1993. *Oiokorashisa-no-yukue* [Directions for masculinities]. Tokyo: Shinyo-sha.

Jolly, Margaret 1997. From point Venus to Bali Ha'i: Eroticism and exoticism in representations of the Pacific. In *Sites of desire, economies of pleasure: Sexualities in Asia and the Pacific*, edited by Lenore Manderson and Margaret Jolly, 99–122. Chicago: University of Chicago Press.

Kandiyoti, Deniz. 1994. The paradoxes of masculinity: Some thoughts on segregated societies. In *Dislocating masculinity: Comparative ethnographies*, edited by Andrea Cornwall and Nancy Lindisfarne, 197–213. London: Routledge.

Kaufman, Michael. 1997. *Working with men and boys to challenge sexism and end men's violence.* Paper presented at UNESCO expert group meeting on Male Roles and Masculinities in the Perspective of a Culture of Peace, September, Oslo.

Kimmel, Michael S. 1987. Rethinking "masculinity": New directions in research. In *Changing men: New directions in research on men and masculinity*, edited by Michael S. Kimmel, 9–24. Newbury Park, CA: Sage.

————. 1996. *Manhood in America: A cultural history*. New York: Free Press

Kimmel, Michael S., and Thomas P. Mosmiller, eds. 1992. *Against the tide: Pro-feminist men in the United States*, 1776–1990, a documentary history. Boston: Beacon.

Kindler, Heinz. 1993. *Maske(r)ade: Jungen- und Maennerarbeit fuer die Pratis* [Work with youth and men]. Neuling: Schwaebisch Gmuend und Tuebingen.

Kinmonth, Earl H. 1981. *The self-made man in Meiji Japanese thought: From Samurai to salary man*. Berkeley: University of California Press.

Kipling, Rudyard. [1901] 1987. *Kim*. London: Penguin.

Klein, Alan M. 1993. *Little big men: Bodybuilding sub culture and gender construction*. Albany: State University of New York Press.

Klein, Uta. 1997. *Our best boys: The making of masculinity in Israeli society*. Paper presented at UNESCO expert group meeting on Male Roles and Masculinities in the Perspectives of a Culture of Peace, September, Oslo.

Lewes, Kenneth. 1988. *The psychoanalytic theory of male homosexuality*. New York: Simon & Schuster.

MacDonald, Robert H. 1994. *The language of empire: Myths and metaphors of popular imperialism*. 1880–1918. Manchester, U.K.: Manchester University Press.

McElhinny, Bonnie. 1994. An economy of affect: Objectivity, masculinity and the gendering of police work. In *Dislocating masculinity: Comparative ethnographies*, edited by Andrea Cornwall and Nancy Lindisfarne, 159–71. London: Routledge.

McKay, Jim, and Debbie Huber. 1992: Anchoring media images of technology and sport. *Women's Studies International Forum* 15 (2): 205–18.

Messerschmidt, James W. 1997. *Crime as structured action: Gender, race, class, and crime in the making.* Thousand Oaks, CA: Sage.

Messner, Michael A. 1992. *Power at play: Sports and the problem of masculinity.* Boston: Beacon.

———. 1997. *The politics of masculinities: Men in movements.* Thousand Oaks, CA: Sage.

Metz-Goeckel, Sigrid, and Ursula Mueller. 1986. *Der Mann: Die Brigitte-Studie* [The male] Beltz: Weinheim & Basel.

Mies, Maria. 1986. *Patriarchy and accumulation on a world scale: Women in the international division of labour.* London: Zed.

Moodie, T. Dunbar. 1994. *Going for gold: Men, mines, and migration.* Johannesburg: Witwatersand University Press.

Morrell, Robert. 1994. Boys, gangs, and the making of masculinity in the White secondary schools of Natal, 1880-1930. *Masculinities* 2 (2): 56–82.

———. ed. 1996. *Political economy and identities in KwaZulu-Natal: Historical and social perspectives.* Durban, Natal: Indicator Press.

Nakamura, Akira. 1994. *Watashi-no Danseigaku* [My men's studies]. Tokyo: Kindaibugei-sha.

Oftung, Knut, ed. 1994. *Menns bilder og bilder av menn* [Images of men]. Oslo: Likestillingsradet.

Phillips, Jock. 1987. *A man's country? The image of the Pakeha male, a history.* Auckland: Penguin.

Poynting, S., G. Noble, and P. Tabar. 1997. *"Intersections" of masculinity and ethnicity: A study of male Lebanese immigrant youth in Western Sydney.* Paper presented at the conference Masculinities: Renegotiating Genders, June, University of Wollongong, Australia.

Roper, Michael. 1991. Yesterday's model: Product fetishism and the British company man, 1945–85. In *Manful assertions: Masculinities in Britain since 1800,* edited by Michael Roper and John Tosh, 190–211. London: Routledge.

Schwalbe, Michael. 1996. *Unlocking the iron cage: The men's movement, gender politics, and the American culture.* New York: Oxford University Press.

Segal, Lynne. 1997. *Slow motion: Changing masculinities, changing men.* 2d ed. London: Virago.

Seidler, Victor J. 1991. *Achilles heel reader: Men, sexual politics and socialism.* London: Routledge.

Shire, Chenjerai. 1994. Men don't go to the moon: Language, space and masculinities in Zimbabwe. In *Dislocating masculinity: Comparative ethnographies,* edited by Andrea Cornwall and Nancy Lindisfame, 147–58. London: Routledge.

Simpson, Amelia. 1993. *Xuxa: The mega-marketing of a gender, race and modernity.* Philadelphia: Temple University Press.

Sinha, Mrinalini. 1995. *Colonial masculinity: The manly Englishman and the effeminate Bengali in the late nineteenth century.* Manchester, U.K.: Manchester University Press.

Taylor, Debbie. 1985. Women: An analysis. In *Women: A world report,* 1–98. London: Methuen.

Theberge, Nancy. 1991. Reflections on the body in the sociology of sport. *Quest* 43: 123–34.

Thorne, Barrie. 1993. *Gender play: Girls and boys in school.* New Brunswick. NJ: Rutgers University Press.

Tillner, Georg. 1997. *Masculinity and xenophobia. Paper presented at UNESCO meeting on Male Roles and Masculinities in the Perspective of a Culture of Peace.* September, Oslo.

Tomsen, Stephen. 1997. A top night: Social protest, masculinity and the culture of drinking violence. *British Journal of Criminology* 37 (1): 90–103.

Tosh, John. 1991. Domesticity and manliness in the Victorian middle class: The family of Edward White Benson. In *Manful assertions: Masculinities in Britain since 1800*, edited by Michael Roper and John Tosh, 44–73. London: Routledge.

United Nations Educational, Scientific and Cultural Organization (UNESCO). 1997. *Male roles and masculinities in the perspective of a culture of peace: Report of expert group meeting, Oslo, 24–28 September 1997.* Paris: Women and a Culture of Peace Programme, Culture of Peace Unit, UNESCO.

Walby, Sylvia. 1990. *Theorizing patriarchy.* Oxford, U.K.: Blackwell.

Walker, James C. 1988. *Louts and legends: Male youth culture in an inner-city school.* Sydney: Allen & Unwin.

Wallerstein, Immanuel. 1974. *The modern world-system: Capitalist agriculture and the origins of the European world-economy in the sixteenth century.* New York: Academic Press.

Whitson, David. 1990. Sport in the social construction of masculinity. In *Sport, men, and the gender order: Critical feminist perspectives*, edited by Michael A. Messner and Donald F. Sabo, 19–29. Champaign, IL: Human Kinetics Books.

Widersprueche. 1995. Special Issue: Maennlichkeiten. Vol. 56/57.

Williams, Walter L. 1986. *The spirit and the flesh: Sexual diversity in American Indian culture.* Boston: Beacon.

Xaba, Thokozani. 1997. Masculinity in a transitional society: The rise and fall of the "young lions." Paper presented at the conference Masculinities in Southern Africa, June, University of Natal-Durban, Durban.

REFLECTION QUESTIONS FOR CHAPTER 7

1. What is wrong with a gender-neutral analysis of globalization?

2. How does research on maids, nannies, and sex workers help us see globalization in a new light?

3. What does Sassen mean by "survival circuits" and how are they crucial to building new economies and expanding existing ones?

4. How does Connell's discussion of "hegemonic masculinity" illustrate his argument that although globalization implies homogenization, it produces many unequal forms of masculinity?

5. Does globalization reproduce the worst tendencies of gender inequality or can it provide opportunities for women to leave the worst excesses of patriarchy behind? Provide examples.

8

The Globalization of Terror

INTRODUCTION

As nations become increasingly connected, they are increasingly vulnerable to terrorist attacks by individuals and organizations.[1] This vulnerability was revealed by the attacks on September 11, 2001, by al-Qaeda on the World Trade Center and the Pentagon killing about 3,000 Americans. In the past terrorism, which is defined "as a methodology of using violence to gain political objectives,"[2] was more or less localized. With the tools of globalization, however, terrorists can now commandeer airplanes and fly into buildings, blow up airplanes (e.g., planting a bomb in Pan Am 747 in 1988, killing 270), and use suicide bombers in a small boat (e.g., the attack on the USS *Cole* in a Yemen harbor in 2000). With globalization there is the potential to invade cyberspace ("cyber-terrorism") by shutting down systems ("weapons of mass disruption") such as providing false signals to, for example, halt the movement of oil and natural gas transported through pipelines or shut down the air traffic control system. Another possible form of terrorism is to make a relatively small nuclear weapon (the technology is widely known), transport it by private plane or boat into another country or to store the bomb in a container that is shipped through commercial shipping, to be detonated by remote control in the harbor of destination. There is also the possibility of biological and chemical attacks ("bioterrorism") by terrorists. Infectious diseases such as anthrax, typhus, Ebola, smallpox, and botulism, and toxic chemicals such as cyanide, nerve gas, and sarin, could be used to sicken and kill people or to destroy livestock and crops.

The current wave of terrorism has changed the nature of war. Previously, wars were fought over land, resources, and ideology, but always nation against nation. But the terrorists of the twenty-first century do not represent a nation and they are not intent on occupying territories. They do not have battleships and airfields to be targeted. Instead of an organized army, they are loosely organized through small groups with embedded "cells" to carry out violence. The U.S. State Department in 2007 counted forty-one foreign terrorist organizations with bases in at least twenty-five nations and the Palestinian territories. Many of these, but not all, are fundamentalist Islamic groups.[3]

The first essay, by law professor Amy Chua, examines the roots of terrorism. Using the example of the Philippines, she shows how poverty, indignity, hopelessness, and grievance fuel terrorist acts by powerless ethnic Filipinos against the affluent Chinese minority. The poor view the Chinese as exploiters, foreign intruders, with inexplicable wealth, and intolerable superiority. Chua extrapolates from the Philippines case an explanation for transnational terrorism. It lies, she says, "in the relationship—and increasingly the explosive collision—among the three most powerful forces operating in the world today: markets, democracy, and ethnic hatred." For her, the "global spread of free-market democracy—at least in its current, raw, for-export form—has been a principal aggravating cause of ethnic violence throughout the non-Western world."

The next reading, by Sajid Huq, provides a compelling picture of the car bomb as a defining tactic in the era of global urban terrorism. Car bombs reflect "asymmetric" warfare in which the weaker power gains physical and symbolic power.

Finally, Edward S. Herman and David Peterson direct our attention toward the threat of global *state* terrorism. Instead of focusing on Bin Laden and his network of terrorist cells, the authors show how highly militarized superpowers, most notably the United States, are engaged in really large-scale killing. The U.S. Code defines terrorism as "any activity . . . dangerous to human life . . . intended to intimidate or coerce a civilian population." They then show how the United States is a terrorist nation.

ENDNOTES

1. Much of this introduction is from D. Stanley Eitzen and Maxine Baca Zinn. 2003. *Social Problems*, 9th ed. Boston: Allyn & Bacon, Chapter 18.
2. William Greider, 2004. "Under the Banner of the 'War' on Terror." *The Nation* (June 21), p. 11.
3. U.S. Department of State. 2007. Country Reports on Terrorism (April 30). Online: http://www.state.gov/s/ct/rls/rm/07/83999.htm.

27

Globalizing Hate

AMY CHUA

One beautiful blue morning in 1994, my mother phoned me from California. In a hushed voice, she told me that my Aunt Leona, my father's twin sister, had been murdered in her home in the Philippines, her throat slit by her chauffeur. My mother broke the news in our native Hokkien Chinese dialect. But "murder" she said in English, as if to wall off the act from the family through language.

The murder of a relative is horrible for anyone, anywhere. My father's grief was impenetrable; to this day, he has not broken his silence on the subject. For the rest of the family, though, there was added disgrace. For traditional Chinese, luck is a moral attribute, and a lucky person would never be murdered. Like having a birth defect, or marrying a non-Chinese, being murdered is shameful.

My three younger sisters and I were very fond of Aunt Leona, who was petite and quirky and had never married. Like many wealthy Filipino Chinese, she had all kinds of bank accounts in Honolulu, San Francisco, and Chicago. She visited us in the United States regularly. She and my father—Leona and Leone—were close, as only twins can be. Having no children of her own, she doted on her nieces and showered us with trinkets. As we grew older, the trinkets became treasures. On my tenth birthday, she gave me ten small diamonds, wrapped in toilet paper. My aunt loved diamonds and bought them up by the dozen, concealing them in empty Elizabeth Arden face moisturizer jars, some right on her bathroom shelf. She liked accumulating things. When we ate at McDonald's, she stuffed her Gucci purse with free ketchups.

According to the police report, my Aunt Leona, "a 58-year-old single woman," was killed in her living room with "a butcher's knife" at approximately 8 P.M. Two of her maids who were questioned confessed that Nilo Abique, my aunt's chauffeur, had planned and executed the murder with their knowledge and assistance.

"A few hours before the actual killing, respondent was seen sharpening the knife allegedly used in the crime." After the killing, "respondent [Abique] joined the two witnesses and told them that their employer was dead. At that time, he was wearing a pair of bloodied white gloves and was still holding a knife, also

SOURCE: From *World On Fire: How Exporting Free Market Democracy Breeds Ethnic Hatred and Global Instability* by Amy Chua, copyright © 2003, 2004 by Amy Chua. Used by permission of Doubleday, a division of Random House, Inc.

with traces of blood." But Abique, the report went on to say, had "disappeared" with the warrant for his arrest outstanding. The two maids were released.

After the funeral, I asked one of my uncles whether there had been any further developments in the murder investigation. He replied tersely that the killer had not been found. His wife explained that the Manila police had essentially closed the case. Why were they not more shocked that my aunt had been killed by people who worked for her, lived with her? Or that the maids had been released? When I pressed my uncle, he was brusque. "That's the way things are here," he said. "This is the Philippines—not America."

My uncle was not simply being callous. As it turns out, my aunt's death is part of a common pattern. Hundreds of Chinese in the Philippines are kidnapped every year, almost invariably by ethnic Filipinos. Many victims, often children, are brutally murdered, even after ransom is paid. Other Chinese, like my aunt, are killed without a kidnapping, usually in connection with a robbery.

Nor is it unusual that my aunt's killer was never apprehended. Police in the Philippines, all poor ethnic Filipino themselves, are notoriously unmotivated in these cases. Asked by a Western journalist why it is so frequently the Chinese who are targeted, one grinning Filipino policeman explained, "They have more money."

My family is part of the Philippines' tiny but entrepreneurial, economically powerful Chinese minority. Just 1 percent of the population, Chinese Filipinos control as much as 60 percent of the private economy, including the country's four major airlines and almost all its banks, hotels, shopping malls, and conglomerates. My own relatives in Manila, who run a plastics conglomerate, are only "third-tier" Chinese tycoons. Still, they own swaths of prime real estate and several vacation homes. They also have safe deposit boxes full of gold bars, each the size of a Snickers bar. My Aunt Leona FedExed me a similar bar as a law school graduation present.

Since my aunt's murder, one childhood memory keeps haunting me. I was eight, visiting from the United States, and staying at my family's splendid hacienda-style house in Manila. It was before dawn, still dark when I went to the kitchen for a drink. But I must have gone down an extra flight of stairs because I literally stumbled onto six male bodies.

I had found the male servants' quarters. My family's house-boys, gardeners, and chauffeurs—I sometimes imagine that Nilo Abique was among them—were sleeping on mats on a dirt floor. The place stank of sweat and urine. I was horrified.

Later that day, I mentioned the incident to my Aunt Leona, who laughed affectionately and explained that the servants—there were perhaps 20 living on the premises, all ethnic Filipino—were fortunate to be working for our family. If not for their positions, they would be living among rats and open sewers, without even a roof over their heads. A Filipino maid then walked in with a bowl of food for my aunt's Pekingese dog. The Filipinos, my aunt continued—in Chinese, but not caring whether the maid understood—were lazy and unintelligent, and didn't really want to do much else. If they didn't like working for us, they were free to leave any time. After all, they were employees, not slaves.

According to the World Bank, UNICEF, and official statistics of the Philippines, nearly two-thirds of the Philippines' 80 million ethnic Filipinos live on less than $2 a day, 40 percent spent their entire lives in temporary shelters, and 70 percent of all rural Filipinos own no land. Almost a third have no access to sanitation.

But that is not the worst of it. Poverty alone never is. Poverty by itself does not make people kill. To poverty must be added indignity, hopelessness, and grievance.

In the Philippines, millions of ethnic Filipinos work for Chinese; almost no Chinese work for Filipinos. The Chinese dominate industry and commerce at every level of society. Global markets intensify this dominance: When foreign investors do business in the Philippines, they deal almost exclusively with Chinese. Apart from a handful of corrupt politicians and a few aristocratic Spanish mestizo families, all of the Philippines' billionaires are Chinese. By contrast, all menial jobs in the Philippines are filled by Filipinos. All peasants, domestic servants, and squatters are Filipinos. Outside Manila, thousands of ethnic Filipinos lived on or around the Payatas garbage dump: a 12-block-wide mountain of fermenting refuse known as The Promised Land. By scavenging through rotting food and animal carcasses, squatters eked out a living. In July 2000, as a result of accumulating methane gas, the garbage mountain imploded and collapsed, smothering more than 100 people, many of them young children.

When I asked an uncle about the Payatas explosion, he was annoyed. "Why does everyone want to talk about that? It's the worst thing for foreign investment."

I wasn't surprised. My relatives live literally walled off from the Filipino masses, in a posh, all-Chinese residential enclave, on streets named Harvard, Yale, and Princeton. Armed, private security forces guard every entry point.

Each time I think of Nilo Abique—he was 6'2" and my aunt was 4'11"—I well up with a hatred and revulsion so intense, it is actually consoling. But over time, I have also had glimpses of how the Chinese must look to the vast majority of Filipinos, to someone like Abique: as exploiters, as foreign intruders their wealth inexplicable, their superiority intolerable. I will never forget the entry in the police report for Abique's "motive for murder": not robbery, despite the jewels and money the chauffeur was said to have taken, but just one word: "Revenge."

There is a connection between my aunt's killing and the waves of global violence and mass murder that we read about with mounting frequency. It lies in the relationship—and increasingly the explosive collision—among the three most powerful forces operating in the world today: markets, democracy, and ethnic hatred.

After the fall of the Berlin Wall, a common economic and political consensus emerged, not only in the West, but to a considerable extent around the world. Markets and democracy, working hand in hand, would transform the world into a community of modernized, peace-loving nations. In the process, ethnic hatred, religious zealotry, and other "backward" aspects of underdevelopment would be swept away. The sobering lesson of the last 20 years, however, is that the global spread of free-market democracy—at least in its current, raw, for-export form—has been a principal aggravating cause of ethnic violence throughout the non-Western world.

The reason has to do with a phenomenon—pervasive outside the West, yet rarely acknowledged—indeed often viewed as taboo—that turns free-market democracy into an engine of ethnic conflagration. The phenomenon is that of *market-dominant minorities*: ethnic minorities who—for widely varying reasons ranging from entrepreneurialism to a history of apartheid or colonial oppression—can be expected under market conditions to economically dominate the "indigenous" majorities around them, at least in the near to mid-term future.

Examples of market-dominant minorities include the Chinese, not just in the Philippines, but throughout Southeast Asia. Most recently, in Myanmar, ethnic Chinese have literally taken over the economies of Mandalay and Yangon. Whites are a market-dominant minority in South Africa and Zimbabwe—and, in a more complicated sense in Bolivia, Ecuador, Guatemala, and much of Latin America. Indians are a market-dominant minority in East Africa, Fiji, and parts of the Caribbean, as are Lebanese in West Africa and Jews in post-Communist Russia. Ibo are a market-dominant minority in Nigeria as were Croats in the former Yugoslavia and Tutsi in pre-genocide Rwanda.

In countries with a market-dominant minority, markets and democracy will tend to favor not just different people or different classes, but different ethnic groups. Markets magnify the often astounding wealth of the market-dominant minority while democracy increases the political power of the impoverished "indigenous" majority.

In such circumstances, where the rich aren't just rich—but belong to a resented, "outsider" ethnic group—the pursuit of free-market democracy often becomes an engine of catastrophic ethno-nationalism, pitting a poor "indigenous" majority, easily aroused and manipulated by opportunistic politicians, against a hated ethnic minority.

Consider Indonesia: Free-market policies in the 1980s and 1990s led to a situation in which the country's 3 percent Chinese minority controlled 70 percent of the country's private economy. The introduction of democracy in 1998—hailed with euphoria in the United States—produced a violent backlash against both the Chinese and markets. Some 5,000 shops and homes of ethnic Chinese were burned and looted, 2,000 people died, and 150 Chinese women were gang-raped. Free and fair elections in the midst of all this gave rise to ethnic scapegoating by demagogic politicians, along with calls for confiscation of Chinese assets and a "People's Economy" that would return Indonesia's wealth to the country's "true owners," the *pribumi* (indigenous Indonesian) majority. The wealthiest Chinese left the country, along with $40 billion to $100 billion of Chinese-controlled capital, plunging the country into an economic crisis from which it has still not recovered.

Indonesia is part of a much larger global problem: Whenever free-market democracy is pursued in the presence of a market-dominant minority, the result is not peace and prosperity but tremendous instability and some form of backlash—even mass slaughter. Sept. 11 brought this same dynamic home to the United States.

While Americans are not an ethnic minority, the world now sees us as a kind of global market-dominant minority, wielding outrageously disproportionate economic power relative to our numbers. With just 4 percent of the world's population, the U.S. is the principal engine and beneficiary of global capitalism. We are also seen as "almighty," "exploitative," and "able to control the world" by the world's poor, whether through military power or through the IMF-implemented austerity measures forced on developing populations. As a result, the United States has become the object of the same kind of mass popular, demagogue-fueled resentment that afflicts so many other market-dominant minorities around the world.

For the last 20 years, the United States has been promoting throughout the non-Western world a bare-knuckled, laissez-faire brand of capitalism abandoned by every Western nation—including the United States—long ago. At the same time, it has been using most of the developing world, with the conspicuous exception of the Middle East, to hold immediate majority-rule elections—"overnight democracy"—whereas Western democracies evolved much more gradually.

The U.S. attempt now underway in Iraq to install free-market democracy raises grave concerns. Like the former Yugoslavia, Iraq's ethnic dynamics are extremely complex—including long-suppressed hatreds among Kurds, Shiites, and Sunnis—and cross-cutting desires for revenge, especially against the brutal Baathist regime and its allies. Post-invasion chaos has made predictions impossible, but many fear that overnight elections could create a fundamentalist Islamic state that is opposed to free markets, to Washington, and to individual liberties, especially for women.

Moreover, because the U.S. is the world's most powerful and resented market-dominant minority, every move it makes with respect to Iraq will be scrutinized by hostile eyes. The best strategy for the U.S. may be the same one that market-dominant minorities everywhere would be well-advised to pursue: cooperate openly and fairly to advance a broad public interest, and support a government that ensures that the country's resources and wealth—in the Iraqi case, oil—benefit *all* the people.

28

The Car-Bomb
Terror's Globalisation

SAJID HUQ

A week before the second anniversary of the London bombs of 7 July 2005, the city had another taste of the fear terrorism can inflict with the discovery on 29 June 2007 of two Mercedes cars packed with gas cylinders, petrol and nails parked in city-centre locations. The failure of this attempt to inflict great loss of life was followed by another unsuccessful attack at Glasgow airport the next day, also using a Mercedes that drove at speed towards the airport entrance but became stuck in the attempt to enter the building.

The second narrow escape from mass slaughter at the hands of—it appears—a radical Islamist cell is an uncomfortable reminder of the continuing potency of this ideology and of the motivation of its adherents. The politics and background of the members of this group responsible continue to be intensively discussed in the media, in particular the nature of any links with the wider al-Qaida movement. Yet alongside this concern, the promiscuous technology used in the attempted operation is significant in its own right. The London and Glasgow incidents reveal also what might be called the continuing globalisation of terror tactics in the first decade of the 21st century, in which the car-bomb has come to acquire a prominent place.

BEIRUT, BAGHDAD, BEYOND

The security agencies of states fighting the "war on terror" characterise car-bombs as "vehicle-borne improvised explosive devices" (VBIED), a key tactic in what the Pentagon calls "fourth-generation" or "open-source" warfare. Indeed, car-bombs can be very powerful weapons that can destroy entire buildings, even (if several are detonated together) entire cityscapes; but their political and media impact is even greater than the physical.

In his excellent book on the phenomenon, *Buda's Wagon: A Brief History of the Car Bomb,* the historian Mike Davis traces the history of the car-bomb: from the use of a horse-drawn vehicle as a weapon by an Italian anarchist (Mario

SOURCE: Sajid Huq, "The Car-Bomb: Terror's Globalisation," *Open Democracy* (June 7, 2007). Online: http://www.opendemocracy.net/conflicts/democracy_terror/car_bomb.

Buda) that killed forty people and injured 400 in New York's financial district in 1920; through the ultra-right Stern Gang's truck-attack on a British police station in Palestine in Haifa, Palestine in January 1947 that killed four and injured forty-seven; to the era of "national-liberation" wars in Vietnam, Algeria and Northern Ireland. But a key moment in establishing the effectiveness of this tactic—one which points forward to a new era rather than backwards to an old—was in Lebanon on 23 October 1983.

The truck-bomb that obliterated the United States marine barracks at Beirut international airport arguably surpassed in lethal effectivity the combined firepower of the bombers and battleships of the US sixth fleet stationed off the coast of Lebanon at the time. The operation, which killed 241 marines, came six months after the truck-bomb that hit the US embassy on 18 April 1983 that killed sixty-three people (including the CIA's national-intelligence officer for the near east, Robert C. Ames). But it is not just the huge death-toll that makes it a landmark in the history of urban terrorism.

Mike Davis argues that the event's geopolitical repercussions were more severe even than the loss of Saigon in 1975, on the grounds that while the latter belonged to a bygone era of bipolar rivalry, the group responsible—Hizbollah—here etched the blueprint for asymmetric warfare that augured the generation to come (see Mike Davis, "The Poor Man's Air Force," *Truthout*, 11 April 2006).

Hizbollah, formed in the wake of the Israeli invasion of Lebanon in 1982 by the coalescing of a number of Islamic groups (including pro-Khomeini elements nurtured by the Iranian *pasdaran* [Revolutionary Guards]), was long regarded by many as a tentacle of Iran's Islamic revolution. Until the 1983 attacks, the then Ronald Reagan administration, the CIA, and the French intelligence services (representing another target of the group) had paid surprisingly little attention to Hizbollah's emergence. So the pick-up truck that carried 12,000 pounds of explosives through Beirut's traffic and into the US marine compound was a devastating announcement of Hizbollah's arrival and of a form of warfare that—by precipitating the ignominious retreat of US forces from Lebanon—showed its capacity to affect the policy of even the most powerful of governments.

CIA agent Robert Baer described the scene: "The USS *Guadalcanal*, anchored five miles off the coast, shuddered from the tremors. At ground zero, the centre of the seven-story embassy lifted up hundreds of feet into the air, remained suspended for what seemed an eternity, and then collapsed in a cloud of dust, people, splintered furniture, and paper."

If this annihilation of a western military outpost in the middle east helped to reshape the history of urban guerrilla warfare and terror tactics, the impact can be seen most sharply in Iraq today, where the number of car-bombs—directed against US forces but also innocent civilians in shopping areas and members of the sectarian "enemy"—fluctuates but remains consistently high. Mike Davis estimates the number of car-bombs detonated in Iraq between June 2003 and June 2006 at 578. He also argues that the very social geography that made the invasion so easy for American forces—a largely desert landscape of heavily urbanised oases connected by a complex highway system—makes it a perfect playground for car-bombers (also often suicide-bombers).

Beyond Iraq, vehicular bombs punctuate a new age of globalised urban warfare which utilise the advantage for the weaker party of "asymmetric warfare." A lethal, inexpensive weapon can inflict great damage and act as an enormous psychological boost to the perpetrating group; it also closes the "asymmetry" of the struggle in perception, if not in reality. The propaganda effect here is of inestimable value to insurgents, for car-bombs are also quintessential terrorist "spectacle"—seducing an image-hungry media, scandalising the public, provoking politicians.

The car-bomb is a defining tactic of the era of global urban terrorism. Just as political assassination and suicide-bombing have spread by example and technology-transfer across the world, so the car-bomb is becoming less and less a monopoly of any particular group—notwithstanding the fact that certain networks (at present, al-Qaida or *Sunni* insurgents in Iraq) seem to "brand" it as one of their special tactics. Even where it doesn't work, as in London and Glasgow on 29–30 June 2007, it still frightens. There may be only the smallest of chances of any individual citizen falling victim to this anonymous threat, but its symbolic power has everyone in its grip.

29

The Threat of Global State Terrorism

Retail vs Wholesale Terror

EDWARD S. HERMAN

DAVID PETERSON

We are living in a very dangerous time, but for reasons almost exactly the opposite of those conventionally accepted. The consensus view in the United States right now is that the danger lies in the terror threat from Bin Laden and his network, and perhaps other terrorists hostile to the West. But Bin Laden and his network, though evidently formidable terrorists, cannot compete in terrorizing with states, and especially with a highly militarized superpower. His is a "retail" terror network, like the IRA or Cuban refugee terrorist network: It has no helicopter gunships, no offensive missiles, no "daisy cutters," no nuclear weapons, and although its death-dealing on September 11 was remarkable (although down from the initially estimated 6,000 or more to below 3,900), it was unique for a non-governmental terrorist organization.

Really large-scale killing and torture to terrorize—"wholesale" terrorism—has been implemented by states, not by non-state terrorists. The reason people aren't aware of this is that states define terrorism and identify the terrorists, and they naturally exempt themselves as always "retaliating" and engaging in "counter-terror" even when their own actions are an exact fit to their own definitions. And their mainstream media always follow the official lead. The U.S. Code definition—"any activity . . . dangerous to human life . . . intended to intimidate or coerce a civilian population . . . [or] to influence the policy of a government by intimidation"—surely fits U.S. policy toward Iraq, where the incessant bombings and "sanctions of mass destruction" have been designed to intimidate the Iraqi people and influence Iraqi government policy. This serious terrorism has been killing more children per month than the total casualty figure for the September 11 terrorist attacks, but in this country it is Iraq that, if not terrorizing, is a terrorist threat getting what it deserves. This distorted perspective is made possible by a mainstream media that serves state policy by focusing attention on Hussein's efforts to develop "weapons of mass destruction," while keeping pictures of dying Iraqi children out of sight.

SOURCE: Edward S. Herman and David Peterson, "The Threat of Global State Terrorism," *Z Magazine* 15 (January 2002), pp. 30–34. Reprinted with permission from the authors.

As another illustrative case, Israel has been using torture on an administrative basis for at least 25 years; a feat no retail (non-state) terrorist could duplicate. This, and the U.S. policy toward Iraq, are wholesale terrorist operations, carried out on a large scale over an extended period of time, as only the institutions of state terrorism are capable of doing. As the 1984 Alfonsin National Commission on the Disappeared explained after reviewing the record of the deposed military regime of Argentina, which had tortured and killed thousands in over 300 detention centers from 1976 to 1983, that regime's (wholesale) terrorism was "infinitely worse" than the (retail) terrorism it was combating.

The real danger to world peace and security arising out of the events of September 11 lies in the responsive wholesale terrorism that will result—and already is resulting—from the resurgent aggressiveness of the United States, with its excessive military power, its global interests that can be served by a forward military policy, its self-righteousness and habituation to getting its way, and the absence of any country or group of countries able to contain it. This country is also especially dangerous by virtue of its being perhaps the most religiously fundamentalist in the world (ranging from the Christian Right and its various militialike sects to the blind patriotic fervor in the wake of September 11 to belief in close encounters of the third kind, angels, and End Times); and with a population that, with the help of the mainstream media, can be brought to approve or ignore any level of external violence that the leadership deems useful. We may recall that the United States is the only country that has used nuclear weapons and has threatened their further use many times. Its employment of chemical weapons more than competes with Saddam Hussein's use in the 1980s, one of the U.S. legacies being some 500,000 Vietnamese children with serious birth abnormalities left from a decade of U.S. chemical warfare in the 1960s.

The September 11 bombing was a windfall for the Bush administration and military-industrial complex, so much to their advantage that theories have been circulating suggesting that the U.S. leadership engineered, or at least failed to interfere with, the bombings. We don't accept the purported evidence for this, but we do believe that after the initial shock at their failure to protect U.S. citizens from attack, the leadership realized that this was what they had been waiting for as a substitute for the Soviet Threat to justify a new projection of U.S. power. In fact, the "war against terrorism" may prove to be more serviceable as a tool for managing the public than the Soviet Threat, given its open-ended and nebulous character.

The Soviet Threat gave the United States a Cold War propaganda cover to justify its support of numerous military dictators and other goons of convenience who would serve U.S. economic and political interests. Thus, in the name of fighting both Soviet "expansionism" and "terrorism" the United States supported terrorist states that engaged in really serious terrorism, combatting a lesser (retail) terrorism that was frequently a response to that state terrorism. One document produced by the Catholic Church in Latin America in 1977, made the telling observation that the military regimes needed to employ terror because the ruthless economic policies that they encouraged, their "development model," which featured helping foreign transnationals by giving them a "favorable climate of investment" (i.e., crushing labor unions), "creates a revolution that did not previously exist." It is hardly a coincidence that "liberation theology," with its "theology from the underside of history" and its "preferential option for the poor" (Gustavo Gutierrez), was born out of the turmoil and victimization of this era of U.S.-sponsored counter-revolutionary violence.

In the earlier period the United States got away with claims that it was opposed to and was fighting terrorism, while it was actually supporting "infinitely worse" terrorisms. The mainstream media allowed the government to define terrorism and name the terrorists; so, for example, the *New York Times* regularly referred to the retail terrorism in Argentina as "terrorism," but never called the infinitely worse state terrorism in that country by its right name. And the *Times*—and the rest of the mainstream media—rarely discussed the ugly details of Argentinian state terrorism, never related it to any development model, and failed to express indignation over it. Also, they never referred to the Nicaraguan contras or Savimbi's UNITA as terrorists or the United States as a sponsor of terrorism for giving them support.

In the Cold War years, also, the media never questioned the alleged objectives of U.S. interventions. If the U.S. government claimed back in the early 1950s that it was overthrowing the elected government of Guatemala for fear of Soviet control and to stop the spread of communism, the media never doubted this; they never suggested that this was a fraudulent cover for the desire to protect the United Fruit Company, to dispose of an annoyingly reformist and independent government, and resulted from an arrogantly imperialistic government's refusal to brook any opposition in its backyard. The media served then as uncritical propagandists for the "war against communism," featuring the alleged threats and focusing heavily on the progress of that notorious intervention. They made the destruction of a democratic government and introduction of a police state into a noble venture that saved the United States from a wholly fabricated threat.

Sound familiar? It should, as the media are doing the same job of protecting state actions today. If their government says that what it is doing in Afghanistan is a "war against terrorism," that is what the media label it. If the Administration hints at extending the war on terrorism to Iraq as one of its state sponsors, the media talk about this only in terms of strategy, whether allies will go along, and possible repercussions. They never suggest that the attack on Afghanistan was itself an act of terrorism, or beyond that, an act of aggression done in straightforward violation of the UN Charter and international law. They never suggest that Iraq has been a victim of very serious state-sponsored terrorism for more than a decade, in which 23 million Iraqis have served as hostages to be starved into rebellion. Never. Although what this country does may fit the official U.S. definition of terrorism with precision, the supposedly free and independent media exempt its actions from the label as a matter of course.

As they did back in 1950–1954 in reference to Guatemala, the mainstream media focus on U.S. claims regarding enemy maneuvers and sinister plans (back then, Red infiltration; today, the location and tricks of Bin Laden and Al-Qaeda); the planning and military activities of the forces supported by the United States (back then, the "contra" army invading Guatemala from

Somoza's Nicaragua; today the military successes of the bombing and "coalition" fighting on the ground in Afghanistan); who is winning and losing in the fighting and diplomatic maneuvering. There was no discussion in the earlier years of objectives other than that supposed "war against communism"—such as the welfare of United Fruit, or the U.S. objection to any social democratic reforms or independent state in its backyard—just as today the media will not discuss the Bush administration's broader agenda—gaining access to and control over the Caspian Basin's enormous oil and natural gas resources, or using antiterrorism as the rationale for going after any global target, or to help create a moral environment that will serve to advance its domestic programs.

Just as the Cold War provided a cover for U.S. support of a "real terror network," so now the "war against terrorism" is providing a cover for a similar and rapid gravitation to contemporary goons of convenience like Russia's Putin, Pakistan's Musharraff, and Uzbekistan's Karimov. Putin is a major wholesale terrorist, whose political career has been built on terrorizing Chechnya; Musharraff is a military dictator who previously was closely allied with the Taliban; and Karimov is another holdover dictator from the Soviet era, whose only virtue is a willingness to serve the "war on terrorism." Just as the media back in 1954 never discussed the fact that that first generation contra invasion of Guatemala, allegedly to "free" Guatemala, was being organized in Somoza's "unfree" Nicaragua, nor questioned U.S. support of that dictator, so today the media never ask the obvious question: How can a new order of democracy be created by supporting and consolidating the power of dictators and wholesale terrorists?

The "war against terrorism" has given a freer hand to terrorist governments that are "with us," like Russia's but also that of Israel, whose leaders quickly recognized their improved political position after September 11 and greatly intensified their violence in the occupied territories. China has also joined the fight against terrorism, and is expected to "use the international war against terror for a new crackdown on the Turkic-speaking Uighurs," and "arrests in the region have increased significantly" since September 11 ("China using terror war against separatists," UPI, October 11, 2001). The new "war" has encouraged governments across the globe to ask for military support from the United States to fight their own "terrorists," and the Bush administration has already come through with aid to the Philippines and Indonesia in these local struggles. So it looks very much as if insurgents anywhere, if they don't happen to be supported by Washington as "freedom fighters," will be transformed into targets of the new "war against terrorism," now to be fought on a global basis. Whereas in the Cold War years these insurgents were tied to Moscow in preparation for supporting states like Argentina, which would then crush them; now they will be branded "foreign terrorist organizations" or linked to Bin Laden, or perhaps that won't even be necessary in the New World Orders—just call them terrorists, flash pictures of the victims of the World Trade Center, and bomb them.

In the earlier years, also, as the government wanted the public mobilized to the frightful threat posed by the disarmed Guatemala, the media beat a steady and incessant drum, day in and day out. Similarly, since September 11, the Bush administration wanting the public frightened and mobilized to support its new and open-ended war, the media have provided incessant and frightening—as well as hugely biased—coverage of "A Nation Challenged," as the *New York Times*'s daily section would have it, or "At War With Terror," in the *Philadelphia Inquirer*'s regular account. The public is led to believe that the Pitiful Giant has had its back against the ropes in its struggle against retail terror, a truly frightening situation; whereas in the earlier case, a social democratic government threatening United Fruit and U.S. prerogatives, but linked to Moscow, provided the media with grist for creating public panic, and justifying U.S. aggression.

In the earlier case, after the elected government of Guatemala was overthrown in June 1954, and was replaced by a puppet that proceeded to dismantle all the human rights and social gains brought by democracy, media attention to Guatemala disappeared, and it stayed invisible as a counterinsurgency state, built on wholesale terror, took over and has remained in place for almost half a century. The media helped overthrow the democratic government, and in the years that followed they kept the public unaware that under U.S. auspices, with U.S. funding, training, Green Beret participation in counterinsurgency campaigns, and diplomatic support, a terror state was built, aided, and protected (for details, Michael McClintock, *The American Connection: State Terror and Popular Resistance in Guatemala* [Zed, 1985]). The same pattern was observable in the case of Nicaragua in the 1980s: huge media attention to the Sandinista government's "threat of a good example" that followed U.S. support of the Somoza dictatorship for 45 years; then after the ouster of the Sandinistas, with the crucial aid of U.S. direct and sponsored terrorism, the media once again lapsed into silence.

This media practice allows the United States to carry out a hit-and-run policy, without any serious public cost to its leadership, as the public is kept in the dark about the fact that this country has "run" following its extended and devastating "hit," because media attention falls to close to zero.

This should clue us in on the likely developments in Afghanistan after this fearsome military challenge is met—and the United States and its antiterrorist "coalition" can celebrate another victory in which they created a desert and called it peace. There is a great deal of talk now of "nation-building" and modernizing Afghanistan, but that is now, when the establishment needs to fend off suggestions that it is better at killing and starving people than it is at spreading democracy and development that helps them. But Vietnam, Guatemala, Nicaragua, Kosovo, and many other cases, teach us that there will be no nation-building at all, although building oil and natural gas pipelines and military bases is another matter.

Once the great military victory is achieved, budget priorities will hardly extend out to Afghanistan, any more than they did to other victims of imperial violence. Official attention will disappear and the media can be counted on to shift their focus elsewhere. Call it a law of the free press, which falls in line whenever duty calls and boldly follows the flag and priorities of the elite and government establishment. If these call for nation destruction, and then a silent exit, so be it.

REFLECTION QUESTIONS FOR CHAPTER 8

1. Chua's essay shows how free markets and democracy, as pushed by the most powerful nations and transnational corporations, actually encourage ethnic terrorism. What is her rationale? Provide critique.

2. Why does Huq argue that the car bomb provides a psychological boost to the weaker party in globalized urban warfare? Do you agree that the symbolic power of the car bomb exceeds the physical damages it inflicts?

3. Do you agree or disagree with Herman and Peterson's assertion that the United States is a terrorist nation? If you disagree, then how do you deal with their supporting evidence?

9

The Globalization of Social Problems

INTRODUCTION

Globalization as represented by transnational flows of people, information, and commerce has intensified social problems globally and locally. There are crime networks that control the global trade in illicit drugs, the black market in guns and explosives, and also traffic in sex workers, sweatshop workers, and domestic servants. Transnational corporations and banks are sometimes involved in money laundering, which evades taxes and indirectly supports criminal activities. Transnational corporations may also exploit workers and degrade the environment in low-wage and developing countries. Of the many social problems exacerbated by globalization, we focus here on three: transnational crime, worker exploitation, and environmental degradation.

For transnational crime, we have selected three readings. The first, by *BusinessWeek,* reports how the Internet makes any illegal activity—drugs, gambling, terrorism, and child pornography—more accessible than ever. The second, by investigative reporter Mark Schapiro, uncovers the tobacco industry's multibillion-dollar global smuggling network. The third, by journalist Lucy Komisar, describes how the world's biggest banks and transnational corporations have set up a shadowy system to secretly move trillions of dollars—a system that can be exploited by tax evaders, drug runners, and even terrorists.

The next articles direct attention to the "New Slavery,"[1] which includes more than 27 million slaves in the world today. The "New Slavery," just as slavery in other times, means the loss of freedom, the exploitation of people for profit, and the control of slaves through violence or its threat. But today's forms of slavery also differ from the past. Typically, slavery is no longer a lifelong condition, as the slave is freed/discarded after he or she is no longer useful or has paid off a debt. Second, sometimes individuals and families become slaves by choice—a choice forced by extreme poverty. Often the poor must place themselves in bondage to pay off a debt. The slave must work for the slaveholder until the slaveholder decides the debt is repaid. This is problematic since many slaveholders use false accounting or charge very high interest, making repayment forever out of reach. The articles here show two forms of the "new slavery" resulting from globalization. Nicholas Stein tells the sordid story of how foreign workers (e.g., a Filipino who wants to work in a Taiwan factory) must pay an exorbitant amount to labor brokers for travel and a job, binding them to their jobs, in a form of indentured servitude. Factories that make products for Nike, Motorola, Ericsson, and other transnational corporations rely on overseas contract workers who are in debt bondage. The second article, by Andrew Cockburn, focuses on sex slaves, prostitutes who are bought and sold and who are then debt slaves who, for their freedom, must pay the amount their new owners paid for them.

The final set of articles deal with environmental degradation, offering differing perspectives on global warming. Kit Batten, Kari Manlove, and Nat Gryll illustrate the many ways in which global warming induces migration with worldwide social and economic consequences. According to the authors, climate change disasters are already a bigger cause of population displacement than war and persecution. Dick Field's essay, instead, focuses on how migration *produces* global warming, as migrants become high-energy consumers in the world's urban centers.

ENDNOTES

1. Ken Bales, 1999. *Disposable People: New Slavery in the Global Economy.* Berkeley: University of California Press.

30

The Underground Web

IRA SAGER
BEN ELGIN
PETER ELSTROM
FAITH KEENAN
PALLAVI GOGOI

It's the kind of call everyone dreads. For Kristen Bonnett, the daughter of NASCAR race driver Neil Bonnett, it came on Feb. 11, 1994—the day her father crashed during a practice run at the Daytona International Speedway. A few hours later, he died. Bonnett was devastated, but she got on with her life. Then, seven years later, came a second call. This time, it was a reporter asking for comment on autopsy photos of her father that were posted on the Internet. Shocked, she quickly got online. "Forty-eight thumbnail pictures, basically of my Dad on the table, butt-naked, gutted like a deer, were staring me directly in the face," says Bonnett. Now, when she thinks of her father, she pictures him lying atop an autopsy table.

Warning: You are about to enter the dark side of the Internet. It's a place where crime is rampant and every twisted urge can be satisfied. Thousands of virtual streets are lined with casinos, porn shops, and drug dealers. Scam artists and terrorists skulk behind seemingly lawful Web sites. And cops wander through once in a while, mostly looking lost. It's the Strip in Las Vegas, the Red Light district in Amsterdam, and New York's Times Square at its worst, all rolled into one—and all easily accessible from your living room couch.

Indeed, the very nature of the Web is what makes it such a playground for hoodlums. Its instant, affordable, far-flung reach has fostered frictionless commerce and frictionless crime. Fraudsters can tap into an international audience from anyplace in the world and—thanks to the Net's anonymity—hide their activities for months, years, forever. And they can do it for less than it costs in the physical world: $200 buys an e-mail list with the names of thousands of potential dupes. "The Web dramatically lowers transaction costs. Mostly, we think of that as a good thing," says Erik Brynjolfsson, professor of management at Massachusetts Institute of Technology's

Center for eBusiness. "But it makes it difficult to control many of the activities we want to control."

That has spawned a bustling Underground Web that's growing at an alarming rate. Black-market activity conducted online will reach an estimated $36.5 billion this year—about the same as the $39.3 billion U.S. consumers will spend on the legitimate Internet this year, according to researcher comScore Media Metrix. Today, illegal online gambling is the eighth-largest business on the Internet. Complaints about child porn in cyberspace have grown sixfold since 1998. And of the total number of fraud complaints being received by the government, 70% occur on the Internet. "North of 70% of all e-commerce is based on some socially unacceptable if not outright illegal activity" says Jeffery Hunker, dean of the H. John Heinz III School of Public Policy at Carnegie Mellon University, who helped craft cybersecurity policy in the Clinton Administration.

And that doesn't even factor in terrorism. Law-enforcement officials say terrorists are using the Internet for communication, research, recruitment, and fund-raising. The men involved in the September 11 attacks plotted and coordinated by trading e-mails from locations as innocuous as the public library. Even now, security experts say al Qaeda is trying to use the Web to plan more attacks. Computers analyzed by law-enforcement officials indicate that the terrorist group researched the U.S. telephone, electric, and water systems online, learning, for example, how digital switches operate those systems. "What keeps me awake at night is a physical attack in combination with some sort of cyberattack that would disrupt the abilities of our 911 systems," says Ronald L. Dick, head of the FBI's National Infrastructure Protection Center.

Not all threats are so overt. The Underground Web, if unchecked, has the potential to undermine the values of society. It enables—even encourages—ordinary citizens to break the law. People who wouldn't even jaywalk find themselves bombarded with offers to place bets at offshore casinos or order drugs online. For many, the offers are hard to resist. There's no need for surreptitious rendezvous in back alleys. It's antiseptic crime. "The Internet breaks down inhibitions to violate the law because the risks are much lower," says Kevin A. Delli-Colli, who heads the U.S. Customs Service CyberSmuggling Center in Fairfax, Va. "You can contact the seller anonymously, click on the product, and it's in your house."

To understand the depth of the problem, a team of five *BusinessWeek* reporters spent four months visiting the seedy side of the Internet. We sat beside gamblers as they placed bets on illegal gaming sites, interviewed people who bought drugs online, and talked with those who have lost loved ones because of cybercrimes. One of them was Barbara Perrin, a Long Island teacher who watched her 22-year-old son die after he bought on the Web a drug banned for bodybuilding. "My heart is broken into a million pieces," she says.

So far, the government's efforts to police the Underground Web have done little to stop its growth. Our reporters found more than 100 Web sites that appear to be engaged in a wide range of illegal or restricted activities. Italian switchblade seller AB Coltellerie, for instance, lists merchandise in U.S. dollars on its www.switchblades.it site. A *BusinessWeek* reporter contacted the site's online customer support to ask if they would ship switchblades to California. It's illegal to own one of these weapons with a blade in excess of two inches in California, which would make the majority of the site's inventory illegal in the state. Customer support's prompt response: "We take orders from California. Seizures are about 10% of total airmail shipments. To improve chances, you'd better choose express carrier as shipping method."

What's more, we found that even legitimate businesses enable Web outlaws. Mainstream sites such as Yahoo!, MSN, and Google help steer U.S. customers to gambling sites. They accept advertising from online casinos and display these ads to viewers in the U.S.—including an easy one-click link to place a bet. The practice is so widespread that the online-gaming industry has emerged as the fifth-biggest buyer of Web ads—$2.5 billion last year, according to comScore Media Metrix. "There are definitely some legality questions" surrounding this practice, says I. Nelson Rose, a professor at Whittier Law School. Microsoft Corp., which owns MSN, declined comment. A spokesperson for Google says it accepts ads from online casinos but says that policy could change. Yahoo says it will stop running gambling ads at the end of the third quarter. AOL Time Warner does not accept gambling ads on AOL but does on its Web properties such as Netscape and MapQuest.

Banks lend a hand, too, by processing the payments of customers in the U.S. who are gambling online illegally. Only under pressure from state attorneys general have some banks started to cut off credit lines to gamblers. "Online gambling poses real enforcement difficulties for us," says New York Attorney General Eliot Spitzer, who helped get a June 14 agreement with Citicorp that requires the credit-card issuer to decline payments for online-gambling transactions. The message is getting through: On July 8, auction giant eBay Inc., which agreed to pay $1.5 billion to acquire online-payment processor PayPal, said it will cancel PayPal's gaming business because of the "uncertain legal situation" surrounding it. In July, PayPal received two federal grand jury subpoenas concerning its processing of online gambling transactions.

Not all Underground Web activity is outright dishonest. Some is just plain vile. Anyone with a cause, no matter how weird, can have a Web site or chat room open to the world. Bonnett's autopsy photos were posted as a protest against race-car driving. And Deathndementia.com tries to appeal to rubber-neckers by displaying gory accident photos and offering links to 2,000 sites, including Celebrity Morgue.

The Underground Web is bigger, broader, scarier, and more damaging than most people realize. Here's why:

GAMBLING

For Debi Baptiste, an addiction to online gambling proved to be more than she could handle. After she lost thousands of dollars playing video poker in bars near her Portland (Ore.) home, she and her husband, John, moved to San Jose,

Calif., in 1999 for a fresh start. But when the family bought a home computer, Debi, 40 at the time, logged on to the Internet and began gambling at offshore Web sites—losing more than $50,000. John discovered what she was doing and changed the computer's password to lock her out. So she started staying late at her executive-secretary job to wager from the office.

Her gambling drove the Baptistes' relationship past its breaking point. On Oct. 5, 2000, John left divorce papers on the kitchen table before going to work. That morning, Debi swallowed 40 Vicodin tablets, went into the garage, and sat in the driver's seat of her car. Putting the divorce papers on the dashboard, alongside pictures of her two stepdaughters and her dog, she turned on the car. John found Debi in the exhaust-choked garage hours later. "I still loved her. I would have stuck with her." he says. "When I brought a computer into my house, little did I know I also brought a slot machine into my house."

Type "casino" into any Internet search engine, and hundreds of gambling sites surface. If you spend any time on the Web, you're almost certain to run across advertisements trying to lure you to visit a gambling site. With names like Prestige Casino, River Belle, and Aces High, these online casinos try to convey all the pizzazz of the Las Vegas Strip.

And people are betting Vegas-size bankrolls. The amount of money pouring into Internet casinos has skyrocketed from $2.2 billion in 2000 to $4.1 billion this year, according to researcher Christiansen Capital Advisors LLC. That's about 5% of the size of the legal U.S. gambling industry. Bear, Stearns & Co. estimates that 1,500 gambling sites have sprouted across the globe, more than double the 650 casinos on the Internet two years ago. They're covering bets from approximately 4.5 million people worldwide, slightly over half of them from the U.S.

Strict laws in the U.S. prohibiting online gambling are proving about as powerful a deterrent as Prohibition was to drinking in the 1920s. Most forms of online gambling are banned by a patchwork of federal and state laws, save for state-by-state exceptions for things such as lotteries or horse-betting. Yet at least 80% of the online gambling done in the U.S. is illegal, estimates Bear Stearns analyst Jason N. Ader.

Law-enforcement authorities can't do much about it because online casinos typically set up their headquarters in countries such as Britain, Australia, or Costa Rica, where Internet gaming is legal. U.S. Justice Dept. officials say they can flex their authority if a casino owner travels to U.S. soil, operates through a U.S. bank, or sets up offices inside the country. So far, there have been few convictions.

BusinessWeek reporters tagged along with gamblers to get an inside look at online wagering. Take Kenny (not his real name), a 29-year-old financial analyst who bets weekly on sports at Sportingbet USA, a site owned by a British company and operated in Costa Rica. The site offers wagers on everything from professional baseball to NASCAR races. Kenny chooses baseball. After entering a password, Kenny looks over a betting sheet, enters a dollar amount, and clicks the box next to the team he wants to bet on. He had previously given the site his charge-card number. In less than two minutes, Kenny has $50 riding on the underdog New York Mets vs. the New York Yankees. The Mets win, and he pockets a tidy $59 profit. "It's all really very easy," he says.

Maybe too easy. Without the time-consuming effort of traveling to a casino, the pressure on problem gamblers such as Debi Baptiste is hard to resist. Indeed, the always-available gambling fix has led to more addicted gamblers while presenting a steep challenge to their recovery, say health-care workers. The California Council on Problem Gambling says 20 callers to its help line last year pinpointed Internet gambling as their downfall—up from virtually none in past years. "The Internet is making the problem a thousand times worse because of its accessibility and increased ability to hide the problem behavior," says Eric Geffner, a clinical psychologist in Southern California who works with gambling addicts.

DRUGS

The easy availability of drugs on the Web proved deadly for Eric Perrin. An avid bodybuilder, Perrin bought some dinitrophenol, or DNP, over the Net last summer because it was supposed to help him lose weight and get better muscle definition. While DNP is promoted on some fitness Web sites, it's illegal to sell for human consumption. The chemical is legal only for use in industrial applications such as a coating on railroad ties to kill fungus. In humans, DNP can shut down the liver, kidneys, and central nervous system. Last August, Perrin took DNP for several days. As his body temperature began to rise and his heart started to race, his mother, Barbara, grew concerned. "He told me, 'Don't worry, Mom, I'll be all right,'" she says. "He was in a lot of pain." Eric died on Aug. 6 at a hospital near his home in Baldwin, N.Y. He was 22.

While the local U.S. Attorney is prosecuting the man who allegedly sold Eric Perrin the DNP, Barbara Perrin thinks the dealer isn't the real culprit. She places most of the blame on the Internet and Elite Fitness, a New York company that runs the Web site where her son read about the supposed benefits of DNP and got in touch with the dealer. She is convinced that without the Web, her son would be alive today. "DNP is not something you find easily," she says. Without the Internet, "Eric may have gotten steroids, but not DNP."

Even today, Elite Fitness provides what appears to be a forum for people to meet who are interested in drugs. With a quick search of the site, *Business Week* found dozens of postings from bodybuilders promoting the benefits of DNP, explaining how to use the drug, and downplaying its health risks. After one visitor asked on an electronic bulletin board why people die from taking DNP, one of the site's moderators responded by writing: "Get your fluids, and you'll [b]e A-O.K." Another moderator posted ground rules for members to communicate in private so they could share information about "sources." And members write that the best way to check out a source for restricted drugs is to e-mail a moderator. Paul Willingham, a partner at New York's Caliber Design Inc., which owns Elite Fitness, says the site simply provides a vehicle for bodybuilders to talk about any subject. "We don't provide a forum to buy and sell drugs," he says. "We're building a community for discussing physical fitness."

Willingham, like the moderator on Elite, argues that DNP is safe. He says that DNP is only dangerous if it is combined with other drugs, such as Ecstasy or speed. Dr. Thomas Manning, the chief toxicologist at the Nassau County Medical Examiner's office, says that Perrin had no other chemicals in his body at the time of death.

Drug trafficking over the Internet is rampant. Bodybuilding drugs are plentiful. You can find recipes for making methamphetamines, Ecstasy, and the notorious date-rape drug GHB as well as links to buy the chemicals needed to make them. Pot? Simply go to Marijuana.com, and there's an advertisement for "Top-quality Marijuana seeds delivered discreetly worldwide." Says Kansas Attorney General Carla J. Stovall: "There are really no limitations to what you can get over the Internet."

Illegal drugs aren't even the big problem. The most explosive kind of drug dealing on the Internet is selling prescription drugs without a prescription. Rogue pharmacies have been set up throughout the U.S. and abroad and are blanketing the Internet with offers for all sorts of drugs. The most popular is the sexual aid Viagra—available from many sites without an in-person doctor's exam, even though Viagra requires a medical exam in most states. The National Association of Boards of Pharmacy estimates that the number of "instant" online pharmacies, which send out prescription drugs with no doctor exam, has ballooned, to about 400 from fewer than 30 in 1999.

And it's not all as innocent as trying to get Viagra without an embarrassing doctor's visit. Painkillers are among the most popular drugs sold by rogue pharmacies because some, in large doses, can give users a high similar to heroin. In March, a federal grand jury in Texas indicted three doctors and several other people for running Pillbox Pharmacy, an Internet store that sold the painkiller hydrocodone and other drugs to patients who were never examined. One of the doctors involved pleaded guilty, and the others are awaiting trial. "This is a very popular drug— and very addictive," says Jerry Ellis, the Drug Enforcement Administration manager who headed the investigation. Pillbox sold at least $7.7 million worth of drugs and attracted 5,000 customers in 2000 and 2001 before being shut down.

In some cases, it's innocent people who get hurt. Dr. Pietr Hitzig used the Net to solicit patients, claiming he could treat just about anything—cirrhosis, obesity, even Gulf War syndrome. The Baltimore doctor lured in more than 1,000 patients in the mid-1990s, charging each $1,500 and up for a combination of the diet drug phentermine and other controlled substances. The DEA found that Hitzig's treatments were causing psychosis and other problems in patients, and last November he was sentenced to 45 months in prison for illegal distribution of controlled substances. "He attracted people who weren't seeking drugs. They were looking for help," says Cathy Gallagher, a DEA supervisor in Baltimore.

CHILD PORN

In May 2000, Russians Sergey Garbko and Vsevolod Solntsev-Elbe created a booming international business overnight: selling child pornography via the Internet. Their Blue Orchid site attracted customers who were willing to pay up to $300 for videos made in the 1980s. Then, to get something fresh to offer their clientele, the two Muscovites hired an acquaintance, Victor Razumov, to make new videos of himself having forced sex with a 15-year-old boy.

Not long after, Russian authorities investigating Blue Orchid discovered that the English-language site was running on a computer in the U.S. Moscow police called in U.S. Customs, and investigators set up a sting. On Mar. 2, 2001, Garbko and Solntsev-Elbe were busted. Police have made 16 arrests in the U.S. and Russia. Razumov is serving seven years for rape. Because Russia has no child-porn laws, Garbko and Solntsev-Elbe received only six-month sentences.

The Internet has brought new life to the child-porn trade. That's maddening for officials who thought they had nearly wiped it out with tougher laws. Anyone caught possessing, making, or distributing child porn in the U.S. can get up to 15 years in jail. But the anonymity of the Web and the difficulty of finding and shutting down sites around the world helps pornographers and their customers escape the law's clutches. Instead of having to scrounge for material in red-light districts, child-porn offenders can meet thousands of others like themselves online—buying or sharing porn without much fear of arrest. "We're seeing all new people becoming involved who have no prior police contact," says Peter D. Banks, director of training at the National Center for Missing & Exploited Children. "The ability to be anonymous has put them over the edge."

Law-enforcement officials estimate that there are thousands of child-porn sites. And the number is growing. The National Center's Cyber Tipline says it received 21,611 complaints about such sites in 2001, up from 16,724 the year before and 3,267 in 1998. Last year, the FBI made 514 arrests for online child porn, up from 68 in 1996.

As the number of sites rises, so do fears that more children are being sexually abused. Some online porn rings require members to post new photos to join, since many of the photos circulating on the Net are old.

Finding child porn online is shockingly easy. A Web search by *BusinessWeek* reporters for "little lolas" or "little boys" turned up five sites with hundreds of pictures of naked children. These sites claim they remain within the law because the photos they post do not show sex acts. But law-enforcement officials say that about 80% of the sites that show nude pictures of children also feature kiddie porn or provide links to child-porn sites. A search of the domain registry for these sites showed owners in Russia, Britain, Grand Cayman, and Tonga. None of the sites returned e-mail questions or phone calls.

It may get even harder to stop child porn. On Apr. 16, the Supreme Court ruled that it's unlawful to ban virtual child porn—computer-generated photos—

because no child is involved or harmed in making them. The case was brought by a trade association for adult-movie makers, which objected to the law on First Amendment grounds. The government says the ruling will make it tougher to prosecute cases because it will be difficult to tell virtual porn from the real thing. "We're going to be forced to prove that every picture is a real child," says U.S. Customs' Delli-Colli.

MONEY SCAMS

Cheryl Muzingo's travails started with a phone call from Discover Financial Services two years ago. Someone had used her name and Social Security number to apply for 16 credit cards online—and two had been approved. The swindler racked up $11,000 in bills for tickets to Disneyland, cash advances at casinos, and visits to a nail salon. The applications were online, so there was no paper trail. And the police wouldn't investigate, claiming the crime was outside their jurisdiction. The 37-year-old accountant from Henderson, Nev., did some sleuthing and found that Joanessa Warner, who worked at her company's travel agency, had stolen her data. Last October, Warner was sentenced to three years' probation and had to repay $9,550 to one credit-card company. Today, Muzingo fears her identity will be stolen again. "It has been horrible," she says.

Identity theft, stock manipulation, stolen credit cards—the Internet makes all these scams easier than ever before. Frauds that used to take days or weeks to cook up because they required office space, phones, and the postal service are done in minutes. In that time, they reach millions of potential marks. "The Internet has altered the playing field for scam artists," says John Reed Stark, chief of Internet enforcement at the Securities & Exchange Commission.

BusinessWeek estimates that financial fraud on the Net costs businesses and consumers $22 billion annually, based on law-enforcement and analyst projections. Online identity theft led to losses of $12 billion last year, according to the Identity Theft Resource Center, a San Diego nonprofit group that helps victims. Meanwhile, the SEC prosecuted cases last year in which investors lost $1.5 billion from Internet stock-manipulation schemes. Since the government estimates that only 1 in 10 Web cases is reported, actual losses could easily top $10 billion.

The scams aren't hard to find. A *BusinessWeek* reporter visited Yahoo's discussion boards and was directed to private discussion groups, known as Internet Relay Chats (IRC). From there, it was easy to glide into DALnet, a hangout for dealers in stolen credit-card numbers, obtained by hacking into systems of Internet merchants. A visit to any of the chat rooms—#thecc, #thacc, #cchome, or #shell_root—revealed hackers buying and selling credit-card numbers for 50¢ to $1 each. At the #thecc chat room, there's a plea from user xsythehell: "Need Discover cards, msg me for a deal." Within seconds, a note pops up from user kamusapa with a few Discover card numbers.

The con artists are careful not to get ripped off themselves, and no transaction takes place without first checking the validity of the numbers. On the message board is a program that checks the 16-digit number against the same database on-line merchants use to verify credit-card numbers. Several numbers quickly pop up as invalid. Then, a rash of numbers that are also being shopped go through as valid. After that, the transaction becomes private—the buyer and seller use instant messenger software to contact each other and set up payment and delivery of the credit-card numbers.

Falling prey to ripoff artists is surprisingly easy. In the course of researching this article, the credit-card number of one of the writers of this story was pinched and used to try to buy 30 Intel Pentium 4 chips at solutions4sure.com, a subsidiary of Office Depot Inc. Red flags went up because the order called for delivery to an address in Atlanta, though the cardholder lives in New York. When the Internet retailer called to verify the $6,700 order, it was squelched. It's not known how the information and phone number were obtained.

WHAT TO DO

Every 44 seconds, an unsavory act is committed on the Internet. The potential for sordid activities is as vast as the Web. There are no borders to patrol, and no single law-enforcement agency is authorized to clean up cyberspace.

What can be done? First, there needs to be a better understanding among law-enforcement officials, legislators, and citizens that the Underground Web is a serious problem. High-profile sectors such as terrorism and child porn are getting the funding for investigators and the necessary technology to weed out wrongdoing. But other areas, including gambling and drugs, get only modest attention from officials. When it comes to drugs, politicians think citizens care more about sales on the streets than on the Net. Yet online drug peddling can be worse. "A doctor prescribing drugs over the Internet can reach many, many more people than a street-level drug dealer," says Robert McCampbell, a U.S. Attorney in Oklahoma who has prosecuted Net drug sales.

One consistent problem is balkanization. Too many cops are stuck in a game of jurisdictional roulette. Internet financial fraud, for example, can be investigated by the FBI, Secret Service, Justice Dept., SEC, or the Federal Trade Commission. If it's international, then the Customs Service can weigh in. The resulting competition and confusion among agencies works to the advantage of criminals.

One solution is to make clear who is responsible for policing the Web. The Secret Service could take overall responsibility for financial fraud on the Internet since it has a lot of experience fighting cybercrime. The FBI would be the logical choice for online gambling and child porn. The DEA could target Internet drug dealing. The agency already has set up a special Net investigations unit, though it hasn't begun operating yet.

After establishing who will fight crime on the Web, there are a number of state models that cops could follow to attack the problems. California, for example, has made progress in stopping identity theft. Because a lot of thieves get credit-card data from paper receipts, the state requires all credit-card receipts to include only the last five card numbers. California also requires police to take reports from victims, something many local police forces are reluctant to do since they view ID theft as out of their jurisdiction. Expanding the California approach nationwide may prove effective.

When it comes to drug sales, Kentucky has one of the most advanced systems in the country. Pharmacies in most states don't share data. Kentucky, however, has built an integrated computer system that tracks drug sales from all pharmacies in the state. The technology allows doctors or pharmacists to see in an instant whether a patient has a drug problem—and it lets regulators see whether a doctor or pharmacist is prescribing unusual quantities of drugs. "If we could clone Kentucky, we would," says Kate Malliarakis, a branch chief at the Office of National Drug Control Policy.

Many states, led by Nevada, have tried to crack down on spam, but there are no federal laws against mass, unsolicited e-mail. Such legislation is important since spam is one of the chief ways fraudsters market their scams. To cut down on spam, heavy fines should be imposed. One bill before Congress—Controlling the Assault of Non-Solicited Pornography & Marketing (CANSPAM)—comes the closest. The bill would let the FTC penalize senders of unsolicited e-mail and require valid "remove me" options on all messages.

Nothing will work, however, without putting teeth into the laws that are already on the books. The best deterrent may be a clear message that both the supplier and the buyer of illegal goods will face stiff penalties if they're caught. That hasn't been so in the past, but it could be changing. On May 17, one of the leaders of an international Internet ring that pirated software, movies, and games was sentenced to 46 months in prison, one of the longest sentences for theft of intellectual property. Customs agents who monitor chat rooms where pirates hang out say there was shock over such stiff sentences. "People were saying, 'I'm getting out of the game.'" says Customs Agent Allan Doody.

Congress appears ready to get tough. On July 15, the U.S. House of Representatives approved The Cyber Security Enhancement Act, which promises life sentences for cyber attacks that recklessly endanger human life. Today, the maximum prison term is 10 years.

There will always be a seedy side to the Internet, just as there is one to every city. Cleaning up the Net will take vigilance and a slew of legal and public actions. For now, though, the Web has too many dark and dangerous corners and too little law and order.

31

Big Tobacco

MARK SCHAPIRO

Tobacco is one of the most globalized industries on the planet. More cigarettes are traded than any other single product, some trillion "sticks," as they're known in the business, passing international borders each year. As a result, American brands have been propelled into every corner of the world, with just four companies controlling 70 percent of the global market. Marlboro, Kool, Kent: They have become as omnipresent around the world as they are here in the United States. With declining sales in this country, foreign markets have become increasingly critical to the tobacco companies' financial health: The top U.S. tobacco firms now earn more from cigarettes sold abroad than in the United States. How they got there is a tale that leads straight into a global underground of smugglers and money launderers who have played a key role in facilitating the tobacco companies' entry into foreign markets.

A six-month investigation by *The Nation*, the Center for Investigative Reporting, and the PBS newsmagazine show *NOW With Bill Moyers* (which airs its investigative report on April 19), has unpeeled the many layers of a complex distribution system of a multibillion-dollar trade in smuggled cigarettes. Twenty-five percent of exported cigarettes, according to the World Health Organization, are smuggled. Smuggling has enabled multinational tobacco companies to increase sales volume dramatically by evading local tariffs and competing head to head with domestic producers, thereby helping to establish internationally recognizable brands.

The smuggling has landed the tobacco companies in U.S. court. Lawsuits filed by European and Canadian governments and Colombian state governments against Philip Morris and British American Tobacco (BAT, Brown & Williamson's British-based parent company) have highlighted the companies' alleged links to smugglers and money launderers. Documents released as a result of the historic $200 billion-plus settlement with U.S. state attorneys general in 1998 also provide a glimpse into the way the companies devised advertising and distribution strategies that helped fuel the market for smuggled cigarettes. The companies stand accused of violating the Racketeer Influenced and Corrupt Organizations Act (RICO), of defrauding governments of hundreds of millions

SOURCE: Reprinted with permission from the May 6, 2002, issue of *The Nation*. For subscription information, call 1-800-333-8536. Portions of each week's *Nation* magazine can be accessed at http://www.thenation.com.

in tax revenues and of hiding and ultimately taking the illicit profits back to the United States, which constitutes money laundering.

As the cases were unfolding just one month after the September 11 terrorist attacks, the tobacco companies—with support from the White House—fought back in the U.S. Congress, where they took advantage of the nation's distraction to win changes in the USA Patriot Act in a brazen effort to shield themselves from liability. But their headache has not gone away. The cases are still winding through the courts, and the companies' attempts to evade accountability are the focus of growing international outrage.

UNDERGROUND IN COLOMBIA

I went to Colombia, a country infamous for smuggling exports to the United States, to see how the flip side of that equation—smuggling from the United States to Colombia—worked for more than a decade.

The journey from the main tobacco hubs in the United States to Colombia has been a circuitous one, a route designed for ease of smuggling rather than ease of transport. From the modern, state-of-the-art ports of Wilmington, North Carolina, and Miami, huge cranes lift pastel-colored containers loaded with 10,000 kilos of cigarettes apiece onto cargo ships with the routine rhythms of oversized insects. Transiting through the free zones of Panama and Aruba, by the end of their journey at Colombia's La Guajira port of Portette, they might as well have traveled back in time. The bawdy port is what one anthropologist who studied the region calls a "phantom town"—it's not included on maps of the country and has had, until recently, few connections to the official structures of the Colombian government.

The hot, sparsely populated province of La Guajira, which sticks out of Colombia's Caribbean coast like a thumb, is home to one of the country's strongest indigenous tribes, the Way'uu. Last winter, when I visited, torrential rains knocked out the bridge between Santa Marta and Barranquilla, making coastal travel impossible. The road through the Sierra Nevada mountain range between Riohacha—the provincial capital on the coast—and Valledupar, the closest major city in the interior, runs through territory disputed by the ELN guerrilla movement and right-wing paramilitary groups. For parts of the year, the only way into La Guajira is by air or boat.

Colombia's other Caribbean ports, Santa Marta and Barranquilla, host sophisticated trucking and railroad depots right on the docks that are designed to facilitate the movement of large quantities of duty-paid cargo into the Colombian interior. Portette has no such facilities. It is a port designed, quite literally, for smugglers—and it's here that the schooners and creaky old ships from throughout the hemisphere pulled into Colombian waters with their crates of Johnnie Walker and Old Parr, and name-brand sound systems and electronic appliances, bales of textiles and those telltale cartons of Philip Morris's Marlboro and Brown & Williamson's Kool.

From Portette, trucks travel for two hours over a single rutted, mostly dirt, road to the town of Maicao, a dusty outpost of weatherbeaten shop-fronts and mud-splattered stucco buildings. In Maicao, young men are perched on stools along the side of the road amid plastic containers full of gasoline—skimmed from Venezuelan tankers and sold at a tax-free discount. Above them, a sign looms in peeling blue and yellow paint: Welcome to the Commercial Hub of Colombia. Maicao has for decades been the primary transit center in La Guajira for contraband headed for Colombian markets. The sight of brand-name whiskies, stereos, shampoos, car parts and cigarettes provides a jarring contrast to the muddy streets and crumbling kiosks where many of these products are sold.

Maicao's 70,000 inhabitants are divided between the Way'uu and a population of Middle Eastern immigrants—Colombians who emigrated from Lebanon, Syria and elsewhere in the Middle East. Historically, the Way'uu and the Turkos—as those with Middle Eastern roots are known—have divided the contraband trade between them. The mostly Muslim Turkos trade in textiles, appliances and other consumer products, leaving the vices of alcohol and cigarettes to the Way'uu.

But the Way'uu do not perceive themselves as criminals in any sense of the word. Since Colombia passed a new Constitution in 1991, decentralizing federal power, the tribe has been in charge of most of La Guajira; the bulk of the state is a *reserva indigena*, in which they enjoy a limited form of autonomy. From the Way'uu perspective, they are merely traders—their main economic activity for centuries.

"*Asi es la vida*," says Francisca Sierra, a Way'uu community leader and trader in Maicao, shrugging her shoulders as she explains the tribe's longtime role as renowned smugglers. That history predates even the formation of modern-day Colombia, which revolted against the Spanish in 1814. It was the Way'uu and ranchers in Santander province who helped spark that rebellion, when they refused to pay taxes on cigarettes and coffee imposed by the Spanish—Colombia's own Boston Tea Party. For 300 years, the Way'uu have facilitated the entrance of foreign products into Colombia below the noses of the national authorities.

In the last decades of the twentieth century, the Way'uu of La Guajira became a critical link in a chain of commercial relationships stretching from the tobacco farms of the southeast United States to corporate boardrooms in Louisville, New York and London, to tax havens like Aruba and Panama, and on into the interior of Colombia. Maicao itself is part of a special free trade zone, but once goods leave that zone, they become contraband. The Way'uu were the ones who unloaded the ships in Portette and then drove the trucks south out of Maicao into the interior, providing Philip Morris and BAT a detour around the tariffs that once made Colombia one of the more restricted markets in Latin America. The Federation of Colombian Departments, representing the country's state governments, estimates that the cigarette contraband cost them more than $500 million in tax revenues over ten years—revenues that would have paid for social projects like education and healthcare, including treatment of the health effects of smoking.

Statistics compiled by Roberto Steiner, an economist and director of the Center for the Study of Economic Development at the University of the Andes in Bogotá, indicate how smuggling served the tobacco companies' long-term interests. The boom in cigarette smuggling into Colombia in the 1990s, according to Steiner, coincided closely with Philip Morris's emergence as the dominant player in Colombia's cigarette market. As the companies sold tax-free cigarettes at prices comparable to those of Colombia's homegrown brands, smokers in Bogotá, Cali, Medellin and elsewhere throughout the country became accustomed to "prestigious" imports like Philip Morris's Marlboro, Brown & Williamson's Kool and BAT's Kent. From 1984 to 1993, says Steiner, the number of cigarettes illegally imported into the country quadrupled. Meanwhile, domestic cigarette producers' share of total cigarette sales dropped from an 85 percent market share in 1984 to just 30 percent in 1993. Colombia used to have a thriving domestic tobacco industry, but since 1984 the amount of hectares devoted to tobacco crops has plummeted. As the domestic cigarette industry imploded, many tobacco farmers made the shift to Colombia's far more famous addictive crop, coca.

A comparison of Colombian tobacco imports with U.S. tobacco exports reveals just how many contraband cigarettes were being shipped southward from the United States. According to the U.S. Department of Agriculture's Economic Research Service, $21.6 million worth of cigarettes—1.06 billion sticks—were exported from the United States to Colombia in 1996. In that same year, the Colombian Department of National Statistics (DANE) officially recorded $10.7 million worth of cigarettes—just over 800 million sticks—as having been legally imported into the country from the United States. That discrepancy between exports and imports appeared through most of the 1990s.

Internal company documents dating back to 1991, made available as a result of the 1998 states' settlement and introduced as evidence in the Colombian lawsuit, reveal how Philip Morris and BAT were battling for market share during this time—the same period in which the overwhelming bulk of cigarette smuggling to Colombia occurred. The record, for example, of a January 14, 1992, meeting in Miami held by BAT executives representing the company's wholly owned subsidiaries in the United States (Brown & Williamson), Brazil (Souza Cruz) and Venezuela (Bigott), under the heading "Colombian Group Meeting Minutes," shows officials discussing cigarette marketing in Colombia, indicating the per-pack, no-tax price in pesos in 1991, a year in which the company had negligible legal cigarette exports to the country. The minutes noted that the company would begin selling "duty paid"—i.e., legally imported cigarettes on which taxes are paid—in the coming year, 1992.

A document covering roughly the same period from the Philip Morris International division, titled "LATIN AMERICA REGION Strategic Plan," provides a listing of prices for its "duty-free" customers in La Guajira and Aruba for the years 1991–93. During this time and into the late 1990s, Philip Morris was advertising heavily and maintaining an office in Bogotá, while the company's legal imports amounted to, as Steiner put it, "close to zero."

In fact, both BAT and Philip Morris were deploying mass advertising and discount marketing, and were providing favorable financing terms to their distributors in the battle for market share, when their sales were almost entirely illegal. Documents from both companies reveal the intense competition and propose measures such as discounts to wholesalers, contests and free gifts to outflank each other. "Plans for 1992," Brown & Williamson minutes from a 1991 meeting state, "are to offer a 5% free goods incentive in Maicao and in the San Andresitos to expand distribution in Bogotá and Medellin." ("San Andresitos" is a colloquial reference to the kiosks that abound in Colombian cities selling smuggled goods; the name came originally from the Colombian island of San Andres, located off the east coast of Nicaragua, which itself has served as a key smuggling center.)

Until recently, few tobacco-industry insiders were willing to talk about the companies' role in smuggling. But in February, Alex Solagnier, a twenty-year veteran with BAT's primary Colombian distributor for cigarettes, an Aruba-based company called ROMAR, went public. Solagnier worked as a marketing and finance manager and finally as the company's chief financial officer, until he was fired after a business dispute with his superior in 1999; his chief responsibility had been selling BAT brands in Colombia. My *NOW* co-producer Oriana Zill and I were the first American journalists to speak with Solagnier, whom we filmed at his home in Aruba.

Solagnier says that BAT was integrally involved in setting the pricing, organizing distribution routes and marketing of cigarettes to the company's distributors at a time when, he says, "95 percent of it [BAT's cigarettes] was contraband." ROMAR itself, Solagnier explained, was set up with financing from BAT, in partnership with an Aruban businessman, Roy Harms, specifically to sell to the Colombian market when the bulk of BAT imports were smuggled into the country. (After a lengthy legal battle in which Solagnier and Harms traded accusations about financial mismanagement, an Aruban court ordered Harms to pay Solagnier more than $400,000 in severance pay, a figure for which Harms was reimbursed by BAT's London headquarters.)

Solagnier explained that during the years 1994–96, most of BAT's cigarettes were sold by ROMAR in Maicao "on consignment," meaning that while ROMAR handled the distribution, the cigarettes were owned by BAT when they were sold in Maicao. He recalls going on trips with BAT officials to assess the placement of the cigarettes, to determine their credit needs and to assess local demand. After studying the preferences of Colombian smokers, Solagnier says that BAT even designed a special cigarette package for its Belmont brand, which was produced by the company's Venezuelan subsidiary, Bigott, with a hinge lid on a hard-box pack, distinguished from the soft packs sold in Venezuela. BAT was promoting Belmont as competition for Philip Morris products. "They knew that all these cigarettes were being smuggled," he says.

Solagnier also explains that in the early 1990s, BAT and Philip Morris discovered the benefits of selling at least a small portion of their cigarettes legally, with full duties paid. Thus, reference to distinctions between "DP" (Duty Paid) and "DNP" (Duty Not Paid) begin to appear in both companies' internal documents. On April 16, 1992, a fax sent from BAT's British headquarters to its branch office in Venezuela indicated the company's growing sensitivity to attention being paid to contraband. In the memo, the executive asks whether the company could continue "with DP and DNP in parallel *and* be seen as a clean and ethical company at the same time." [underline in original] "Can we really do all this *and* continue DNP," he adds.

Translation: The company was interested in whether it would be beneficial to pursue legal imports along with its existing illegal imports. The answer, as shown by company documents and import statistics, was: yes. Both BAT and Philip Morris gradually began increasing the number of legitimately imported cigarettes—while the flood of smuggled cigarettes continued. Solagnier says that this dual system came to be known as the "umbrella"—a system of providing legal cover for advertising and marketing a product the bulk of which continued to be smuggled.

And the advertising was, by the mid-1990s, everywhere. In magazines, at sporting events, on billboards, ads for American cigarettes seemed more abundant than they are here in the United States, where ever-tighter restrictions have been placed on the tobacco companies' ability to advertise. At the same time, the companies launched a particularly cynical ploy—pressuring Colombian government officials to lower taxes on cigarettes as a means of reducing the incentive for smuggling. José Manuel Arias, director of the Colombian Federation of Departments, says that representatives of Philip Morris and BAT lobbied the state and national governments to lower Colombia's relatively hefty taxes on cigarette imports.

Their efforts paid off. According to Dr. Diego Roselli, a professor of pediatrics at Javieriana University in Bogotá and former chairman of the Colombian Council Against Cancer, the country saw a drop in cigarette tariffs from 125 percent to 45 percent in the mid-1990s. But the dramatic tax cut had a negligible impact on smuggling. Cigarettes continued to pour through the smuggling pipeline, selling for a little over a dollar a pack, just a quarter more than cigarettes produced in Colombia. At the same time, the tobacco companies gradually increased the amount of legal imports, where the margins were slimmer, but where they now enjoyed a lower tax rate and still obtained critical cover for advertising and other marketing activities.

TURNING OFF THE TAP

The week in late November when I arrived in La Guajira, trouble was brewing. The government had initiated a crackdown on contraband: The previous weekend, Maicao traders attacked the warehouse of Colombian customs (DIAN) in the town, looting it of all the goods that the DIAN had confiscated in the previous weeks, including cartons of cigarettes. The director of DIAN, Ricardo Ramirez Acuna, would later explain that an "arrangement" had been struck in which the companies agreed to assist the customs service in insuring that their cigarettes traveled through legal channels.

While Colombian officials see this as good news, they also say that it is a strong indication of how deeply the companies have been involved in the smuggling enterprise. When they decided to turn off the tap, off it went.

As a result, however, the Way'uu, long accustomed to being the transport mules of the contraband business, now feel betrayed by Philip Morris. For the first time, they were willing to speak publicly about the longtime relationship they had with Philip Morris during more than a decade of boom times, fueled partly by the cigarette company's nicotine contraband.

"We feel betrayed by Philip Morris because the Way'uu were the ones to bring the Marlboro cigarettes from the Caribbean islands into Colombia," asserts Alvaro Iguaran, a Way'uu lawyer and legal adviser. "Philip Morris sent their cigarettes through Maicao. . . . The Way'uu's were the ones who distributed the cigarettes and showed them to the rest of the country. Once the market was established, now they leave us and go elsewhere."

With the crackdown on smuggling, unemployment among the Way'uu in La Guajira has jumped 20 percent. "Philip Morris should build us a hospital and some schools," argues Iguaran, who doesn't want to wait for the lawsuit to be resolved. "They should do this on their own, and not just because of this legal case!"

Iguaran's plea is echoed in the comments of Ingrid Betancourt, a former congresswoman and senator running as an independent for president on an anticorruption platform. "Philip Morris pushed enormous quantities of cigarettes through Maicao into all of Latin America," she told me last November in Bogotá. (Betancourt was kidnapped by the FARC guerrillas in February and remains in custody.) "If the Way'uu don't do contraband, they starve. . . . Philip Morris has poured millions of dollars into a new NGO they created to promote the culture, dances, folklore of Colombia. Fine. But what about the Way'uu?"

Reflecting on BAT and Philip Morris's deal with the Colombian authorities, Alex Solagnier comments: "They know they got caught. . . . Now they want to cooperate to combat something they initiated and organized. They invented it. And the question is not what they're going to do now, but what did they do to create this problem?"

DRUG-MONEY LAUNDROMAT

In addition to cultivating the dependence of the Way'uu, the tobacco companies helped lubricate corrupt political and financial empires in Colombia. "You could say that Philip Morris has been influencing the political parties in La Guajira and Colombia for decades," comments Lucho Gomez, former mayor of Riohacha, the La Guajira state capital. Gomez has been the nemesis of an entrenched political machine run by a former senator from the state, Santo Lopesierra, a veteran political boss notorious for his connection to smugglers and commonly known in Colombia as "the Marlboro Man."

The Colombian newsmagazine *Semana* reported that representatives of Philip Morris's Colombian distributors, the Aruba-based Mansur family, met with Ernesto Samper during his 1994 presidential campaign and gave him more than $500,000. The Conservative Party of current President Andres Pastrana has ties to the industry too: Among several top officials with links to Philip Morris is the company's longtime lobbyist and attorney, Martha Lucia Ramirez, now minister of foreign trade.

Despite strong opposition from many in the Colombian political and economic elite, in the late 1990s the DIAN conducted an investigation and concluded that the boom in smuggling was tied to vast amounts of cash being generated by drug sales in the United States. The U.S. Drug Enforcement Administration shared the Colombians' concern, as did other U.S. law enforcement agencies. "In our undercover operations," Edward Guillen, chief of financial operations at the DEA's Washington headquarters, told us at *NOW*, "we started to find that what we initially might have thought were straw corporations . . . were actually involved in genuine commerce, actually buying goods, be they television sets or cigarettes . . . and then those goods were ultimately smuggled into Colombia."

Carlos Ronderos was minister of foreign trade from 1994 to 1998, during Samper's presidency. At a time when the U.S. government was launching an offensive against the Colombian government for the smuggling northward of cocaine, Ronderos began pressuring the U.S. government to rein in Philip Morris's southbound smuggling enterprise. Ronderos, interviewed in Bogotá, recalled a meeting in 1998 that he arranged with then-U.S. Ambassador Myles Frechette, the U.S. front man on the drug war: "I told him that you can't ask Colombia to stop the flow of cocaine if you are not willing to stop the flow of cigarettes and other goods used to launder the money from the sale of that cocaine." Frechette, according to Ronderos, rejected his plea, responding that "'it was purely a customs problem for Colombia.' And I felt like saying, 'OK, well drugs are just a customs problem for the United States.'"

He didn't say that, but after Frechette's rebuff, Ronderos went straight to Washington with his complaint. The Washington representative for the Colombian Trade Office, Carlos Acevedo—who is now working as one of the lead attorneys on the Colombian lawsuit—invited U.S. money-laundering experts to review the government's files in Bogotá. In February 1998—at the height of the Clinton Administration's efforts to isolate the Samper government—a five-person team, including specialists in money laundering from the Treasury Department's Financial Crimes Enforcement unit (FinCen), the IRS and Customs came to Bogotá to investigate the allegations.

Al James, a top FinCen agent at the time in charge of money-laundering investigations and chief organizer of the U.S. inquiry, has fond memories of that trip. "Ronderos was a real gentleman," he commented in a telephone interview. "He opened up everything to us, both the good and the bad stuff. We worked real well together." The joint investigation began to put into high relief a critical aspect of the narcotics trade: the means by which narco-dollars from the United States were channeled into the purchase of U.S. goods such as cigarettes. Those goods were transferred through Caribbean tax havens and ultimately sold to Colombian consumers for pesos as part of a complex money-laundering chain that came to be known as the Black Market Peso Exchange. James became chairman of a multiagency task force known as the Black Market Peso Exchange Working Group. "We began to understand," says James, "that what they were calling contraband smuggling was actually the other side of narcotics money laundering."

During his trip to Bogotá, James met with top Philip Morris executives to express his concerns about money laundering. "I warned them when we were in Colombia," he says. The officials told him that they had nothing to do with the cigarettes once they reached Colombian shores. James had similar meetings with other companies back in the States, informing them of the potential use of their products for drug-money-laundering purposes. Most, he says, responded by taking precautionary measures and instituting tighter surveillance of their sales operations. But not, says James, Philip Morris: "The evidence [of smuggling] started to seem pretty clear to me. Philip Morris had a 'legitimate' sales office in Bogotá. But they were losing millions of dollars if you looked at their legal sales. They spent more on advertising than they were making out of legitimate cigarette sales. They told me they were spending the ad money to sustain the legitimate sales. Bullshit!" Phillip Morris refused to respond to these allegations directly, but in an e-mailed statement declared, "Philip Morris does not condone money laundering; nor do our business practices facilitate it." The company states that it has instituted "know your customer" policies suggested by U.S. law enforcement and has stopped accepting cash or third party check payments, which could be used for laundering drug money.

TAKING IT TO COURT

U.S. law enforcement has little leverage over U.S. corporations overseas, and no legal action was taken against the cigarette companies. In May 2000, frustrated by the continuing flow of smuggled cigarettes into the country, twenty-two Colombian states and the city of Bogotá filed a lawsuit against Philip Morris and BAT in New York federal court, alleging various violations of U.S. law, including fraud, smuggling, money laundering and contraventions of the RICO act. The suit accuses the companies of "orchestrating and profiting from" the smuggling of cigarettes on a massive scale. It alleges that the companies were involved in shipping and distributing cigarettes that evaded customs duties and other taxes; that they disguised and moved the ill-gotten profits back to the United States, which constituted money laundering; and that cigarette smuggling was used in the laundering of Colombian drug profits. It was, says José Manuel Arias and other Colombian officials, the publicity from the lawsuit that prompted the cigarette companies to strike the deal with DIAN and stanch the flow of cigarettes passing illegally into Maicao. (Putting further pressure on the companies, last year the Colombian Congress passed a law mandating that their advertising expenditures cannot exceed the amount of their legal imports.)

According to Arias, Philip Morris lobbied every Colombian governor against signing on to the lawsuit, to little avail. But the company's lobbying of the national government did bear fruit: President Pastrana refused to sign on the national government, even though millions of dollars yearly in customs duties were allegedly diverted from the national treasury. (As it happened, back in Washington, Philip Morris emerged as one of the few nonmilitary companies to lobby heavily on behalf of Plan Colombia when it was winding its way through Congress.)

Six months after the Colombian filing, the European Union and ten European governments sued Philip Morris and RJ Reynolds on essentially the same grounds. Last August, after a federal court judge ruled that the EU was an inappropriate body to bring the suit, it was refiled by the ten European countries, including France, Germany, Spain, Belgium, the Netherlands, Greece and Italy.

While technically distinct, the Colombian and European cases are being argued in parallel. The cases are the first in which a phalanx of foreign governments are pitted against a trio of corporate powerhouses, and will be a significant test of whether U.S. corporations can be held accountable when they run afoul of U.S. and foreign law in their overseas operations. At a time when the fates of national economies are ever more intertwined, the cases promise to establish important precedents in the realm of international law governing corporations. "We are looking," says attorney Carlos Acevedo, "for the legal system to embrace the challenges of the modern globalized economy, in which production and distribution facilities have been flung far and wide across the globe."

In the long run, the companies could face repercussions from this legal offensive that are even more severe than the historic $200 billion–plus settlement with the U.S. states of four years ago. That legal crusade hinged on the companies' knowledge of the harmful effects of cigarettes. This time, the companies could face not only hundreds of millions in damages but criminal charges for smuggling and money laundering.

Thus far, the plaintiffs have suffered some setbacks. On February 19 a district court judge found in favor of the companies' argument that a common-law precedent dating from the eighteenth century, known as the "revenue rule," prevents U.S. courts from adjudicating disputes over uncollected foreign taxes. On March 25 the Europeans and Colombians announced their intention to appeal that decision. The judge didn't block them from pursuing the case on the money-laundering allegations, which they intend to do in a separate filing. But the plaintiffs' prospects would now be considerably brighter if the tobacco companies had not engineered an audacious reshaping of the USA Patriot Act to prevent them from acquiring a potent new legal tool.

PATRIOT GAMES

Roused into action by the terrorist attacks on September 11, Congress rushed to tighten U.S. laws governing money laundering and smuggling and to require transparency among financial institutions in order to strike at the means by which terrorists generate funds through illicit financial enterprises. In the original House version of the Patriot Act, introduced on October 3, and then known as the Financial Anti-Terrorism Act, Section 107(b) expanded the definition of money laundering to include "fraud or any scheme to defraud against a foreign government or foreign government entity, if such conduct would constitute a violation of this title if it were committed in interstate commerce in the United States." The Justice Department had asked for that section to strengthen its hand in pursuing legal prosecutions for money laundering—but the section would also have established the jurisdiction of U.S. courts over precisely the sort of activity of which the tobacco companies now stand accused.

At the time, the tobacco companies were facing legal assaults on several fronts. The government of Canada was preparing to appeal a lower-court decision that threw out its case accusing RJ Reynolds of evading $1 billion in taxes by smuggling cigarettes into Canada. And the Colombian and European RICO cases were on the docket at federal district court in New York. The provision would have provided clear legal standing to the plaintiffs in those lawsuits. But on October 11, with the country still reeling from the attacks one month before, GOP Representative Michael Oxley of Ohio, chairman of the House Financial Services Committee, undermined the Justice Department's original request and removed the provision before the committee hearing. He undertook that maneuver at the behest of the White House, according to a Congressional source close to the negotiations.

"The tobacco companies didn't care that in striking that provision they might have opened the American people to greater risk of a terrorist attack and funding terrorist groups that might attack our own people," comments Congressman Henry Waxman, a leading antagonist of the tobacco industry in Congress. "They wanted to make sure that that provision would not have been interpreted to give standing to these foreign countries." Philip Morris does not dispute the latter point, but in a letter insisted the changes were supported by "the business community at large," which has long been concerned with such matters of foreign liability, and vigorously denied that the change would hamper "the government's ability to bring suits to combat terrorism."

According to one Waxman staffer, in Congress the tobacco companies took a "belt and suspender approach to the Patriot Act in an effort to insulate their international operations from legal challenges in U.S. courts. The "belt"—Section 107(b)—was what Oxley removed from the act. The "suspenders" came late at night on October 16, when Chairman Oxley inserted a provision in the bill after it had been debated and approved by the full committee. Oxley's addendum specifically blocked any expansion of jurisdiction for U.S. courts to hear civil claims for damages from foreign nations seeking compensation for violations by U.S. corporations of foreign tax laws. The measure had the support of the White House and the top Republican leadership, including House majority leader Tom DeLay, according to a report by the Center for Public Integrity's International Consortium of Investigative Journalists. On the morning of the 17th—an infamous day, as the anthrax scare jumped from the Senate to the House, where members prepared to evacuate—the Financial Anti-Terrorist Act passed overwhelmingly in the House, including the new provision that would get the companies off the hook.

But staff aides to Waxman and Massachusetts Democrat Martin Meehan caught wind of the change. They alerted the Campaign for Tobacco Free Kids, which concluded in a memo that the only relevance of the provision was to "the . . . currently pending lawsuits . . . brought by Canada, the European Union and several Latin American countries . . . against major U.S. cigarette companies. . . . And the only future lawsuits likely to be affected would be similar lawsuits directed at the U.S. cigarette companies' involvement with international cigarette smuggling."

When the bill reached the Senate, there was outrage at what Senator Patrick Leahy described as the attempted "carve-out of tobacco companies from RICO liability for foreign excise taxes." Senator Paul Sarbanes, chairman of the Banking Committee, removed the offending passage from the bill. On the day it passed the Senate, Massachusetts Senator John Kerry, a longtime advocate of tighter money-laundering laws, introduced a statement into the Congressional Record clarifying that the law could be used to pursue the sort of legal challenges now faced by the tobacco companies: "It is the intent of the legislature that our allies will have unimpeded access to our courts and the use of our laws if they are the victims of smuggling, fraud, money laundering or terrorism." On October 26, the Patriot Act was signed by President Bush without Oxley's language.

While the "suspenders" in the tobacco industry's offensive were gone, however, the "belt" remained—a narrowed definition of money laundering that denied future and current plaintiffs against the tobacco industry an important legal instrument. "What was left out," says an infuriated Waxman aide, "was far more important than what was not put in." As Richard Daynard, director of the Tobacco Litigation Center at Northeastern University, says, "The bill as originally drafted would have made the tobacco companies a lot more vulnerable to the [money laundering] charges."

As in Colombia, those who argued on behalf of the tobacco industry were also major recipients of the industry's largesse: A report by the Campaign for Tobacco Free Kids reveals that Republicans received 82 percent of the more than $18 million that the tobacco industry has poured into political campaigns since 1997. Oxley himself has received $34,300 from the tobacco industry since 1999, both for his political campaigns and his PAC, Leadership 2000, and he held a party at the 2000 Republican convention that was paid for partly by Philip Morris.

THE GLOBALIZATION OF SMUGGLING

With an annual turnover of some $400 billion, tobacco is one of the world's largest industries. Across the globe, there are stories similar to that of the Way'uu and Philip Morris: "mules" who have helped put four tobacco companies in control of 70 percent of the world market. Smuggling insulates the companies from national controls to limit cigarette consumption—which the World Health Organization warns will cause another 10 million deaths by 2030. Wherever smuggling occurs, the pattern, says Luk Joossens, a consultant to the International Union Against Cancer and member of the Belgian delegation to the WHO, is the same: "If you have high tariffs or a state [tobacco] monopoly, they smuggle to get into the market, weaken the state monopolies, and lead the market into the hands of the multinationals."

Smuggling has also become a big-time criminal activity. The European Union's Anti-Fraud Office, which has investigated cigarette smuggling in conjunction with the national police forces of Spain, Italy, the Netherlands and elsewhere, claims that organized crime is increasingly a major player in what has become a multibillion-dollar business. In Montenegro last December, the Parliament held a series of explosive hearings on allegations raised by a Croatian newsweekly, *Nacional*, that Montenegro's President, Milo Djukanovic, has ties to cigarette smugglers linked to the Italian Mafia. In August the Iranian health ministry released statistics indicating that up to two-thirds of all cigarettes in the country had been smuggled. As part of its legal complaint, the European Union introduced evidence in February indicating that the profits from smuggling have gone to finance terrorist groups in Iraq and elsewhere. Here in the United States, four Arab immigrants confessed in early March to sending the profits from cigarette smuggling back to Hezbollah contacts in Lebanon; another fourteen people will be going on trial in Charlotte, North Carolina, this spring on the same charges, which now, according to the FBI, include "aiding and abetting a terrorist organization."

The World Health Organization has come to see smuggling as a major public health issue, asserting that it incapacitates one of government's best weapons for lowering tobacco consumption: high taxes. The WHO puts forth a simple calculation: More smuggling equals cheaper cigarettes equals more smokers, which means more smoking-related illnesses and deaths. According to the World Bank, if the price of cigarettes were to increase just 10 percent—which could be mandated through taxation—an estimated 40 million people would quit smoking worldwide.

At a meeting in Geneva March 18–23, representatives from WHO's 191 member states began finalizing plans for a Framework Convention on Tobacco Control, which would become the first international public health treaty. Proposals include measures to combat smuggling by requiring that tobacco companies mark each cigarette pack with a clear electronic code identifying its origins and destination; licensing all parties involved in cigarette distribution; and eliminating duty-free sales, which is a primary means of skimming off tax-free cigarettes into national markets.

The tobacco industry has been fighting those provisions, as well as others proposed by the WHO. Representative Waxman charges that the Bush Administration's delegation has been trying to weaken the organization's effort to limit tobacco consumption and contraband. In a letter to President Bush last November, Waxman accused his negotiators of embracing ten out of eleven changes to the convention proposed by Philip Morris, which had expressed opposition to the strongest of the anti-smuggling measures as well as controls on cigarette advertising, and even a proposal insuring that health warning labels appear in the language of the country of destination. A follow-up meeting on the issues raised in Geneva will be held in New York in July under the aegis of the Bureau of Alcohol, Tobacco and Firearms.

The tobacco companies refused repeatedly to be interviewed for this article. A spokesman for BAT, David Betteridge in London, said that the company would not comment on anything relating to smuggling, due to an ongoing investigation by the British Department of Trade and Industry. In December the company stated publicly that it would "apply even more stringent criteria" to its international distribution system to counter smuggling, and shortly thereafter BAT issued a statement that it was revising its projected earnings downward for the coming year. Philip Morris's director for public communication, John Sorrells, e-mailed me a statement on March 22, which reads, in part, that "Philip Morris does not condone, facilitate or support the smuggling of cigarettes and cooperates with governments in their efforts to prevent an illegal trade in the products we manufacture. We have taken significant steps, both internally and in cooperation with foreign governments, to prevent the smuggling of our products." The company also indicated that it now agrees with several measures proposed by the World Health Organization and foreign governments "to prevent cigarette smuggling," including "licensing of distribution chains" and "marking of duty-free products" intended to make it easier to track contraband cigarettes.

Clearly, the companies are uneasy about the lawsuits still winding their way through the U.S. court system. At a court hearing on the case last January, Philip Morris's attorney, Irvin Nathan of Arnold & Porter, expressed the company's dismay. "We are a public corporation," Nathan stated to the court. "It is unfair to us to have to be engaged in discussions about terrorism and money laundering. To have to put that in our disclosure documents is misleading to our shareholders."

As for the Way'uu, Alvaro Iguaran says: "The biggest social debt Philip Morris has is with us, the Way'uu. We showed Colombia and Venezuela that Marlboros existed . . . they used us because we opened their markets. And after their markets were opened, they didn't need us anymore. They owe us a lot of money."

32

Explosive Revelation$

LUCY KOMISAR

In the tax haven of Luxembourg, a little-known outfit called Clearstream handles billions of dollars a year in stock and bond transfers for banks, investment companies and multinational corporations. But a former top official of this "clearinghouse" says Clearstream operates a secret bookkeeping system that allows its clients to hide the money that moves through their accounts.

In these days of global markets, individuals and companies may be buying stocks, bonds or derivatives from a seller who is halfway across the world. Clearinghouse like Clearstream keep track of the "paperwork" for the transactions. Banks with accounts in the clearinghouse use a debit and credit system and, at the end of the day, the accounts (minus "handling fees," of course) are totaled up. The clearinghouse doesn't actually send money anywhere, it just debits and credits its members' accounts. It's all very efficient. But the money involved is massive. Clearstream handles more than 80 million transactions a year, and claims to have securities on deposit valued at $6.5 trillion.

It's also an excellent mechanism for laundering drug money or hiding income from the tax collector. Banks are supposed to be subject to local government oversight. But many of Clearstream's members have real or "virtual" subsidiaries in offshore tax havens, where records are secret and investigators can't trace transactions. And Clearstream, which keeps the central records of financial trades, doesn't get even the cursory regulation that applies to offshore banks. On top of that, it deliberately has put in place a system to hide many of its clients' transactions from any authorities who might come looking.

According to former insiders:

- Clearstream has a double system of accounting, with secret, non-published accounts that banks and big corporations use to make transfers they don't want listed on the official books.

- Though it is legally limited to dealing with financial institutions, Clearstream gives secret accounts to multinational corporations so they can move stocks and money free from outside scrutiny.

SOURCE: Lucy Komisar, "Explosive Revelation$," from *In These Times* (April 2002), pp. 18–22. This article is reprinted with permission from *In These Times magazine* (2004), and is available at www.inthesetimes.com.

- Clearstream carried an account for a notoriously criminal Russian bank for several years after the bank had officially "collapsed," and clearinghouse accounts camouflaged the destinations of transfers to Colombian banks.

- Clearstream operates a computer program that erases the traces of trades on request from its members.

- Clearstream was used to try to hide a dubious arms deal between French authorities and the Taiwanese military.

Many of these charges were first made in a controversial book called *Révélation$,* written by Denis Robert, a French journalist, and Ernest Backes, a former top official at the clearinghouse who helped design and install the computer system that facilitated the undisclosed accounts. The book's impact was explosive. Six European judges called it "the black box" of illicit international financial flows. Top Clearstream officials were fired. The scandal made headlines in big European newspapers; TV networks broadcast specials; the French National Assembly's financial crimes committee held a hearing. Luxembourg authorities ordered an investigation, and then they effected a cover-up. Yet *Révélation$* remains unpublished and relatively unknown in the United States.

A bearded, heavyset man in his mid-fifties, Backes spoke with *In These Times* in Neuchâtel, Switzerland, where he'd gone to attend a conference on international crime, and explained how he'd started fighting "organized crime in banking."

Ernest Backes was born in 1946 in Trier, Germany. (As he likes to joke, "There were two important people born in Trier; the other is Karl Marx.") His father was a Luxembourg metal worker, his mother a German nurse. From 14, he worked on an assembly line to pay for school and joined the Young Catholic Workers. After a job in the Luxembourg civil service, he was hired in 1971 by Clearstream's predecessor Cedel (short for "central delivery" office), set up the year before by a consortium of 66 international banks. Backes helped design and install Cedel's computerized accounting system in the '70s.

Cedel and its main competitor, Brussels-based Euroclear, were started to manage transfers of "eurodollars," U.S. currency kept in banks outside the United States. According to Barbara Garson's book *Money Makes the World Go Around,* eurodollars were invented in the '50s by the Chinese and the Soviets so they would not have to put their assets in banks where the U.S. government could seize them. But others saw value in eurodollars, and they began to be traded for other currencies. Some banks attracted eurodollars with higher interest than was being paid in America, and U.S. corporations and individuals began using the accounts to avoid laws on domestic banks. The euromoney market was born. (By the '90s, the Federal Reserve estimated that about two-thirds of U.S. currency was held abroad as eurodollars.)

Cedel and Euroclear eventually expanded into handling transfers of stock titles and other financial instruments. Their clients needed a system that would guarantee the creditworthiness of their trading partners and keep records of the trades. The clearinghouses provided speed, discretion, and a system that didn't make the records of their deals and profits readily accessible to outsiders. Every few months, a list of members' codes was distributed. For transfers, members just entered the codes, and Clearstream handled the deals with no further inquiries.

In 1975, several big Italian and German banks wanted to centralize their accounting and didn't want other members of Cedel to send transfers through their numerous individual branches. The Cedel council of administration—its board of directors—authorized banks with multiple subsidiaries not to put all their accounts on the lists. Backes and Gerard Soisson, then Cedel's general manager, set up a system of non-published accounts. A bank would send a transfer to the code of the headquarters bank, which would send it on to the non-published account of its subsidiary. The bank would regulate this operation internally.

Soisson authorized each non-published account, which would be known only by some insiders, including the auditors and members of the council of administration. As Cedel's literature to clients explained: "As a general rule, the principal account of each client is published: the existence of the account, as well as its name and number, are published. . . . On demand, and at the discretion of Cedel Bank, the client can open a non-published account. The non-published accounts don't figure in any printed document and their name is not mentioned in any report."

Requests for non-published accounts came from some banks that weren't eligible, but Soisson turned them down.

By 1980, Backes had become Cedel's No. 3 official, in charge of relations with clients. But he was fired in May 1983. Backes says the reason given for his sacking was an argument with an English banker, a friend of the CEO. "I think I was fired was because I knew too much about the Ambrosiano scandal," Backes says.

Banco Ambrosiano was once the second most important private bank in Italy, with the Vatican as a principal shareholder and loan recipient. The bank laundered drug- and arms-trafficking money for the Italian and American mafias and, in the '80s, channeled Vatican money to the Contras in Nicaragua and Solidarity in Poland. The corrupt managers also siphoned off funds via fictitious banks to personal shell company accounts in Switzerland, the Bahamas, Panama and other offshore havens. Banco Ambrosiano collapsed in 1982 with a deficit of more than $1 billion. (Unknown to many moviegoers, Banco Ambrosiano inspired a subplot of *The Godfather Part III*.)

Several of those behind the swindle have met untimely ends. Bank chairman Roberto Calvi was found hanged under Blackfriars Bridge in London. Michele Sindona, convicted in 1980 on 65 counts of fraud in the United States, was extradited to Italy in 1984 and sentenced to life in prison; in 1986, he was found dead in his cell, poisoned by cyanide-laced coffee. (Another suspect, Archbishop Paul Marcinkus, the head of the Vatican Bank, now lives in Sun City, Arizona with a Vatican passport; U.S. authorities have ignored a Milan arrest warrant for him.)

Just two months after Backes' dismissal in 1983, Soisson, 48 and healthy, was found dead in Corsica, where he'd gone on vacation. Top Cedel officials had the body returned immediately and buried, with no autopsy, announcing that he had died of a heart attack. His family now suspects he was murdered. "If

Soisson was murdered, it was also related to what he knew about Ambrosiano," Backes says. "When Soisson died, the Ambrosiano affair wasn't yet known as a scandal. [After it was revealed] I realized that Soisson and I had been at the cross-roads. We moved all those transactions known later in the scandal to Lima and other branches. Nobody even knew there was a Banco Ambrosiano branch in Lima and other South American countries."

After leaving Cedel, Backes got a job in the Luxembourg stock market, and later became manager of a butchers' cooperative. But he kept friends inside the clearinghouse and began to collect information and records about Cedel's operations.

With Soisson out of the way, there was nothing to stop the abuse of the system. Whereas Soisson had refused numerous requests to open non-published accounts (from such institutions as Chase Manhattan in New York, Chemical Bank of London and numerous subsidiaries of Citibank), Cedel opened hundreds of non-published accounts in total irregularity—especially after the arrival of CEO André Lussi in 1990. No longer were they just sub-accounts of officially listed accounts, Backes charges. Some were for banks that weren't subsidiaries or even official members of Cedel. At the start of 1995, Cedel had more than 2,200 published accounts. But in reality, according to documents obtained by Backes, Cedel that year managed more than 4,200 accounts, for more than 2,000 clients from 73 countries.

Clearstream was formed in 1999 out of the merger of Cedel and the compensation company of Deustche Börse (the German stock exchange). "No accounts are secret," insists spokesman Graham Cope. "We are controlled by the local authorities . . . who have access to information on all accounts. The term 'secret' is misused again and again. Our customers choose to have unpublished accounts, which simply means—like a telephone number—they choose not to display the name and number in our publications. Customers often have many unpublished accounts, which they use for their own internal management purposes to ensure there is no confusion between their accounts."

But Backes thinks otherwise. "I discovered an increasing number of unpublished accounts," he says. "There were more unpublished than published accounts, and a [large] proportion were not sub-accounts of a principal account, which is what the system was supposedly for. The owners of these accounts were not inscribed on the official list of the clients of the firm."

How does the system work? Backes explains, for example, that a bank with a published account could open an unpublished account for a branch in the Cayman Islands, an offshore tax haven. A drug trafficker easily could have the Cayman branch debit cash from his personal account to buy stocks on Wall street. The transaction would be handled by Clearstream, which would transfer the money electronically to a New York bank that had its own clearinghouse account. Soon the shares could be sold to buy real estate in Chicago with "clean" money. But regulators or investigators, depending only on published accounts, would find it nearly impossible to trace the money. Backes says Clearstream employees joke that the company name means "the river that washes."

While clearinghouse clients may want to keep transactions secret, detailed information on every transfer, including those via non-published accounts, is listed on daily "security statements"—records to prove that the stock or cash has been sent. These statements are stored on microfiche and, under Luxembourg law, must be kept 10 years for commercial enterprises and 15 years for banks. A Clearstream insider gave Backes 10 years worth of these records. "The documents are a mine of information for any financial inquiry," Backes say. "The archives of the clearinghouses can contribute to retracing where funds have gone. The knowledge of the list and the codes relative to non-published accounts, until now guarded secrets, offer immense possibilities."

Backes notes that similar records exist for the other big clearinghouses, Euroclear and Swift, also based in Brussels. "It is possible," he explains, "when one knows the date of an operation and the bank of entry, to reconstitute inside the clearing companies the voyage of the money and stocks or bonds—to follow the tracks."

Révélation$ charges that Cedel/Clearstream further violated its own statutes by setting up unpublished accounts for industrial and commercial companies. With accounts in their own names, companies could avoid passing through banks or exchange agents to use the clearinghouse. They thus skirted mandated due diligence and record-keeping. When Siemens was proposed for membership, Backes says, some Cedel employees protested that this violated Luxembourg law. However, management told them that Siemens' admission had been negotiated at the highest level.

Among the major companies with secret accounts, Backes discovered the Shell Petroleum Group and the Dutch agricultural multinational Unilever, one of whose accounts was associated with Goldman Sachs. On the French TV broadcast "Les dissimulateurs" ("The Deceivers") in March 2000, Clearstream President Lussi simply denied the accounts existed. "Only banks and brokers are eligible for membership," he said, "as it has always been the case. No private company accounts, no commercial or industrial companies."

But his own spokesman contradicts this claim. "Customers of Clearstream can be banks or, exceptionally, corporate clients who have their own treasury departments the size of banks," Cope wrote in an e-mail to *In These Times*. "We cannot accept CEOs of multinationals or terrorists and have strict account-opening procedures to prevent such problems."

By 2000, according to Backes, Clearstream managed about 15,000 accounts (of which half were non-published) for 2,500 clients in 105 countries; most of the investment companies, banks and their subsidiaries are from Western Europe and the United States. Most of the new non-published accounts were in offshore tax havens. The banks with the most non-published accounts are Banque Internationale de Luxembourg (309), Citibank (271) and Barclays (200).

Backes found numerous discrepancies in the lists he obtained of the secret accounts. For example, code No. 70287 on the published list belongs to Citibank NA-Colombia AC in Nassau, and code No. 70292 is that of the Banco Internacional de Colombia Nassau Ltd. But on the non-published list, the numbers both belong to Banco Internacional de Colombia in Bogota.

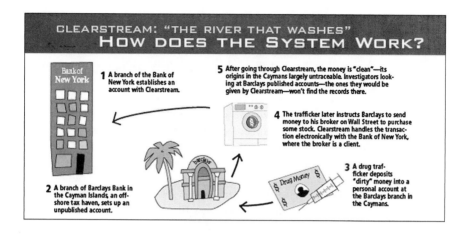

CLEARSTREAM: "THE RIVER THAT WASHES"
HOW DOES THE SYSTEM WORK?

Bank of New York
1 A branch of the Bank of New York establishes an account with Clearstream.

5 After going through Clearstream, the money is "clean"—its origins in the Caymans largely untraceable. Investigators looking at Barclays published accounts—the ones they would be given by Clearstream—won't find the records there.

4 The trafficker later instructs Barclays to send money to his broker on Wall Street to purchase some stock. Clearstream handles the transaction electronically with the Bank of New York, where the broker is a client.

2 A branch of Barclays Bank in the Cayman Islands, an off-shore tax haven, sets up an unpublished account.

3 A drug trafficker deposits "dirty" money into a personal account at the Barclays branch in the Caymans.

Drug Money

There's no mention of Citibank. Based on the published list, members may think they are dealing with two banks in the Bahamas, one of which is a subsidiary of Citibank, but anything sent to these establishments goes directly to the country of cocaine cartels. On the April 2000 Clearstream list, there are 37 Colombian accounts, of which only three are published. (Richard Howe, spokesman for Citicorp in New York, declined repeated requests for comment. Cope declined to talk about any individual customers or accounts, citing Luxembourg banking secrecy laws.)

Clearstream's dealings with Russian banks are another area of concern. Menatep Bank, which had been bought in a rigged auction of Soviet assets and has been linked to numerous international scams, opened its Cedel account (No. 81738) on May 15, 1997, after Lussi visited the bank's president in Moscow and invited him to use the system. It was a non-published account that didn't correspond to any published account, a breach of Clearstream's rules. Menatep further violated the rules because many transfers were of cash, not for settlement of securities. "For the three months in 1997 for which I hold microfiches," Backes says, "only cash transfers were channeled through the Menatep account."

"There were a lot of transfers between Menatep and the Bank of New York," Backes adds. Natasha Gurfinkel Kagalovsky, a former Bank of New York official and the wife of a Menatep vice president, stands accused of helping launder at least $7 billion from Russia. U.S. investigators have attempted to find out if some of the laundered money originated with Menatep, which they believed had looted Russian assets. (The Justice Department declined to comment on the investigation.)

Even though Menatep officially failed in 1998, it oddly remained on the non-published list of accounts for 2000. (Clearstream also lists 36 other Russian accounts, more non-published than published.) Kathleen Hawk, a U.S. spokeswoman for Clearstream, says that was "a mistake." But Cope contradicts her: "Closed accounts remain on our files and systems even though they're non-active because we don't reuse numbers. We keep the records for many years so there is no future confusion from reused numbers."

But Backes explains that there's no systematic rule about delisting canceled accounts. He found that "some that didn't exist any longer were on the list. Others were delisted when they didn't exist. And still other accounts were delisted, when we knew they existed, though the numbers no longer appeared."

Régis Hempel, a computer programmer who worked for Clearstream, says some dormant accounts were activated for special transactions. "Such an account can be opened in the morning, used for a transaction, and closed to appear as delisted in the evening," Backes explains. "Only the guy who gave the order to open it in the morning knows about the transaction. An investigator or auditor would not look at such an account because it doesn't appear on the accounts list."

Hempel also claims that Clearstream erased the records of some transfers. In testimony before the French National Assembly's financial crimes committee last year, he explained that a computer system had been developed to wipe out the traces of transactions in non-published accounts. When a bank wanted to carry out such a transaction, Hempel testified, it simply contacted a Cedel staff person. "We made a 'hard coding' in the program and corrected the instruction that was going to come," he explained. "[An instruction could be] a purchase, a sale, a movement of funds or a security. We made it disappear, or we put it on another account. Then, when all was finished, we put back the old program and removed the exception. It was not seen or known."

He said such requests came every two or three days.

Hempel volunteered to help Luxembourg prosecutor Carlos Zeyen investigate Clearstream. But Hempel says local authorities seem more interested in blocking an investigation than in exercising oversight. Zeyen responded that the inquiry into Hempel's charges hadn't produced any evidence and dismissed claims that Hempel had been prevented from seeing relevant files as "rubbish." In a July 2001 public statement, Zeyen said the investigation would continue.

Luxembourg sources say Zeyen was looking into how Menatep used the system and also into improper ways André Lussi might have gained personally. In January, a French judge took depositions about Menatep corruption. According to Luxembourg journalist Marc Gerges, writing in the local newspaper *Land,* the FBI and the German BKA are also interested in what might be revealed about the role of Menatep in the diversion of IMF funds. Gerges says investigators are also looking to implicate Lussi in suspected financial swindles conducted through holding companies and trusts in the offshore financial havens of Guernsey or Jersey. (Lussi could not be located; his attorney did not respond to phone and e-mail requests for comment.)

The publication of *Révélation$* brought forward others with stories about how Cedel/Clearstream had facilitated corruption. Joël Bûcher, former deputy general director of the Taiwan branch of the bank Société Générale, wrote Zeyen volunteering to testify that SG used the clearinghouse to hide bribes and to launder money. In his deposition for Zeyen—which is cited in Denis Robert's new book on the Clearstream saga, *The Black Box*—Bûcher said he had worked for the bank for 20 years, but quit in 1995 out of disgust at its rampant money-laundering. He said much of that occurred though a Luxembourg affiliate working through non-published accounts at Cedel. "Cedel didn't ask any questions about the origin of funds that would have appeared suspect to any beginner," he told Robert. "[As a result] we directed our clientele with funds of doubtful origin to Luxembourg."

In the early '90s, Bûcher contends, Cedel was used to launder $350 million in illegal "commissions" on a contract for the sale by Thomson-CSF, a French government arms company, of six French frigates to Taiwan. He said that the money, handled by an SG subsidiary, was paid as a registered securities transfer to a "nominee"—a stand-in for the real beneficiary—and that Thomson (now known as Thales) didn't appear in the transaction except in the Cedel archives.

The kickbacks were exposed after the 1993 murder of a naval captain named Yin Ching-feng, who had written a critical report on the purchase and its inflated $2.8 billion price. Bûcher told Taipei authorities that a third of the kickbacks went to Taiwanese generals and politicians, while the rest was pocketed by French officials. Taiwan courts sentenced 13 military officers and 15 arms dealers to between eight months and life in prison for bribery and leaking military secrets.

In March, Bûcher will testify before a French court examining French complicity. "SG is very much implicated," he told *In These Times*. "Taipei police searches found many records of transfers of commissions" relating to the frigates and also to the sale of French Mirage fighter planes. In New York, SG spokesman Jim Galvin denies that the bank had any involvement in the arms deal.

There has been no legal action by the Luxembourg prosecutor based on any of his investigations. However, Clearstream Banking, Lussi and others have filed 10 lawsuits for libel in Luxembourg, France, Belgium and Switzerland against Backes, Robert and their publisher, Les Arenes. The first case, *Clearstream v. Backes* went to court in March in Luxembourg. Another case began its first hearings in Paris a few days later. With no sense of irony, the liquidator of Russia's notorious Menatep Bank is also suing the authors and publishers for damage to its reputation. (Mikhail Khodorkovsky, the Russian oligarch who controlled Menatep, did not respond to a request for comment.)

Backes' knowledge and records make him a valuable investigative partner, and he cooperates with numerous authorities, though he prefers not to say in which countries. But his agenda is larger than that. Backes is lobbying for oversight by an international public body. Unlike banks, Clearstream has no effective outside surveillance. It is audited by KPMG, one of the "big five" international accounting firms, which either has been ignorant of or has overlooked the non-published accounts system. KPMG announced last year it found "no evidence" to support the allegations made in *Révélation$*, though its report was not made public.

Local officials' attempts to defend financial secrecy are not surprising. Luxembourg's multi-billion-dollar financial sector brings in 35 percent of GNP and gives the inhabitants a per capita income of more than $44,000, the highest in the world. (Next on the list are Liechtenstein, Switzerland and Bermuda, all money-laundering centers, with the United States fifth.) For years, local officials have refused to provide bank information to other countries.

But Luxembourg authorities have turned their sights on Backes. Using a March 2001 judicial order based on a complaint made by Lussi before he was fired, police raided Backes' house on September 19 in search of records. He says they seized unimportant documents and diskettes; he keeps the microfiches outside the country as "life insurance." "The raid was organized to impress [others] not to repeat what this dangerous guy Ernest Backes has done," he says. "Those who know me well know I am not at all impressed by such a raid."

33

No Way Out

NICHOLAS STEIN

For the privilege of working 12-hour shifts seven days a week in a factory where she makes plastic casings for Motorola cellphones, Mary, 30, will be in debt for years to come.

Mary already owes every penny she earns. In the circular, crazy logic of the global labor economy, she owes the money precisely because she has a job. And she is bound to her job as a result of her debt. Once it would have been called indentured servitude. Today, in some parts of the world, it's standard hiring practice.

This is how it works: To secure work at the Motorola subcontractor, which is in Taiwan, Mary had to pay $2,400 to a labor broker in her native Philippines. She didn't have that kind of money, so, as is common, she borrowed from a local money lender at an interest rate of 10% per *month*. That payment, however, got her only as far as Taiwan. A second labor broker met Mary at the Taipei airport and informed her of his separate $3,900 fee before delivering her to the new job.

Before she left the Philippines, Mary rejoiced at the $460 a month she would be earning in Taiwan; it was a princely sum, more than five times what she could make doing similar work, if she could even find it, in her own country. But once in Taiwan she began to realize that after the brokers' fees and other deductions, she would be left with almost nothing. Out of her monthly check came $215 to repay the Taiwanese broker, $91 for Taiwanese income tax, $72 for her room and board at the factory dorm, and $86 for a compulsory contribution to a savings bond she will get only if she completes her three-year contract. After 18 months she will have repaid the Taiwanese labor broker. But she still must contend with the Philippine debt and its rapidly compounding interest. "It is very painful for us to have to pay so much of our money," says Mary, who asked *Fortune* to change her name to protect her identity. "But we don't have a choice. We either have to take it or leave it."

There are many forms of debt bondage. As students of American history can attest, we've seen our share on these shores, from coal miners forced to buy overpriced food at the company store to sharecroppers trapped by the money they owe landowners. Even today many illegal Mexican immigrants are working

SOURCE: Nicholas Stein, "No Way Out," January 20, 2003, issue, *Fortune*. Reprinted with permission from *Fortune* and Nicholas Stein. © 1995 *Time* Inc. All rights reserved.

to pay off debts to the so-called coyotes who smuggled them across the Rio Grande.

But unlike coyotes, the Asian labor brokers to whom workers like Mary are indebted operate in the open. Their services are sought by the factories that import foreign workers and sanctioned by the governments that send and receive them. The labor trade they facilitate functions in the name of global competition.

When Motorola, Ericsson, Nike, and other Western companies contract with Asian factories to produce cellphones and modems, sew clothes, or prepare leather for shoes, they are thinking about costs. The globalization of manufacturing has been a tremendous boon to corporations, allowing them to seek the lowest-cost suppliers, wherever they may be. Today the vast majority of suppliers reside in Asia, and the lowest costs on the continent are found in the new manufacturing mecca: China.

What does that last fact have to do with the labor trade? Everything. With its vast pool of cheap labor, China has proved irresistible to Western manufacturers, which have been flocking there since liberalization in the mid-1990s. As a result, factories that operate in countries such as Taiwan, South Korea, and Malaysia, which have higher labor costs, have had to scramble to compete. The solution they devised was to import workers from poor neighbors—Vietnam, Thailand, the Philippines—and sign them up for two- or three-year contracts. The cost savings are real: In Taiwan, for example, native factory workers earn between $600 and $850 a month, while their foreign co-workers get the minimum wage of $460 and are ineligible for raises and promotions. Moreover, because they need the job to pay off labor brokers, overseas workers are less likely to complain about long hours or abusive supervisors. The logic has caught on: The number of foreign contract workers in Taiwan alone has doubled, to 316,000, in the past seven years.

Five years ago the chief labor issue for American companies like Nike, Liz Claiborne, and Gap was the sweatshop conditions in suppliers' factories. In response to protests and boycotts, U.S. companies began to demand that factories meet basic health and safety standards, providing workers with facemasks, bathroom breaks, and well-ventilated workspaces. But the debt bondage ensnaring many foreign workers in those factories has not yet hit the radar of most big corporations. It will.

During a trip to Taiwan and the Philippines in December, *Fortune* visited four factories in the apparel and high-tech sectors, spoke with half-a-dozen labor brokers, and interviewed more than 50 overseas contract workers from Vietnam, Thailand, and the Philippines. All the workers reported paying broker fees similar to Mary's, and many had suffered other abuses as well. In theory, engaging foreign contract workers is a solution that should benefit all parties: Poor countries reduce their unemployment, wealthy countries get cheaper labor, and the workers earn far more abroad than they could at home. In practice, however, the labor brokers have every incentive and opportunity to gouge the workers under their control.

Li Tung International, one of Taiwan's largest labor brokerages, occupies the fifth floor of a low-rise office building on the industrial outskirts of Taipei. The firm's general manager, Eric Chiang, has been in the business for ten years and has a fleet of luxury automobiles to show for it, including a Porsche Boxster, a Lotus Elise, and a chauffeur-driven BMW sedan. Sitting in his well-appointed office one December afternoon, surrounded by fine art and antiques, Chiang acknowledges that other Taiwanese labor brokers may take advantage of foreign workers. But he insists that his firm charges only the legal limit. "We sign contracts with the Thai and Philippine governments," he says. "It is impossible that we charge higher than what the law requires." For a worker on a three-year contract, that limit is $1,725, about 10% of a worker's gross pay, collected in monthly installments.

The foreign workers under contract to Li Tung, however, tell a different story. *Fortune* spoke to more than a dozen, some in the company of their factory managers and others in the privacy of their dorms. All cited payments far *in excess of the legal limit*. At a tannery that provides leather to Nike for its athletic shoes, a young worker from Thailand says he is paying $2,100 for his three-year contract. At the plastics factory where Mary works producing parts destined for Motorola, General Motors, Mercedes-Benz, and others, a Thai worker says he is paying $2,900 on a two-year contract. And at a garment factory that supplies brassieres to the underwear giant Wacoal, a Filipina worker tells *Fortune* that she paid $2,900, all of it in the first 15 months, which meant monthly payments four times the legal limit. A Li Tung spokesman would not confirm these fees but says many workers return to the agency for a second contract, so they must find the terms acceptable.

Overcharging, it turns out, is an accepted business practice for labor brokers, and some are not afraid to admit it. In Manila a labor broker from the D.A. Rodrigo agency said that her firm charges Taiwan-bound workers a $1,600 fee, equivalent to more than three months' salary even though Philippine law prohibits a broker's fee from exceeding one month's wages. The laws are simply not enforced. In fact, the broker admitted overcharging during an interview that took place while she was waiting to file papers inside the Philippine Overseas Employment Administration, the government branch charged with protecting the interests of workers who go abroad. "One month's salary is not enough to maintain a business like this," says Salvador Curameng, a Manila labor broker who presides over the profession's trade association, the Asian Recruitment Council. "Once you dip your fingers into this market, you are committing to do something illegal."

Governments are willing to look the other way because of what they get in return: The labor trade means jobs and capital will stay in their countries and not get shipped to China. Nations that import labor also tailor their laws to keep local factories happy. To hold turnover to a minimum, governments allow factories to retain workers' passports, impose curfews, and deduct compulsory savings bonds—or "run-away insurance"—which workers get back only when they have completed their contract. In South Korea, which limits foreign laborers to a single three-year visit, workers are considered trainees their first two years, so they are exempt from most of the country's labor laws, including minimum wage and overtime. In an effort to aid Taiwan's slumping manufacturing sector, the government last year passed a law allowing factories to charge foreign workers room and board.

The contracts are meant to be short term. Once they have finished, importing nations are eager to ensure that the workers won't find a way to stay. Almost all allow factories to administer pregnancy tests to female workers before they arrive. If a worker gets pregnant in Malaysia, the factory can terminate her contract and force her to cover the cost of her return airfare. Though Taiwan recently changed its law to allow pregnant workers to stay, in practice they are typically given the choice of abortion or deportation. Foreign workers who think marriage to a Taiwanese national is the route to permanent residency are out of luck: Marriage is grounds for immediate deportation.

Manila's C-5 expressway runs straight through Ecopa 3, a cramped and crowded neighborhood of rickety wood-and-tin shanties. Beneath the highway's overpass, 34-year-old Edwin lives with his brother and sister in a dark two-room dwelling. The stench of raw sewage seeps in from the narrow streets. Edwin has attempted to work overseas twice. He first left home in 1996 for a job in a Taiwanese factory, testing circuitboards for Sony computers. The experience proved disastrous. The factory was having financial troubles and eventually shut down. Edwin was shuttled to another factory owned by the same company but never got the four months' salary he was owed from the first job—nor the $2,000 in run-away insurance that had been deducted from his pay. When he returned to Manila, all Edwin had to show for three years of work was $1,100.

The second time Edwin sought work in Taiwan, in 2000, the experience was worse. His Philippine labor broker absconded with his $1,000 placement fee and never even got him a job. A subsequent class-action suit claimed that she had bilked 286 others out of fees ranging from $500 to $1,700. Two years later the case remains unresolved.

Stories like Edwin's are not uncommon. Stranded in a foreign country with no knowledge of the local language or labor laws, lacking government protection, and restrained by debt, foreign contract workers are especially vulnerable to mistreatment. "Some of the worst abuses we've ever seen have been in factories with foreign contract workers," says Heather White, executive director of Verité, an Amherst, Mass., nonprofit organization that has audited more than 1,000 factories on behalf of large corporations.

At the Hope Workers' Center, a shelter for foreign laborers in Taiwan, Father Peter O'Neill tends a wall of filing cabinets filled with complaints against employers. Most workers complain about unpaid wages, though there are also instances of forced labor, physical abuse, and even rape. In one case Vietnamese workers who made clothes for an Ann Taylor Loft and Dockers supplier charged that their employer withheld their passports and forced them to sign a contract in Chinese that they couldn't read. "We have workers come to the center who have been in Taiwan for six months," says O'Neill, "and have never seen Taiwanese money."

Felicidad Revolledo was working in Taiwan assembling modems for an Ericsson and Motorola subcontractor, when the factory simply cut her wages in half and suspended overtime pay for four months. When she refused to work more overtime until she got paid, she says the factory forced her to resign. Revolledo was lucky; she didn't owe any money in the Philippines. "A lot of my co-workers still working at the factory can't fight for their rights," says Revolledo, who left Taiwan in December, seven months before the end of her contract. "They are afraid to be sent back before they pay off their loans."

Now home in the Philippines, Revolledo hasn't been able to find another job. She's thinking about borrowing money to pay a labor broker to get back to Taiwan. "I cry a lot," she says. "But I think next time it will be better."

Once you've visited a place like Escopa 3, you immediately understand why people are desperate enough to seek work as an overseas contract worker again and again. "The choice is overseas employment or unemployment and poverty," says Mary Lou Alcid, the executive director of the Kanlungan Center for Migrant Workers in Manila. "Workers' expectations are high that they will be able to make big amounts of money—that the bad experiences will happen to someone else." Dr. Mariano Gagui, a psychiatrist who has treated many contract workers after their return to Escopa 3, says people are ashamed to talk about their ordeal. "A lot get traumatized," he says. "And you are not going to admit that going abroad is why the life of your family has disintegrated, why your husband is with another woman, and why your kids are on drugs."

The plight of overseas contract workers is just starting to enter the consciousness of big corporations. A year and a half ago Gap launched a new division of its global compliance department devoted solely to monitoring the treatment of foreign contract workers by its subcontractors abroad. "The presence of foreign workers is now one of the core issues we look at when we evaluate a new supplier—just like product quality or safety standards," says Dan Henkle, the apparel giant's vice president of global compliance. If a factory does employ foreign workers, Gap requires that factory management let the workers control their own travel documents and wages. Management must also agree to assume the workers' debt and travel costs if they choose to leave. "We want workers to feel they can leave at any time for any reason," says Henkle. Even with those rules in place, Gap remains wary of the possibilities for abuse: Only 5% of the company's production comes from factories that use overseas contract workers.

Gap's active response is unusual. "Companies are only beginning to understand this issue," says Doug Cahn, vice president of Reebok's human rights programs. "Although there are instances where some progress has been made, generally the abuses facing migrant laborers remain widespread." As a result, Reebok pays particular attention to subcontractors who use those workers.

Because of intense criticism over human rights abuses in the apparel and footwear industry over the past half-decade, companies like Gap, Reebok, and Nike are generally alert to labor issues. Many now monitor factories, and judging from *Fortune's* visit to the Yng Hsing tannery in Taiwan, which last year supplied leather for a million pairs of Nike Air Jordan basketball shoes, physical working conditions have improved as a result. "After we started working with Nike, we had to change our philosophy," says Philip Lo, the tannery's vice general manager. "They have strict requests about how you treat safety, health, attitude, environment." Nike is so sensitive to potential criticism that when the company learned of *Fortune's* visit to Yng Hsing, it immediately informed the tannery that unless it passed a hastily arranged inspection, it would be removed from Nike's supplier list.

The debt burdens and abuses of foreign contract workers, however, don't get the same level of attention as factory conditions. While Nike is aware of the abuses often faced by the workers, the company has no policy specifically governing supplier behavior on the issue—beyond a general statement in its code of conduct prohibiting forced labor. And this is a company whose brand was famously slammed in the late 1990s over the conduct of some of its suppliers. Nike's audit of Yng Hsing, which the tannery passed, did not even address the crushing debt loads carried by some of its Thai workers. "We checked out Yng Hsing," says a senior Nike official, "and all their employment contracts conform to Taiwanese law." (Nike declined to comment for attribution because of a pending lawsuit on another matter.)

In the high-tech sector, there appears to be even less attention paid to potential abuses. "I wasn't aware of this at all," said Pia Gideon, Ericsson's vice president of external relations, when first informed about the experiences of Revolledo and other foreign workers at a Taiwanese factory producing the Swedish company's modems. "I've never heard any indication about this situation in Taiwan." During a subsequent conversation, Gideon made a distinction between Ericsson's direct contractors, which it audits periodically, and subcontractors, which it expects to do "self-assessments" of the conditions in their factories. "I don't think you can have a law that says if you don't do this or that we will punish you," says Gideon. "Business doesn't work that way. We want our subcontractors to act a certain way, and we have to trust them." (Motorola representatives declined to be interviewed but issued a written statement saying the company "has a strict policy of adherence to the laws and labor practices in the countries where it operates, in addition to a rigorous code of conduct.")

Fortune visited factories and spoke with workers who make products for Nike, Motorola, and Ericsson, but they are not the only companies whose subcontractors rely on overseas contract workers. And with so many independent monitors now assessing labor rights and working conditions in manufacturing plants, it's hard to believe they could be completely ignorant of debt bondage in their supplier companies. "Five years ago, clients could say, 'I didn't know,'" says Verité's White. "There were no monitors. There was no awareness. They can't say that anymore. They have to acknowledge what is going on."

Companies' acknowledgment of the debt burden—and their doing something about it—is the best hope for the factory workers. And it's not an altogether quixotic one. After all, when companies believed their brands were at risk because of the sweatshop issue, they took action. The alternative is bleak. People like Mary and Edwin will continue to go to Taiwan or other countries as long as jobs remain scarce at home. "There are so many problems if I stay in the Philippines," says Edwin. "So what's more if I go abroad?" And when they do—barring pressure from Western corporations—they will continue to owe nearly everything they earn to the labor brokers.

34

21st Century Slaves

ANDREW COCKBURN

Sherwood Castle, headquarters to Milorad Milakovic, the former railway official who rose to become a notorious slave trafficker in Bosnia, looms beside the main road just outside the northwest Bosnian town of Prijedor. Under stucco battlements, the entrance is guarded by well-muscled, heavily tattooed young men, while off to one side Milakovic's trio of pet Siberian tigers prowl their caged compound.

I arrived there alone one gray spring morning—alone because no local guide or translator dared accompany me—and found my burly 54-year-old host waiting for me at a table set for lunch beside a glassed-in aquamarine swimming pool.

The master of Sherwood has never been shy about his business. He once asked a dauntless human rights activist who has publicly detailed his record of buying women for his brothels in Prijedor: "Is it a crime to sell women? They sell footballers, don't they?"

Milakovic threatened to kill the activist for her outspokenness, but to me he sang a softer tune. Over a poolside luncheon of seafood salad and steak, we discussed the stream of young women fleeing the shattered economies of their home countries in the former Soviet bloc. Milakovic said he was eager to promote his scheme to legalize prostitution in Bosnia—"to stop the selling of people, because each of those girls is someone's child."

One such child is a nearsighted, chain-smoking blonde named Victoria, at 20 a veteran of the international slave trade. For three years of her life she was among the estimated 27 million men, women, and children in the world who are enslaved—physically confined or restrained and forced to work, or controlled through violence, or in some way treated as property.

Victoria's odyssey began when she was 17, fresh out of school in Chisinau, the decayed capital of the former Soviet republic of Moldova. "There was no work, no money," she explained simply. So when a friend—"at least I thought he was a friend"—suggested he could help her get a job in a factory in Turkey, she jumped at the idea and took up his offer to drive her there, through Romania. "But when I realized we had driven west, to the border with Serbia, I knew something was wrong."

It was too late. At the border she was handed over to a group of Serb men, who produced a new passport saying she was 18. They led her on foot into

SOURCE: Andrew Cockburn, "21st Century Slaves," *National Geographic Magazine* (September 2003). Reprinted with permission from the National Geographic Society.

Serbia and raped her, telling her that she would be killed if she resisted. Then they sent her under guard to Bosnia, the Balkan republic being rebuilt under a torrent of international aid after its years of genocidal civil war.

Victoria was now a piece of property and, as such, was bought and sold by different brothel owners ten times over the next two years for an average price of $1,500. Finally, four months pregnant and fearful of a forced abortion, she escaped. I found her hiding in the Bosnian city of Mostar, sheltered by a group of Bosnian women.

In a soft monotone she recited the names of clubs and bars in various towns where she had to dance seminaked, look cheerful, and have sex with any customer who wanted her for the price of a few packs of cigarettes. "The clubs were all awful, although the Artemdia, in Banja Luka, was the worst—all the customers were cops," she recalled.

Victoria was a debt slave. Payment for her services went straight to her owner of the moment to cover her "debt"—the amount he had paid to buy her from her previous owner. She was held in servitude unless or until the money she owed to whomever controlled her had been recovered, at which point she would be sold again and would begin to work off the purchase price paid by her new owner. Although slavery in its traditional form survives in many parts of the world, debt slavery of this kind, with variations, is the most common form of servitude today.

According to Milorad Milakovic, such a system is perfectly aboveboard. "There is the problem of expense in bringing a girl here," he had explained to me. "The plane, transport, hotels along the way, as well as food. That girl must work to get that money back."

In November 2000 the UN-sponsored International Police Task Force (IPTF) raided Milakovic's nightclub-brothels in Prijedor, liberating 34 young women who told stories of servitude similar to Victoria's. "We had to dance, drink a lot, and go to our rooms with anyone," said one. "We were eating once a day and sleeping five to six hours. If we would not do what we were told, guards would beat us."

Following the IPTF raids, Milakovic complained to the press that the now liberated women had cost a lot of money to buy, that he would have to buy more, and that he wanted compensation. He also spoke openly about the cozy relations he had enjoyed with the IPTF peacekeepers, many of whom had been his customers.

But there were no influential friends to protect him in May this year, when local police finally raided Sherwood Castle and arrested Milakovic for trafficking in humans and possessing slaves.

We think of slavery as something that is over and done with, and our images of it tend to be grounded in the 19th century: black field hands in chains. "In those days slavery thrived on a shortage of person power," explains Mike Dottridge, former director of Anti-Slavery International, founded in 1839 to carry on the campaign that had already abolished slavery in the British Empire. The average slave in 1850, according to the research of slavery expert Kevin Bales, sold for around $40,000 in today's money.

I visited Dottridge at the organization's headquarters in a small building in Stockwell, a nondescript district in south London. "Back then," said Dottridge, "black people were kidnapped and; forced to work as slaves. Today vulnerable people are lured into debt slavery in the expectation of a better life. There are so many of them because there are so many desperate people in the world."

The offices are festooned with images of contemporary slavery—forced labor in West Africa, five- and six-year-old Pakistani children delivered to the Persian Gulf to serve as jockeys on racing camels, Thai child prostitutes. File cabinets bulge with reports: Brazilian slave gangs hacking at the Amazon rain forest to make charcoal for the steel industry, farm laborers in India bound to landlords by debt they have inherited from their parents and will pass on to their children.

The buying and selling of people is a profitable business because, while globalization has made it easier to move goods and money around the world, people who want to move to where jobs [are] face ever more stringent restrictions on legal migration.

Almost invariably those who cannot migrate legally or pay fees up front to be smuggled across borders end up in the hands of trafficking mafias. "Alien smuggling [bringing in illegal aliens who then find paying jobs] and human trafficking [where people end up enslaved or sold by the traffickers] operate exactly the same way, using the same routes," said a veteran field agent from the U.S. Immigration and Naturalization Service (INS). "The only difference is what happens to people at the other end." As the fees people must pay for transport rise in step with tightening border controls, illegal immigrants are ever more likely to end up in debt to the traffickers who have moved them—and are forced to work off their obligations as slaves.

It's dangerous for outsiders to show too close an interest in how these trafficking mafias work (a point that had occurred to me at Sherwood Castle), but in Athens I found a man who has made the study of slave trafficking his specialty and lives to tell the tale.

In 1990 Grigoris Lazos, a sociology professor at Panteion University, embarked on what he thought would be straightforward research on prostitution in Greece. Bright and intense, he resolved to go straight to the source, the prostitutes themselves. Through them he eventually made contact with the people who had enslaved them. Over the course of a decade—and in the face of intense disapproval from his professional colleagues—Lazos gained access to trafficking operations from the inside and was able to paint a clear picture of the interplay between prostitution and slavery in his country.

"You should note the difference between a small trafficking gang and a large network, which uses the Internet and bank accounts," he said. "Any bar owner or group of bar owners in Greece can send someone up to southern Bulgaria to buy women for cash. The cost of a girl in that area is $1,000, or, if you negotiate, you might be able to get two for $1,000. Best to try on a Monday for cheap prices, because most trafficking happens at the weekends. Mondays are slow, so you can get the leftovers."

"A network on the other hand," he continued, "has the ability to bargain and complete financial transactions from a distance. Simply call Moscow, ask for women, and they will be sent to Romania and from there on through Bulgaria to Greece. The parties don't even have to know each other. The importer simply says, 'I want so-and-so many first quality women, so-and-so many second quality, so-and-so many third quality.'"

Flicking through his exhaustive files, the professor rattled off the cold data of human trade. "Between 1990 and 2000 the total amount earned in Greece from trafficked women, that is to say those who were forced into this kind of prostitution, was 5.5 billion dollars. Voluntary prostitutes, those who were working of their own accord and are mostly Greek women, earned 1.5 billion dollars."

The efficiency and scope of the Greek traffickers' operations studied by Lazos is by no means unique. In Trieste, the gateway from the Balkans into northern Italy, investigators from the local anti-mafia commission tracked the activities of Josip Loncaric, a former taxi driver from Zagreb, Croatia.

By the time Loncaric was finally arrested in 2000 he owned airlines in Albania and Macedonia and was involved in moving thousands of people destined for work not only in prostitution but in any menial task requiring cheap labor in the prosperous world of the European Union. His Chinese wife, who was also his business partner, provided a link to criminal Chinese triads with which Loncaric did profitable business smuggling Chinese as well as Kurds, Iraqis, Iranians, and any other afflicted people willing to mortgage themselves in hopes of a better future. Many of Loncaric's Chinese victims found themselves locked up and forced to work 18 hours a day in restaurants or in the famous Italian leather workshops.

Trafficking mafias and smugglers, in the last decade of the 20th century, brought 35,000 people a year into Western Europe through the Trieste area, guiding them at night through the rugged mountains and forests straddling the border with Slovenia. But this is only one of many funnels between poor worlds and rich ones. Thousands of miles away I found another flood of migrants fleeing Central America on their way to El Norte, the United States, where they could ultimately become slaves.

These migrants' homes were ravaged by the wars of the 1980s and '90s and reduced to further ruin by a succession of natural and man-made disasters. Hurricane Mitch pounded Honduras and Nicaragua in 1998; afterward the number of homeless street children in Central America jumped by 20 percent. El Salvador was hit by a 7.6 earthquake in 2001. Large parts of the region have been without rain for the past three years, and the world price of coffee has crashed, ruining the Central American coffee industry and leaving 600,000 workers unemployed. In Guatemala more than half a million coffee workers face starvation.

Many economists argue that the North American Free Trade Agreement has made its own contribution to the flood of people trying to move north, maintaining that cheap U.S. corn imported into Mexico has effectively driven millions of Mexican peasant corn farmers out of business and off the land. They suggest that for every ton of corn imported into Mexico, two Mexicans migrate to the U.S.

The tiny Guatemalan town of Tecún Umán lies on the bank of the Suchiate River. Here migrants from Central America gather to cross into Mexico on their way north. Those with valid travel documents for Mexico cross the bridge over the river; those without them pay a few cents to be ferried across on rafts made from tractor inner tubes.

No matter where they come from, a great majority of migrants arrive in Tecún Umán penniless, easy prey for the local hoteliers, bar owners, and people smugglers—known as coyotes—who live off the flow of humanity. It is a town where, in the words of one former resident, "everything and everyone is for sale."

Some of the luckier migrants find a temporary safe haven at Casa del Migrante, a walled compound just a few yards from the muddy riverbank. "Every day, morning and night, I give a speech here," says the Casa's director, Father Ademar Barilli, a Brazilian Jesuit who remains surprisingly buoyant despite the surrounding misery. "I talk about the dangers of the trip north and urge them to go back. It's a bad choice to go home, but a worse one to try to go on to the U.S."

Barilli warns migrants about the bosses in Mexico who may take their precious documents and force them into slavery on remote plantations. He tells them about the brothels in Tapachula, the Mexican town across the river, where girls are forced into prostitution. Most, remembering the misery they have left behind, disregard his warnings. As Adriana, a 14-year-old prostitute in a Tapachula bar, exclaimed when asked if she would consider going home to Honduras: "No, there you die of hunger!"

Despite Barilli and Casa del Migrante, Tecún Umán itself is hardly safe. The week before I arrived, a dead coyote had been dumped just outside the gates of the compound with a hundred bullets in his body. "People are killed here because of the traffic in people and babies. There are many mafias involved in the business of this town. *Aqui uno no sale en la noche*—Here you don't go out at night," Barilli said.

As I calculated the amount of daylight left, Barilli explained what local bar owners say to girls from the buses that roll in every day from the south. "They talk about a job working in a restaurant. But the job is in a bar. After the girl has worked for a while just serving drinks, the owner denounces her to the police and gets her arrested because she has no documents. She is jailed; he bails her out. Then he tells her she is in his debt and must work as a prostitute. The debt never ends, so the girl is a slave."

Barilli cited a recent case involving a bar named La Taverna on the highway out of town. The owner, a woman, had duped six girls in this fashion. "Some of them got pregnant, and she sold the babies," he said. Thanks partly to the efforts of a Casa del Migrante lay worker (who afterward went into hiding in response to a flood of very credible death threats), the bar owner was finally arrested and jailed.

Stepped-up security in the wake of 9/11 has made the major obstacle on the road from the south, the border between Mexico and the U.S., more difficult than ever to clear. With heightened control has come a commensurate increase in the price charged by smuggling gangs to take people across: up from an average of about $1,000 a person to $2,000. Survivors of the journey arrive deeply indebted and vulnerable to slavers.

In Immokalee, Florida, I sat in a room full of men and women with the same Maya features I had last seen on the faces of the people in Tecún Umán. Almost all of them were farm laborers, toiling on Florida's vast plantations to pick fruit and vegetables consumed all over the U.S. They were meeting at the headquarters of a farmworker organization, the Coalition of Immokalee Workers (CIW), to discuss ways of improving conditions in their ill-paid occupation. When the rapid-fire Spanish conversation died away, an elderly man picked up a guitar and began to sing about Juan Muñoz, who left Campeche, Mexico, "to seek his fortune in the U.S." but ended up in Lake Placid, Florida, working "as a slave" for a cruel boss who stole all his money.

Blues singers composed similar laments about the miseries of plantation life in the Old South, and we think of those songs as part of our heritage. But this song was not about the past, Juan Muñoz is a real person, a 32-year-old who left his small farm in Campeche because he couldn't earn enough money to feed his family. He made his way across the border to Marana, Arizona, where a coyote promised him a ride all the way to a job picking oranges in Florida. The ride cost $1,000, which Muñoz was told he could pay off over time. On arrival he found he had in fact joined the modern slave economy.

Highway 27 runs through citrus country in the heart of Florida, which supplies 80 percent of U.S. orange juice. The pickers in the fields that line the highway are overwhelmingly immigrants, many undocumented and all poor. They earn an average $7,500 a year for work that is hard and unhealthy, toiling for bosses who contract with growers to supply crews to pick crops. The law generally leaves these people alone so long as they stick to low-paid but necessary work in the fields.

Sweatshop conditions in the fields are almost inevitable, since the corporations that buy the crops have the power to keep the prices they pay low, thus ensuring that wages paid by harvesting companies to pickers stay low too. These conditions lead to a high turnover in the workforce, since anyone with a prospect of alternative work swiftly moves on. Hence the appeal to crew bosses of debt-slave crews, whose stability and docility are assured. That is how Juan Muñoz found himself held captive along with at least 700 others in the well-guarded camps operated by the Ramos family in and around the little town of Lake Placid.

"They had almost all been picked up in Arizona by coyotes who offered to take them to Florida and then sold them to the crew bosses," says Romeo Ramirez, a 21-year-old Guatemalan who went undercover to investigate the Ramoses' operation on behalf of the CIW.

Captives in eight camps in and around Lake Placid were living "four to a room, which stank, sleeping on box springs." Not surprisingly, the workers were terrified of their bosses. "People knew they would be beaten for trying to get away," said Ramirez, citing the rumor about one would-be escapee who "had his knees busted with a hammer and then was thrown out of a car moving 60 miles an hour."

"The workers were paid by the growers every Friday," Ramirez continued, "but then they would all be herded to the Ramoses' stores in Lake Placid and forced to sign over their checks. By the time they had paid for rent and food, their debt was as high as ever." One such store, Natalie's Boutique, is a block from the police station.

In April 2001 a team from the CIW helped four of the captive laborers, including Muñoz, to make a break. Spurred to action by the unequivocal testimony of the escapees, the FBI and INS mounted a raid—although the prominent "INS Deportation Service" sign on the side of the bus accompanying the raiding party gave the crew bosses enough warning to send the workers out into the orange groves around Lake Placid to hide. Nevertheless, the brothers Ramiro and Juan Ramos, along with their cousin José Luis Ramos, were eventually charged with trafficking in slaves, extortion, and possession of firearms. In June 2002 the three Ramoses were convicted on all counts and received prison sentences totaling 34 years and 9 months.

This 21st-century slave operation may have been ignored by the Ramoses' corporate clients; and federal agencies may have been slow to react to prodding by the CIW. But the slave crews were hardly out of sight. The main camp in which the Ramoses confined their victims was just on the edge of town right beside a Ramada Inn. On the other side of the compound a gated community, Lakefront Estates, offered a restful environment for seniors.

"The slaves in Lake Placid were invisible, part of our economy that exists in a parallel universe," points out Laura Germino of the CIW. "People were playing golf at the retirement community, and right behind them was a slave camp. Two worlds, speaking different languages."

The Ramos case was in fact the fifth case of agricultural slavery exposed in Florida in the past six years. All came to light thanks to the CIW, which is currently promoting a boycott of fast-food giant Taco Bell on behalf of tomato pickers. The corporation boasts of its efforts to protect animal welfare in its suppliers' operations. Corporate officials also say they demand compliance with labor laws, but point out that since they cannot monitor suppliers' labor practices continually they rely on law enforcement to ensure compliance.

Slavery and slave trafficking in the U.S. today extend far beyond farm country into almost every area of the economy where cheap labor is at a premium. In 1995 more than 70 Thai women were rescued after laboring for years behind barbed wire in the Los Angeles suburb of El Monte, making clothes for major retailers while federal and state law enforcement repeatedly failed to obtain a proper warrant to search the premises. In June 2001 federal agents in Yakima, Washington, arrested the owners of an ice cream vending company and charged them with using Mexican slaves, working to pay off transportation debts, to sell ice cream on city streets. According to Kevin Bales, there are between 100,000 and 150,000 slaves in the U.S. today.

The Department of State puts the number of people trafficked into the U.S. every year at close to 20,000. Many end up as prostitutes or farm laborers. Some work in nursing homes. Others suffer their servitude alone, domestic slaves confined to private homes.

The passage by Congress in 2000 of the Victims of Trafficking and Violence Protection Act, which protects such slaves against deportation if they testify against their former owners, perhaps has helped dispel some fearfulness. The growth of organizations ready to give help, like the CIW or the Coalition to Abolish Slavery and Trafficking, a southern California group that has assisted more than 200 trafficked people, means that victims are not alone. Public scrutiny in general is rising.

Still, such captives the world over are mostly helpless. They are threatened; they live in fear of deportation; they are cut off from any source of advice or support because they cannot communicate with the outside world. And the harsh fact remains that this parallel universe, as Laura Germino called it, can be a very profitable place to do business. Before sentencing the Ramoses, U.S. District Court Judge K. Michael Moore ordered the confiscation of three million dollars the brothers had earned from their operation, as well as extensive real estate and other property.

Moore also pointed a finger at the agribusiness corporations that hired the Ramoses' picking crews. "It seems," he said, "that there are others at another level in this system of fruit picking—at a higher level—that to some extent are complicit in one way or another in how these activities occur."

A former slave named Julia Gabriel, now a landscape gardener in Florida and a member of CIW, remembers her arrival in the U.S. from Guatemala at the age of 19. She picked cucumbers under armed guard in South Carolina for 12 to 14 hours a day; she saw fellow captives pistol-whipped into unconsciousness. "Maybe this is normal in the U.S.," she thought. Then a friend told her, "no, this is not normal here," so Gabriel found the courage to escape.

"This is meant to be the country to which people come fleeing servitude, not to be cast into servitude when they are here," says Attorney General John Ashcroft. But some historians argue that the infamous trans-Atlantic slave trade that shipped millions of Africans to the New World was abolished only when it had outlived its economic usefulness. Now slave traders from Sherwood Castle to sunny Florida—and at hundreds of points in between—have rediscovered the profitability of buying and selling human beings. Which means that, in the 21st century, slavery is far from gone.

35

Climate Refugees

Global Warming Will Spur Migration

KIT BATTEN

KARI MANLOVE

NAT GRYLL

Debate over comprehensive immigration reform may have stalled last week in the Senate, but there's one key concern that's just warming up: the exacerbating effect that droughts, severe weather, food shortages, disease, and sea level rises will have on migration.

Worldwide environmental, economic, and social consequences from existing atmospheric greenhouse gas concentrations, even if we were to cease emissions today, will drive migration around the globe. Attention to the migration pressures resulting from global warming should therefore be an essential aspect of a long-term U.S. immigration plan. This will not only focus efforts on helping populations adapt to climate change, but also encourage thought on how to alleviate migration pressures.

According to the International Federation of Red Cross, climate change disasters are already a bigger cause of population displacement than war and persecution. Estimates of climate refugees currently range from 25 to 50 million. And this April, global scientific experts and former U.S. military leaders warned in two reports—the Intergovernmental Panel on Climate Change's Fourth Assessment and the CNA Corporation's "National Security and the Threat of Climate Change"—that the effects of global warming are likely to trigger conflict and mass migrations of affected people.

Large numbers of immigrants to the United States currently come from Mexico and the Caribbean, and with increases in storm intensity, stress on natural resources, and rising sea levels—side effects already affecting these regions—immigration levels will only increase. Northern Mexico's severe water shortages will drive immigration into the United States despite the increasingly treacherous border terrain. The damage caused by storms and rising sea levels in the coastal areas of the Caribbean Islands—where 60 percent of the population live—will likewise increase the flow of immigrants from the region and generate political tension.

SOURCE: Kit Batten, Kari Manlove, and Nat Gryll, "Climate Refugees: Global Warming Will Spur Migration," *Center for America Progress* (July 3, 2007). Online: http://www.americanprogress.org/issues/2007/07/climate_debt.html.

The United States cannot ignore the potentially heightened flow of displaced peoples as it continues to discuss immigration reform. Because we shoulder a large portion of the responsibility for the current levels of global warming pollution in the atmosphere, we have a moral responsibility to invest in solutions that will help ourselves and the world—particularly poor countries—adapt and prevent the growing implications of climate change.

The countries least responsible for greenhouse gas emissions are frequently the most vulnerable to global warming's earliest effects. Developing countries bear minimal responsibility for climate change because they have little industry and produce relatively small amounts of pollution. But their populations—often the poorest of the world's people—are more likely to occupy vulnerable locations such as coast lines, flood plains, and steep slopes and live in structures unable to withstand severe weather events. The governments of these poor countries therefore carry the largest burden associated with climate change and are ill-equipped to recover from disasters and meet the basic needs of their citizens.

The United States therefore owes a "climate debt" that it needs to pay back to these poor countries. China may have recently surpassed the United States in terms of overall greenhouse gas emissions, but the United States has still had the largest historical greenhouse gas emissions as well as the greatest per capita emissions. Each American citizen on average produces four times the amount of greenhouse gas as an average Chinese citizen. And, unfortunately, because greenhouse gases can persist in the atmosphere for hundreds of years, the world will be experiencing the negative effects of these disproportionately large U.S. emissions for years to come.

To tackle this debt, we should direct a portion of the revenue generated by an eventual national greenhouse gas cap-and-trade or carbon tax law toward adaptation projects such as desalinization facilities to provide fresh drinking water, city construction away from low-lying areas inundated by rising sea levels, and investment in drought-resistant crop cultivation. This adaptation assistance will help diffuse potential conflicts over scarce resources and offset global migration pressures.

Alleviating this debt by investing in projects to help these countries adapt to climate change will fulfill a moral obligation and is smart international policy. As Congress steps back to decide its next steps for tackling the immigration crisis, it must begin considering how it will deal with even larger numbers of immigrants and take action before the problem worsens.

36

Global Warming
Population Transfers Ignored

DICK FIELD

A few commentators have expressed grave concern that the continued importation of millions of immigrants to Canada, the UK, the USA and northern Europe contributes to an ever increasing urban sprawl and the consequent destruction of vital agricultural lands. Even Canada's huge land mass cannot support the density of human occupation that its size would lead one to believe. Most of Canada's land mass is virtually uninhabitable. Compatible cultural integration is also becoming a cause of concern.

Some ninety percent of immigrants take up residence in the major cities. As a consequence, transportation systems are experiencing increasing gridlock and many of our vital social infrastructures are being overwhelmed. Is it not strange that environmentalists, scientists and politicians worry about the destruction of trees and rain forests and yet, for the most part, ignore the destruction of our shrinking productive lands, parks, green areas and the damage created to our societies in terms of quality of life?

What is even stranger however is that these same leaders, who talk so much about global warming, pay no attention to the contribution to global warming caused by massive transfers of immigrant populations, legal and illegal, from tropical and sub tropical climates to northern temperate and frigid climates. Should the world's leaders not begin to factor in the damage to the planet that these huge south to north population transfers create? Let's at least think about this problem.

Every immigrant from a tropical or semi-tropical climate such as the Caribbean, Central and South America, Africa, the Indian Sub Continent, Hong Kong, southern China, South East Asia, the Burmese/Malay peninsula, the Philippine archipelago and the Middle East requires huge additional amounts of carbon based energy to survive in North American and European climates. This additional consumption of fossil based energy creates huge amounts of carbon dioxide and other chemicals and hence, atmospheric pollution.

In climates where not much more than a tee shirt and shorts plus minimal housing is needed to stay warm, dry and alive, little consumption of energy is

SOURCE: Dick Field, "Global Warming: Population Transfers Ignored," *Canada Free Press* (February 22, 2007). Online: http://www.canadafreepress.com/2007/field022207.htm.

required. However, in cold climates such as Europe, Canada, Russia and most of the United States, the additional consumption of energy for synthetic and natural fibre winter clothing, heated housing, gasoline and diesel powered transportation is enormous.

On a worldwide basis, massive interclimatic migration from south to north, whether through economic migration or political terror, needs to be examined carefully. While much work needs to be done to accurately assess the dimensions of the problem, here are a few numbers that ought to at least give scientists, environmentalists and politicians cause to reexamine their immigration source and population transfer assumptions.

The annual per capita production of CO_2 (Carbon Dioxide) by persons living in Canada is 4.35 tons. In comparison, the per capita production of CO_2 of a person living in tropical Indonesia is a mere .21 tons (1/20th of a person surviving in Canada). In Mexico the average per capita production of CO_2 is 1.01 tons or less than one one-quarter the amount produced by that same person moving to Canada. Even in the highly industrialized but more moderate climate, society of Japan, the production of CO_2 per capita is 2.34 tons or approximately 50% less than the same person's output if required to live and survive in Canada.[1]

Why is this problem being ignored? It is a difficult question to answer but a partial answer may be that the United Nations and its supporting governments, including Canada, have got into the population transfer business as a solution to the overpopulation and economic problems suffered by much of the tropical world.

There is also a myth that because the industrialized western world is "rich" and peaceful, somehow or other it must be responsible for the poverty and political mayhem in large parts of the undeveloped and "poor" third world. The western industrialized world is thereby falsely pressured into engaging in the band-aid solutions of massive immigration and refugee acceptance, no amount of which can ever be massive enough to relieve the population and political pressures of the grossly overpopulated third world.

Governments in northern climates, especially Canada, northern USA and Europe that look to satisfy their immigration needs by importing untold millions of low energy consumers from warm climate countries and thereby turn them into high energy consumers (and polluters) are acting irresponsibly and must reexamine their assumptions. The whole world is being placed in environmental jeopardy by their carelessness.

By rough calculation, taking into account population transfers to the USA, Canada and Europe, there have been approximately 82,000,000 people brought from southern climates in the last 40 or so years. Taking an estimated and very conservative factor of 3 additional tons of CO_2 per person produced by this inter-climatic transfer of population from the south to north is tantamount to adding another 246,000,000 tons of greenhouse gases, annually, to the world's global warming problems.

Canada's South to North contribution has been a population transfer of approximately 4,455,000 people representing an additional 13,650,000 unnecessary tons of annual greenhouse gas pollution to Canada's already sorry record.[2]

What is the alternative to this very damaging economic and political inter-climatic migration with its substantial permanent contribution to the Global Warming? Surely those of us living in the northern part of the industrialized world need to learn to manage our pollution problems much better than we have done in the past.

Surely our leaders must also become more realistic in respect to immigration planning. Immigration and refugee policies should be carefully reexamined with a view to substantially reducing our ever-expanding interclimatic, south to north, population transfers. While year by year additional tonnages may not seem huge, every year adds permanently to the overall problem and 20, 50, 100 years down the road we will have created an enormous unnecessary problem for ourselves and the world. Is it not about time the world's political leaders and our very voluble environmentalists began to address these troublesome issues? Why have they been silent?

What is the alternative to economic and political interclimatic migration? The alternative should be to help those warm climate countries, with large potentially migrant populations, to develop their own population carrying capacities. We should continue to help them stabilize their economic and political systems. Our policy should be to help them in situ. The solution is there in their countries, not in our northern countries.

We might also look at changing our taxation and incentive systems to encourage our resident population to increase their family size sufficiently to replace our own populations from within. Less expansive immigration policies should be considered and certainly from less pollution creating sources than mass transfers from southern to northern climates.

Is it not time to carefully examine this much overlooked aspect of contributing factors to the Global Warming problem?

ENDNOTES

1. Statistics dated from 2001 environmental studies supplied by Dr. Richard Hummel, Professor Emeritus, University of Toronto.

2. Population Statistics (approximated) from current UK, France, Germany, Spain, Italy, Belgium, Netherlands, Austria, USA, Canada, and other European countries.

REFLECTION QUESTIONS FOR CHAPTER 9

1. The industrial nations burn a disproportionate amount of the fossil fuels that contribute to the greenhouse effect that produces global warming. Should these nations voluntarily reduce these damaging emissions? How does such an act square with capitalism?

2. Almost a third of the world's population live on less than $1 a day, half of the World Bank's definition of poverty. Most of them live in rural areas and the shanty towns in developing societies. Their poverty is a primary source of the "New Slaves" (sex workers, domestic workers, and sweatshop workers). What can be done about this enormous problem? What is the responsibility of the prosperous nations? Transnational corporations? Nations of the developing world?

3. How does the treatment of global warming provided by Field differ from that of Batten and colleagues? Why are both perspectives important in understanding the global dimensions of climate change?

10

Changing Global Structures
Resistance and Social Movements

INTRODUCTION

The articles included in this book address issues and questions regarding the various manifestations of globalization: Does the presence of transnational corporations in the developing world raise the standard of living for the poor in those countries? Will transnational corporations rule the world? Are the political agencies of globalization—for example, the World Trade Organization and the International Monetary Fund—increasing the wealth gap between the rich and poor nations? Is globalization increasing the threats to the environment? Is the future a homogenized Westernized world? For many individuals and groups the answers to these questions are in the negative. Thus, in opposition to the effects of globalization, they have organized to resist globalization or to change its directions. In doing so, they are engaged in human agency by joining social movements to resist, challenge, and to change the direction of globalization. These two phenomena—human agency and social movements—have the potential to change the course of history. In effect, social transformations do not always occur from the top (where the political and economic power is located) but sometimes from the bottom up (by the relatively powerless who join with others to form a social movement).

Social structures constrain what we do, but they do not determine what we do.[1] While these structures are powerful, human beings are not totally controlled. We are not passive actors. Individuals acting alone or with others can shape, resist, challenge, and sometimes change the social structures that impinge on them. These actions constitute human agency.[2]

Individuals seeking change typically join with others for greater power to become part of a social movement. "Social movements are agencies of social

295

transformation that emerge in response to certain social changes and conditions."[3] For our purposes many have become part of social movements to resist global corporations, free trade, and transnational political/economic organizations.

> Millions of workers, consumers, environmentalists, religious activists, farmers, and women around the world are demanding their fair share of the fruits of the global economy. Their strategies are diverse. Some attempt to slow down aspects of globalization, while others aim to reshape its path in ways that promote democracy, equity, and sustainability. Campaigns operate on the local, national, and international levels.[4]

The first reading is an excerpt from Jeremy Brecher, Tim Costello, and Brendan Smith's *Globalization from Below*. The authors address a wide range of efforts to stop, slow down, or change the course of globalization. They provide the blueprint, the rationale, and a brief history of how and why people have and are organizing at the grassroots to affect change "from the bottom up." The second part of their contribution provides an understanding of how social movements arise and how they have the potential to overcome the politically and economically powerful.

The second reading, by Manning Marable, turns our attention to global apartheid and the need for transnational forms of resistance. He first describes the racialized division of resources, wealth, and power that separates Europe, North America, and Japan from Black, indigenous, undocumented, immigrant, and poor people across the planet. He shows that the ever-widening racial inequalities are produced by larger global systems of labor exploitation and political disenfranchisement. Therefore, the anti-globalization struggle must be a worldwide, pluralistic, anti-racist movement with a central goal of destroying global apartheid and white supremacy.

The next article, by David Bacon, calls for transnational solidarity in labor organizing. He interviews Julia Quiñones, coordinator of the Border Committee of Women Workers in Piedras Negras, Mexico, who makes the point forcefully that just as there are no borders to economic capital, U.S. and Mexican workers must build the same borderless solidarity and support one another.

The fourth piece in this section addresses students' mobilization in global activism. Bhumika Muchhala provides an account of both his own radicalization in dissent movements and the history of the Students Against Sweatshops movement on U.S. campuses. Not only does he render an account of the movement's origin and structure, he reflects on its place in the global struggle.

In the next reading, "The Rise of Food Democracy," Brian Halweil examines power relations in the global food system. From Nebraska to Norway and from Egypt to Zimbabwe, he shows how small farms are responding to reclaim the "sovereignty" of small producers and local markets, with some notable effects at the level of the World Trade Organization and some of the world's largest food companies.

The final reading takes up questions about using the Internet as a feminist tool in the context of global inequalities. The dialogue between Radhika Gajjala and Annapurna Mamidipudi suggests that Internet technologies and cyberfeminism can connect women across the world and bridge differences between poor women in the south and the affluent in the north.

ENDNOTES

1. Anthony Giddens, 1991. *Introduction to Sociology*. New York: W. W. Norton.
2. Stanley Eitzen and Maxine Baca Zinn. 2004. *In Conflict and Order: Understanding Society*, 10th ed. Boston: Allyn & Bacon, Chapter 18.
3. Robin Cohen and Paul Kennedy. 2000. *Global Sociology*. New York: New York University Press, p. 287.
4. Sarah Anderson and John Cavanagh, with Thea Lee. 2000. *Field Guide to the Global Economy*. New York: The New Press, p. 91.

37

Globalization and Social Movements

JEREMY BRECHER
TIM COSTELLO
BRENDAN SMITH

It is often said that globalization is inevitable and that there is no alternative. But, in fact, the new global regime is highly vulnerable. It violates the interests of the great majority of the world's people. It lacks political legitimacy. It is riven with divisions and conflicting interests. It has the normal crisis-prone character of capitalist systems, but few of the compensatory non-market institutions that helped stabilize pre-globalization economies. And it has few means to control its own tendency to destroy the natural environment on which it—and its species—depend. These are the reasons that, as the *Financial Times* wrote, the world had swung "from the triumph of global capitalism to its crisis in less than a decade."[1]

GLOBALIZATION FROM BELOW

Just as the corporate and political elites are reaching across national borders to further their agendas, people at the grassroots are connecting their struggles around the world to impose their needs and interests on the global economy. Globalization from above is generating a worldwide movement of resistance: globalization from below.[2]

Throughout the 20th century, nationally based social movements have placed limits on the downsides of capitalism. Workers and communities won national economic regulation and protections ranging from environmental laws to labor unions and from public investment to progressive taxation.

Globalization outflanked both national movements and national economies. It caused a historic break in the institutions, traditions, and movements that had opposed unfettered capitalism since its inception. Not only Communism, but also social democracy, economic nationalism, trade unionism, and democratic government itself were rolled back by the neoliberal tide—and often found their own foundations crumbling from within in the face of forces they could not understand or control.

SOURCE: Jeremy Brecher, Tim Costello, and Brendan Smith, *Globalization from Below,* South End Press. Reprinted with permission from South End Press. Notes in this article have been renumbered.—ED

Nonetheless, the real problems of a system of unrestrained capitalism did not disappear. Globalization only intensified them. And so the impulses that had generated these counter-movements in the first place began to stir.

Like globalization from above, these counter-movements began from many diverse starting points, ranging from local campaigns against runaway plants to union organizing in poor countries, and from protection of indigenous peoples to resistance to corporate-engineered food. Their participants have come to the issues of globalization by way of many different itineraries. For example:

- Acid rain and global warming do not respect national borders. They have forced environmentalists around the world to recognize global ecological interdependence.[3] At the same time, environmentalists became increasingly conscious that the actions of global corporations and of institutions such as the World Bank destroyed local environments—symbolized by the destruction of the Amazon rain forest and India's Narmada Valley. While some argued that globalizing capitalism would actually promote environmentalism in the third world, environmentalists discovered that it was instead creating an environmental race to the bottom as countries lowered environmental standards to attract corporations. The WTO's anti-environmental rules— symbolized by its decision condemning a U.S. law for the protection of sea turtles—brought the environmental movement into direct confrontation with this central institution of globalization.

- In the 1970s, the world's poorer countries formed the G-77 and initiated a North-South Dialogue with the rich countries to formulate a New International Economic Order. When the rich countries withdrew from this effort in the 1980s and began instead to promote neoliberal policies coordinated through the IMF, World Bank, and WTO, most third world governments went along with their plans, albeit in many cases reluctantly. But networks of third world NGOs continued to develop an alternative agenda and to press it both on their own countries and on international institutions. Third world governments have recently begun to follow their lead. As the rich countries prepared their agenda for the 1999 Seattle WTO extravaganza, poor-country governments began to question whether they had benefitted from globalization. Encouraged by the global citizens' movement to halt any new round of WTO negotiations, third world delegations for the first time refused to go along with the rich countries' proposals until their own concerns were addressed, helping to bring the meeting down in shambles. Early in 2000, the G-77 held its first ever head-of-state-level meeting and proposed an alternative program that included debt relief, increased aid, access to technology, and a shift in economic decision making from the World Bank and IMF to the UN.[4]

- People in rich countries have a long history of compassionate assistance for poor countries—sometimes in alliance with religious proselytizing and colonialism. With the development of the third world debt crisis in the 1980s, however, many people of conscience in the first world became deeply concerned about the effect of crushing debts on third world people and began to demand cancellation of their debt. Many then went on to address the broader question of the devastating "structural adjustment" policies being imposed on the debtor countries by the IMF, World Bank, and rich countries.

- When negotiations started in 1986 for what became the WTO, critics argued that U.S. and other first world proposals would benefit agribusiness and transnational commodity traders, but would drive millions of small farmers in both the North and South off their farms. Advocates for small farmers around the world began holding regular counter-meetings at the negotiations and developed a global network to oppose the proposals. They provided much of the core for international opposition to the emerging WTO. What has been described as "the first really global demonstration," in December 1990, brought farmers from Europe, Japan, North America, Korea, Africa, and Latin America to Brussels—helping force the negotiations into deadlock.[5] Since then, small farmers have been at the forefront of opposition to WTO agricultural policies, efforts to turn seeds into private property, and genetically engineered organisms (GEOs).

- From World War II until the 1960s, the labor movement in the United States was a strong supporter of economic liberalization, both as an expression of its alliance with U.S. international policy and as a means to secure expanding markets for U.S.-made products. Faced with a massive loss of jobs in auto, steel, garment, and other industries in the 1970s, the labor movement increasingly campaigned for tariffs and other barriers to imports designed to "save American jobs." Over the 1990s, globalization made such economic nationalist strategies less and less credible. Organized labor increasingly moved toward demanding reform of the global economy as a whole, symbolized by demands for labor rights and environmental standards in international trade agreements to protect all the world's workers and communities from the race to the bottom. Its participation in the Seattle WTO protests represented a new page in U.S. labor history and was followed by the announcement of a long-term "Campaign for Global Fairness."

- The burgeoning identity-based movements of the late 20th century found that many identities did not respect national borders. The women's movement slogan "sisterhood is powerful" evolved into a consciousness that "sisterhood is global." A growing awareness of the global oppression of women led to a struggle to define women's rights as internationally protected human rights. Events surrounding the UN's 1995 Beijing women's conference brought large numbers of women in the United States to an awareness of the impact of IMF and World Bank–imposed structural adjustment austerity programs on women in poor countries, and their similarity to the implications of welfare reform for poor women in the United States. The fact that the great majority of those exploited in overseas factories were young women led to a growing concern about the global sweatshop.

- From the 1960s on, consumer movements in many countries had enshrined a wide range of protections in national laws and had developed effective legal techniques for imposing a degree of accountability on corporations. Consumer organizations—notably Ralph Nader's Public Citizen—discovered that trade agreements like NAFTA and the WTO were overriding high national standards for such things as food and product safety. They also realized that both neoliberal ideology and competition among countries for investment were tending to lower consumer protection standards all over the world. New consumer issues, such as the right of governments to regulate genetically engineered food, have steadily increased consumer concern over globalization.

- African American communities in the U.S. have been concerned with conditions in Africa from the mid-19th century to the struggle against South African apartheid. But the 1990s saw two specific concerns that brought attention to the global economy. The first was the devastation wreaked on African countries by international debt and the brutal structural adjustment conditionalities the IMF and World Bank imposed on African countries in exchange for helping them roll over their debts. The other was the struggle over the African trade bill (known to its critics as the "NAFTA for Africa" bill) that ostensibly opened U.S. markets to African exports but in fact imposed more stringent structural adjustment–type conditions while doing little to provide desperately needed debt relief. Many African American leaders, including a wide swath of black clergy, became involved in the Jubilee 2000 debt relief campaign and the fight against the "NAFTA for Africa" bill and for an alternative proposed by Rep. Jesse Jackson, Jr.

- Groups in Europe, Japan, and the U.S. that had been involved in support for development and popular movements in third world countries found those countries increasingly used as production platforms by global corporations. They began calling attention to the growth of sweatshops and pressuring companies like the Gap and Nike to establish acceptable labor and human rights conditions in their factories around the world. Their efforts gradually grew into an anti-sweatshop movement with strong labor and religious support and tens of thousands of active participants. In the U.S., college students took up the anti-sweatshop cause on hundreds of campuses, ultimately holding sit-ins on many campuses to force their colleges to ban the use of college logos on products not produced under acceptable labor conditions.

Many other people are following their own itineraries toward globalization from below. Some, such as activists in the human rights movement seeking to protect rights of people globally, or public health advocates trying to control tobacco companies and provide AIDS treatment for poor countries, are just as globalized as those described above. Some, such as activists in the immigrant networks spreading out around the world, are in some ways even more global and are challenging globalization from above by their very way of life. Some, like the tens of millions who have participated in nationally organized mass and general strikes and upheavals, are resisting the effects of globalization from above, even if (so far) they are doing so in a national framework.[6] Far more numerous still are the billions of people who are being adversely affected by globalization from above, but who have not yet found their own way to respond. Ultimately, their itineraries may be the most important of all.

Confluence

From diverse origins and through varied itineraries, these movements now find themselves starting to converge. Many of their participants are recognizing their commonalties and beginning to envision themselves as constructing a common movement.

This convergence is occurring because globalization is creating common interests that transcend both national and interest-group boundaries. As author and activist Vandana Shiva wrote in the wake of the Battle of Seattle,

> When labour joins hands with environmentalists, when farmers from the North and farmers from the South make a common commitment to say "no" to genetically engineered crops, they are not acting as special interests. They are defending the common interests and common rights of all people, everywhere. The divide and rule policy, which has attempted to pit consumers against farmers, the North against the South, labour against environmentalist has failed.[7]

Much of the convergence is negative: [D]ifferent groups find themselves facing the same global corporations, international institutions, and market-driven race to the bottom. But there is also a growing positive convergence around common values of democracy, environmental protection, community, economic justice, equality, and human solidarity.

Participants in this convergence have varied goals, but its unifying goal is to bring about sufficient democratic control over states, markets, and corporations to permit people and the planet to survive and begin to shape a viable future. This is a necessary condition for participants' diverse other goals.

Is this confluence a movement, or is it just a collection of separate movements? Perhaps it can most aptly be described as a movement in the early stages of construction. Within each of its components there are some people who see themselves as part of a global, multi-issue movement and others who do not. Those who do are often networked with their counterparts in other movements and other countries. Their numbers are increasing rapidly and they are playing a growing role within their movements and organizations. They are developing a shared vision. And they see themselves as constructing a common movement. It is this emerging movement that we refer to as globalization from below.

Globalization from below is certainly a movement with contradictions. Its participants have many conflicting interests. It includes many groups that previously defined themselves in part via negative reference to each other. It includes both rigidly institutionalized and wildly unstructured elements.

Globalization from below is developing in ways that help it cope with this diversity. It has embraced diversity as one of its central values, and asserts that cooperation need not presuppose uniformity. Its structure tends to be a network of networks, facilitating cooperation without demanding organizational centralization.

Older orientations toward charitable "us helping them" on the one hand, and narrow self-interest on the other, are still present; but there is also a new recognition of common interests in the face of globalization. Solidarity based on mutuality and common interest increasingly forms the basis for the relationships among different parts of the movement.

The movement is generally multi-issue, and even when participants focus on particular issues, they reflect a broader perspective. As Howard Zinn wrote of the Seattle WTO protests,

> In one crucial way it was a turning point in the history of movements of the recent decades—a departure from the single-issue focus of the Seabrook occupation of 1977, the nuclear-freeze gathering in Central Park in 1982, the great Washington events of the Million-Man March, [and] the Stand for Children [march].[8]

Globalization from below has now established itself as a global opposition, representing the interests of people and the environment worldwide. It has demonstrated that, even when governments around the world are dominated by corporate interests, the world's people can act to pursue their common interests.

Globalization from below grew both out of previous movements and out of their breakdown. There is much to be learned from the historical heritage of centuries of struggle to restrain or replace capitalism, and today's activists often draw on past values and practices in shaping their own. But it would be a mistake to simply treat this new movement as an extension of those that went before—or to attach it to their remnants.[9]

Globalization in all its facets presents new problems that the old movements failed to address. That is part of why they declined so radically. It also presents new opportunities that will be lost if the new wine is simply poured back into the old bottles. Besides, the historic break provides an invaluable opportunity to escape the dead hand of the past and to reground the movement to restrain global capital in the actual needs and conditions of people today.[10]

Globalization from below is now a permanent feature of the globalization epoch. Even if its current expressions were to fail, the movement would rise again, because it is rooted in a deep social reality: the need to control the forces of global capital.

★ ★ ★

THE POWER OF SOCIAL MOVEMENTS
(AND ITS SECRET)

The supporters of globalization from above control most of the world's governments. They control the global corporations and most of the world's wealth. They have a grip on the minds of people all over the world. It seems inconceivable that they can be effectively challenged.

Yet social movements have overcome equal or even greater concentrations of wealth and power in the past. Colonized peoples from North America to India, and Africa to Vietnam, have thrown out imperial powers with many times their wealth and firepower. The abolitionist movement eliminated slavery in most of the world and the civil rights movement eliminated legal segregation in the United States. In recent decades, mass movements have brought down powerful dictatorships from Poland to the Philippines. A coordinated domestic and global movement abolished South African apartheid. To understand how social movements are able to overcome what seem to be overwhelming forces, we need to take a deeper look at the processes underlying such successes.

How Social Movements Arise

Normally, most people follow life strategies based on adapting to the power relations of their world, not on trying to change them. They do so for a varying mix of reasons, including:

- Belief that existing relations are good and right.

- Belief that changing them is impossible.

- Fear that changing them would lead to something worse.

- An ability to meet their own needs and aspirations within existing power relations.

- Belief that existing power relations can and will change for the better.

- Identification with the dominant groups or with a larger whole—for example, a religion or nation.

- Fear of sanctions for violation of social rules or the will of the powerful.[11]

Most institutions and societies have elaborate systems for assuring sufficient consent or acquiescence to allow their key institutions to function. These means of maintaining a preponderance of power—often referred to as "hegemony"—range from education to media, and from elections to violent repression.[12]

Over time, problems with existing social relationships may accumulate, initiating a process of change. These problems usually affect particular social groups —for example, particular communities, nations, classes, racial, ethnic and gender groups, religious and political groupings, and the like. The process may start with some people internally questioning or rejecting some aspects of the status quo. It becomes a social process as people discover that others are having similar experiences, identifying the same problems, asking the same questions, and being tempted to make the same rejections. Then people begin to identify with those others and to interact with them. This turns what might have been an individual and isolating process into a social one.[13]

Seeing that other people share similar experiences, perceptions, and feelings opens a new set of possibilities. Perhaps collectively we can act in ways that have impacts isolated individuals could never dream of having alone. And if we feel this way, perhaps others do, too.

This group formation process constructs new solidarities. Once a conscious-ness of the need for solidarity develops, it becomes impossible to say whether participants' motives are altruistic or selfish, because the interest of the individual and the collective interest are no longer in conflict; they are perceived as one.[14]

This process occurs not only in individuals, but also in groups, organizations, and constituencies.

Thus form social movements.[15]

Why Social Movements Can Be Powerful

The fact that people develop common aspirations doesn't mean that they can realize them. Why are social movements able to change society? The power of existing social relations is based on the active cooperation of some people and the consent and/or acquiescence of others. It is the activity of people—going to work, paying taxes, buying products, obeying government officials, staying off private property—that continually re-creates the power of the powerful.

Bertolt Brecht dramatized this truth in his poem "German War Primer":

General, your tank is a strong vehicle.
It breaks down a forest and crushes a hundred people.
But it has one fault: it needs a driver.[16]

This dependence gives people a potential power over society—but one that can be realized only if they are prepared to reverse their acquiescence.[17] The old American labor song "Solidarity Forever" captures the tie between the rejection of acquiescence and the development of collective power:

They have taken untold millions
that they never toiled to earn
But without our brain and muscle
not a single wheel can turn.
We can break their haughty power,
gain our freedom when we learn
That the union makes us strong.[18]

Social movements can be understood as the collective withdrawal of consent to established institutions.[19] The movement against globalization from above can be understood as the withdrawal of consent from such globalization.

Ideally, democracy provides institutionalized means for all to participate equally in shaping social outcomes. But in the rather common situation in which most people have little effective power over established institutions, even those that claim to be democratic, people can still exercise power through the with-drawal of consent. Indeed, it is a central means through which democratization can be imposed.

Withdrawal of consent can take many forms, such as strikes, boycotts, and civil disobedience. Gene Sharp's *The Methods of Nonviolent Action* lists no fewer than 198 such methods, and no doubt a few have been invented since it was written.[20] Specific social relations create particular forms of consent and its withdrawal. For example, WTO trade rules prohibit city and state selective purchasing laws like the Massachusetts ban on purchases from companies that invest in Burma—making such laws a form of withdrawal of consent from the WTO, in effect an act of governmental civil disobedience.[21] (Several foreign governments threatened to bring charges against the Massachusetts Burma law in the WTO before it was declared unconstitutional by the U.S. Supreme Court in June 2000.)

The World Bank depends on raising funds in the bond market, so critics of the World Bank have organized a campaign against purchase of World Bank bonds, modeled on the successful campaign against investment in apartheid South Africa. Concerted refusal of impoverished debtor countries to continue paying on their debts—for example, through a so-called debtors' cartels—would constitute a powerful form of withdrawal of consent from today's global debt bondage.

Just the threat of withdrawal of consent can be an exercise of power. Ruling groups can be forced to make concessions if the alternative is the undermining of their ultimate power sources.[22] The movement for globalization from below has demonstrated that power repeatedly. For example, the World Bank ended funding for India's Narmada Dam when 900 organizations in 37 countries pledged a campaign to defund the Bank unless it canceled its support. And Monsanto found that global concern about genetically engineered organisms so threatened its interests that it agreed to accept the Cartagena Protocol to the Convention on Biological Diversity, allowing GEOs to be regulated.[23]

At any given time, there is a balance of power among social actors.[24] Except in extreme situations like slavery or military occupation, unequal power is reflected not in an unlimited power of one actor over the other. Rather, it is embedded in the set of rules and practices that are mutually accepted, even though they benefit one far more than the other. When the balance of power is changed, subordinate groups can force change in these rules and practices.

The power of the people is a secret that is repeatedly forgotten, to be rediscovered every time a new social movement arises. The ultimate source of power is not the command of those at the top, but the acquiescence of those at the bottom. This reality is hidden behind the machinations of politicians, business leaders, and politics as usual. The latent power of the people is forgotten both because those in power have every reason to suppress its knowledge and because it seems to conflict with everyday experience in normal times. But when the people rediscover it, power structures tremble.

Linking the Nooks and Crannies

New movements often first appear in small, scattered pockets among those who are unprotected, discriminated against, or less subject to the mechanisms of hegemony. They reflect the specific experiences and traditions of the social groups among which they arise. In periods of rapid social change, such movements are likely to develop in many such milieus and to appear very different from each other as a result. In the case of globalization from below, for example, we have seen significant mobilizations by French chefs concerned about preservation of local food traditions, Indian farmers concerned about corporate control of seeds, and American university students concerned about school clothing made in foreign sweatshops. Even if in theory people ultimately have power through withdrawal of consent, how can such disparate groups ever form a force that can exercise that power?

One common model for social change is the formation of a political party that aims to take over the state, whether by reform or by revolution. This model has always been problematic, since it implied the perpetuation of centralized social control, albeit control exercised in the interest of a different group.[25] However, it faces further difficulties in the era of globalization.

Reform and revolution depend on solving problems by means of state power, however acquired. But globalization has outflanked governments at local and national levels, leaving them largely at the mercy of global markets, corporations, and institutions. Dozens of parties in every part of the world have come to power with pledges to overcome the negative effects of globalization, only to submit in a matter of months to the doctrines of neoliberalism and the "discipline of the market." Nor is there a global state to be taken over.[26]

Fortunately, taking state power is far from the only or even the most important means of large-scale social change. An alternative pathway is examined by historical sociologist Michael Mann in *The Sources of Social Power*.[27] The characteristic way that new solutions to social problems emerge, Mann maintains, is neither through revolution nor reform. Rather, new solutions develop in what he calls "interstitial locations"—nooks and crannies in and around the dominant institutions. Those who were initially marginal then link together in ways that allow them to outflank those institutions and force a reorganization of the status quo.

At certain points, people see existing power institutions as blocking goals that could be attained by cooperation that transcends existing institutions. So people develop new networks that outrun them. Such movements create subversive "invisible connections" across state boundaries and the established channels between them.[28] These interstitial networks translate human goals into organizational means.

If such networks link groups with disparate traditions and experiences, they require the construction of what are variously referred to as shared worldviews, paradigms, visions, frames, or ideologies. Such belief systems unite seemingly disparate human beings by claiming that they have meaningful common properties:

> An ideology will emerge as a powerful, autonomous movement when it can put together in a single explanation and organization a number of aspects of existence that have hitherto been marginal, interstitial to the dominant institutions of power.[29]

The emerging belief system becomes a guide for efforts to transform the world. It defines common values and norms, providing the basis for a common program.[30] When a network draws together people and practices from many formerly marginal social spaces and makes it possible for them to act together, it establishes an independent source of power. Ultimately, new power networks may become strong enough to reorganize the dominant institutional configuration.

The rise of labor and socialist movements in the 19th century and of feminist and environmental movements in the 20th century in many ways fits this model of emergence at the margins, linking, and outflanking.[31] So, ironically, does the emergence of globalization from above. . . .

Self-organization in marginal locations and changing the rules of dominant institutions are intimately linked. The rising European bourgeoisie both created their own market institutions and fought to restructure the political system in ways that would allow markets to develop more freely. Labor movements both organized unions and forced governments to protect labor rights, which in turn made it easier to organize unions.

Over time, movements are likely to receive at least partial support from two other sources. Some institutions, often ones that represent similar constituencies and that themselves originated in earlier social movements but have become rigidified, develop a role of at least ambiguous support. And sectors of the dominant elites support reforms and encourage social movements for a variety of reasons, including the need to gain support for system-reforming initiatives and a desire to win popular backing in intra-elite conflicts.

Social movements may lack the obvious paraphernalia of power: armies, wealth, palaces, temples, and bureaucracies. But by linking from the nooks and crannies, developing a common vision and program, and withdrawing their consent from existing institutions, they can impose norms on states, classes, armies, and other power actors.

The Lilliput Strategy

How do these broad principles of social movement-based change apply to globalization from below? In fact, they describe the very means by which it is being constructed. We call this the Lilliput Strategy, after the tiny Lilliputians in Jonathan Swift's fable *Gulliver's Travels* who captured Gulliver, many times their size, by tying him up with hundreds of threads.

In response to globalization from above, movements are emerging all over the world in social locations that are marginal to the dominant power centers. These are linking up by means of networks that cut across national borders. They are beginning to develop a sense of solidarity, a common belief system, and a common program. They are utilizing these networks to impose new norms on corporations, governments, and international institutions.

The movement for globalization from below is, in fact, becoming an independent power. It was able, for example, to halt negotiations for the Multilateral Agreement on Investment (MAI), to block the proposed "Millennium Round" of the WTO, and to force the adoption of a treaty on genetically engineered products. Its basic strategy is to say to power holders, "Unless you accede to operating within these norms, you will face threats (from us and from others) that will block your objectives and undermine your power."

The threat to established institutions may be specific and targeted withdrawals of support. For example, student anti-sweatshop protestors have made clear that their campuses will be subject to sit-ins and other forms of disruption until their universities agree to ban the use of school logos on products made in sweatshops. Or, to take a very different example, in the midst of the Battle of Seattle, President Bill Clinton, fearing loss of electoral support from the labor movement, endorsed the use of sanctions to enforce international labor rights.[32] The threat may, alternatively, be a more general social breakdown, often expressed as fear of "social unrest."[33]

The slogan "fix it or nix it," which the movement has often applied to the WTO, IMF, and World Bank, embodies such a threat. It implies that the movement (and the people of the world) will block the globalization process unless power holders conform to appropriate global norms. This process constitutes neither revolution nor conventional "within the system" and "by the rules" reform. Rather, it constitutes a shift in the balance of power.

As the movement grows in power, it can force the modification of institutions or the creation of new ones that embody and/or impose these norms as enforceable rules.[34] For example, the treaties on climate change and on genetic engineering force new practices on corporations, governments, and international institutions that implement norms propounded by the environmental and consumer movements. Student anti-sweatshop activists force their universities to join an organization that bans university logos on products made under conditions that violate specified rules regarding labor conditions. The world criminal court, endorsed by many countries under pressure of the global human rights movement, but resisted by the United States, would enforce norms articulated at the Nuremberg war crimes tribunal.

These new rules in turn create growing space for people to address problems that the previous power configuration made insoluble. Global protection of human rights makes it easier for people to organize locally to address social and environmental problems. Global restrictions on fossil fuels that cause global warming, such as a carbon tax, would make it easier for people to develop renewable energy sources locally.

While the media have focused on global extravaganzas like the Battle of Seattle, these are only the tip of the globalization from below iceberg. The Lilliput Strategy primarily involves the building of solidarity among people at the grassroots. For example:

- Under heavy pressure from the World Bank, the Bolivian government sold off the public water system of its third largest city, Cochabamba, to a subsidiary of the San Francisco—based Bechtel Corporation, which promptly doubled the price of water for people's homes. Early in 2000, the people of Cochabamba rebelled, shutting down the city with general strikes and blockades. The government declared a state of siege and a young protester was shot and killed. Word spread all over the world from the remote

Bolivian highlands via the Internet. Hundreds of e-mail messages poured into Bechtel from all over the world demanding that it leave Cochabamba. In the midst of local and global protests, the Bolivian government, which had said that Bechtel must not leave, suddenly reversed itself and signed an accord accepting every demand of the protestors. Meanwhile, a local protest leader was smuggled out of hiding to Washington, DC, where he addressed the April 16 rally against the IMF and World Bank.[35]

- When the Japanese-owned Bridgestone/Firestone (B/F) demanded 12-hour shifts and a 30 percent wage cut for new workers in its American factories, workers struck. B/F fired them all and replaced them with 2,300 strikebreakers. American workers appealed to Bridgestone/Firestone workers around the world for help. Unions around the world organized "Days of Outrage" protests against B/F. In Argentina, a two-hour "general assembly" of all workers at the gates of the B/F plant halted production while 2,000 workers heard American B/F workers describe the company's conduct. In Brazil, Bridgestone workers staged one-hour work stoppages, then "worked like turtles"—the Brazilian phrase for a slowdown. Unions in Belgium, France, Italy, and Spain met with local Bridgestone managements to demand a settlement. U.S. B/F workers went to Japan and met with Japanese unions, many of whom called for the immediate reinstatement of U.S. workers. Five hundred Japanese unionists marched through the streets of Tokyo, supporting B/F workers from the U.S. In the wake of the worldwide campaign, Bridgestone/Firestone unexpectedly agreed to rehire its locked out American workers.[36]

- In April 2000, AIDS activists, unions, and religious groups were poised to begin a lawsuit and picketing campaign denouncing the Pfizer Corporation as an AIDS profiteer for the high price it charges for AIDS drugs in Africa. Pfizer suddenly announced that it would supply the drug fluconazole, used to control AIDS side effects, for free to any South African with AIDS who could not afford it. A few weeks later, U.S., British, Swiss, and German drug companies announced that they would cut prices on the principal AIDS drugs, anti-retrovirals, by 85 to 90 percent. Meanwhile, when South Africa tried to pass a law allowing it to ignore drug patents in health emergencies, the Clinton administration lobbied hard against it and put South Africa on a watch list that is the first step toward trade sanctions. But then, according to the *New York Times,* the Philadelphia branch of Act Up, the gay advocacy group, decided

> to take up South Africa's cause and start heckling Vice President Al Gore, who was in the midst of his primary campaign for the presidency. The banners saying that Mr. Gore was letting Africans die to please American pharmaceutical companies left his campaign chagrined. After media and campaign staff looked into the matter, the administration did an about face and accepted African governments' circumvention of AIDS drug patents.[37]

- Two independent unions, the United Electrical Workers Union (UE) in the United States and the Frente Autentico del Trabajo (FAT) in Mexico, formed an ongoing Strategic Organizing Alliance in the mid-1990s. At General Electric (GE) in Juarez, FAT obtained the first secret ballot election in Mexican labor history, aided by pressure on GE in the United States. The trinational Echlin Workers Alliance was formed to jointly organize Echlin, a large multinational auto parts corporation, in Canada, Mexico, and the U.S. In cooperation with Mexican unions, U.S. unions brought charges under the NAFTA side agreements for the repression of Echlin workers. A rank-and-file activist from FAT traveled to Milwaukee, Wisconsin, to help UE organize foundry workers of Mexican origin. Workers from each country have repeatedly conducted speaking tours organized by those across the border. U.S. workers helped fund and build a Workers' Center in Juarez. And the Cross-Border Mural Project has developed binational teams that have painted murals celebrating international labor solidarity on both sides of the border.[38]

How Movements Go Wrong

It is nowhere guaranteed that any particular social movement will succeed in using its potential power to realize the hopes and aspirations of its participants or to solve the problems that moved them to action in the first place. There are plenty of pitfalls along the way.

Schism: From Catholic and Protestant Christians to Sunni and Shiite Muslims, from Communists and socialists to separatists and integrationists, social movements are notorious for their tendency to split. They can often turn into warring factions whose antagonisms are focused primarily on each other. Splits often occur over concrete issues but then perpetuate themselves even when the original issues are no longer salient.

Repression: Movements can be eliminated, or at least driven underground, by legal and extralegal repression.

Fading out: The concerns that originally drew people into a movement may recede due to changed conditions. An economic upswing or the opening of new lands has often quieted farmer movements. Or constant frustration may simply lead to discouragement and withdrawal.

Leadership domination: In a mild form, the movement evolves into an institution in which initiative and control pass to a bureaucratized leadership and staff, while the members dutifully pay their dues and act only when told to do so by their leaders. In a more virulent form, leaders establish a tyrannical control over members.[39]

Isolation: Movements may become so focused on their own internal life that they are increasingly irrelevant to the experience and concerns of those who are not already members. Such a movement may last a long time as a sect but be largely irrelevant to anyone except its own members.

Cooptation: A movement may gain substantial benefits for its constituency, its members, or its leaders, but do so in such a way that it ceases to be an independent force and instead comes under the control of sections of the elite.

Leadership sell-out: Less subtly, leaders can simply be bought with money, perks, flattery, opportunities for career advancement, or other enticements.

Sectarian disruption: Movements often fall prey to sects that attempt either to capture or to destroy them. Such sects may emerge from within the movement itself or may invade it from without.

★ ★ ★

To succeed, globalization from below must avoid these pitfalls; promote movement formation in diverse social locations; establish effective linkages; develop a sense of solidarity, a common worldview, and a shared program; and utilize the power that lies hidden in the withdrawal of consent.

ENDNOTES

1.   "Das Kapital Revisited," *Financial Times*, August 31, 1998, p. 14.

2.   As far as we have been able to determine, the terms "globalization from above" and "globalization from below" were coined by Richard Falk and first appeared in print in Jeremy Brecher, John Brown Childs, and Jill Cutler eds., Global *Visions: Beyond the New World Order* (Boston: South End Press, 1993). This book provides a view of the development of transnational social movements and common vision prior to 1993; it reflects an awareness of the increasing global interconnectedness in the post–Cold War era, but it puts limited emphasis on the global economic integration that was then gathering steam. For movements in the early 1990s responding specifically to economic globalization, see Chapter 5 of *Global Village or Global Pillage*. See also Richard Falk, *Predatory Globalization: A Critique* (Massachusetts Blackwell: Maiden, 1999), especially Chapter 8.

3.   Jeremy Brecher, "The Opening Shot of the Second Ecological Revolution," *Chicago Tribune*, August 16, 1988.

4.   "Poor Countries Draft Proposal On Poverty," *New York Times*, April 12, 2000. The G-77 currently has 133 member nations.

5.   Mark Ritchie, quoted in *Global Village or Global Pillage*, p. 97.

6.   According to labor journalist Kim Moody,

> In the last couple of years there have been at least two dozen political general strikes in Europe, Latin America, Asia, and North America. This phenomenon began in 1994. There have been more political mass strikes in the last two or three years than at any time in the 20th century.

Kim Moody, "Workers in a Lean World," a speech to the Brecht Forum in New York, New York, November 14, 1997. Broadcast on Alternative Radio (tape and transcript available from http://www.alternativeradio.org).

7. Vandana Shiva, "The Historic Significance of Seattle," December 10, 1999, MAI-NOT Listserve, Public Citizen Global Trade Watch.

8. Howard Zinn, "A Flash of the Possible." *The Progressive* 61:1 (January 2000). Available on-line at https://secure.progressive.org/zinn001.htm.

9. For a portrayal of current struggles as a continuation of historical working class struggles, see Boris Kargarlitsky's recent trilogy *Recasting Marxism*, including *New Realism, New Barbarism: Socialist Theory in the Era of Globalization* (London: Pluto Press, 1999), *The Twilight of Globalization: Property, State and Capitalism* (London: Pluto Press, 1999), and *The Return of Radicalism: Reshaping the Left Institutions* (London: Pluto Press, 2000).

10. It is often pointed out that globalization is creating a capitalism that in significant ways resembles the capitalism that preceded World War I. It could also be observed that globalization from below in some ways resembles the international socialist movement before World War I. Globalization provides an opportunity to reevaluate some of the key features of the post-1914 left, such as its relationship to nationalism and the nation state; the schisms between social democracy, Communism, and anarchism; and the development of organizational forms adapted to the effort to secure state power via reform or revolution.

11. For a fuller discussion of this subject, with extensive references, see Gene Sharp, *The Politics of Nonviolent Action: Part One: Power and Struggle* (Boston: Porter, Sargent, 1973), "Why Do Men Obey?" pp. 16–24.

12. The analysis of "hegemony" is generally associated with the work of Antonio Gramsci. See for example Antonio Gramsci, *The Modern Prince and Other Writings* (New York: International Publishers, 1959).

13. E. P. Thompson describes this process of group formation for the specific case of class: "Class happens when some men, as a result of common experiences (inherited or shared), feel and articulate the identity of their interests as between themselves, and as against other men whose interests are different from (and usually opposed to) theirs." E. P. Thompson, *The Making of the English Working Class* (New York: Vintage, 1996), p. 9.

14. Solidarity can take a number of forms. Peter Waterman defines six meanings of international solidarity:

- Identity = solidarity of common interest and identity
- Substitution = standing in for those incapable of standing up for themselves
- Complementarity = exchange of different needed/desired goods/qualities
- Reciprocity = exchange over time of identical goods/qualities
- Affinity = shared cross-border values, feelings, ideas, identities
- Restitution = acceptance of responsibility for historical wrong

He points out that each of these has certain problems and limitations. For example, identity-based solidarity tends to exclude those who don't share the common identity as defined; substitution can lead to an unequal, patronizing relationship of charity. See Peter Waterman, *Globalization, Social Movements, and the New Internationalisms* (London:

Mansell, 1999). Preliminary text available on-line at http://www.antenna.nl/~water-man/dialogue.html. The process of constructing solidarity is illustrated with numerous labor history examples in Jeremy Breche, *Strike! Revised and Updated Edition* (Cambridge South: End Press Classics, 1999) and analyzed on p. 284.

15.   This highly schematic formulation is based primarily on the study and observation of social movements, combined with theories drawn from many sources, for example, Jean-Paul Sartre, *Critique de la Raison Dialectique [Critique of Dialectial Reason]* (Paris: Gallimard, 1960), and Francesco Alberoni, *Movement and Institution* (New York: Columbia UP, 1984).

16.   Bertolt Brecht, *Deutsche Kriegsfibel* ["German War Primer"], in *Gesammilte Werke* (Berlin: Suhrkamp, 1967), vol. 4, p. 638. The translation by Martin Esslin originally appeared in Jeremy Brecher and Tim Costello, *Common Sense for Hard Times* (New York: Two Continents/Institute for Policy Studies, 1976), p. 240.

17.   In the "acquiescent state," people's relation to each other is mediated via the market or common relations to authority. The process of movement creation and group formation to some degree replaces these with direct relations. Sartre analyzes this as the transition from the "series" to the "group" *(Critique de la Raison Dialectique)*.

18.   Written by Ralph Chaplin.

19.   Gene Sharp, who analyzes hundreds of historical examples of nonviolent action in the three volumes of his *Politics of Nonviolent Action* (Boston: Porter Sargent, 1973), con-cludes that the base of nonviolent action is "the belief that the exercise of power de-pends on the consent of the ruled who, by withdrawing that consent, can control and even destroy the power of their opponent" (*Part One; Power and Struggle*, p. 4). Sharp emphasizes that nonviolent struggle requires indirect strategies that undermine the op-ponents' strength rather than annihilate the opponent. (Of course, the picture is made less simple by the fact that "the ruled" are not a homogeneous group, and those who withdraw consent may be defeated by those who do not.) This analysis does not apply exclusively to nonviolence. Even in war, victory usually results not from physical anni-hilation of the enemy but from the withdrawal of support of the population from the war effort ("loss of morale"), defection of political supporters of the war, withdrawal of allies, and change in policy by ruling groups in response to the presence or threat of these factors.

20.   Sharp, *Part Two: The Methods of Nonviolent Action* (Boston: Porter Sargent, 1973). See also *Part Three: The Dynamics of Nonviolent Action* (Boston: Porter Sargent, 1973).

21.   In constitutional terms, this would be described as a form of nullification.

22.   Of course, an irrational ruler may not be deterred from acting to repress a nonviolent movement by the fact that doing so may undermine his or her own power. But given an irrational ruler, violence is no more guaranteed to be an effective deterrent than nonviolence.

23.   "United States negotiators gave in to a demand from Europe and most of the rest of the world for what is known as the 'precautionary principle'. . . Even Greenpeace, an avowed critic of the technology, issued a statement calling the protocol a 'historic step towards protecting the environment and consumers from the dangers of genetic engineering.'" *St. Louis Post-Dispatch*, January 30, 2000.

The British Environment Minister, Michael Meacher, said: "For the first time countries will have the right to decide whether they want to import GM products or not when there is less than full scientific evidence. It is official that the environment rules aren't subordinate to the trade rules. It's been one hell of a battle."

"This protocol is a campaign victory in that it acknowledges that GMOs [genetically modified organisms] are not the same as other crops and products and they require that special measures be taken," said Miriam Mayer of the Malaysia-based Third World Network. . . . The U.S. State Department declined to specify whether the biotechnology company Monsanto had been consulted over the past few days. A State Department source said: "We understand there is no major problem so far as the company is concerned." *The Observer*, January 30, 2000.

24. As Gramsci put it, "The fact of hegemony undoubtedly presupposes that the interests and strivings of the groups over which the hegemony will be exercised are taken account of, that a certain balance of compromises be formed, that, in other words, the leading group makes some sacrifices" (*Modern Prince*, p. 154).

25. This critique has long been elaborated in the anarchist and libertarian socialist traditions, and has more recently been developed by the New Left of the 1960s, the Green movement, and the Mexican Zapatistas.

26. . . . This is not to argue that states are no longer of significance, or that political parties and contests for government power have not played an important role in the past and might not today or in the future. Rather, it is to deny that social movements can or should be reduced to such a strategy.

27. Michael Mann, *The Sources of Social Power: Volume 1* (Cambridge: Cambridge UP, 1986), Chapter 1, "Societies as Organized Power Networks."

28. Michael Mann, *Sources of Social Power*, p. 522.

29. Michael Mann, *Sources of Social Power*, p. 21.

30. We use the term *values* to refer to criteria for classifications of good and bad or of better and worse. We use the term *norms* for the application of such values to the behavior of particular classes of actors, thereby specifying how they should act.

31. Over time, the labor and socialist movements of course became increasingly focused on national governments and increasingly contained within national frameworks.

32. This threat, strongly resented by many third world governments, contributed to the deadlocking of the WTO negotiations.

33. For example, provoking such general social unrest was an articulated objective of many U.S. opponents of the Vietnam War after other means of halting it had failed and public opinion had swung against it without visible effect on policy.

34. For a similar perspective on how social movements make change through imposing norms, with recent examples and proposals for the future, see Richard Falk, "Humane Governance for the World: Reviving the Quest," in Pieterse ed, *Global Futures*, pp. 23ff. See also *On Humane Governance, Toward a New Global Politics* (University Park, Penn: Pennsylvania State UP, 1995).

35. "Bolivian Water Plan Dropped After Protests Turn Into Melees," *New York Times*, April 11, 2000. For further information on the Cochabamba water struggle, prepared by Jim Schultz, a Cochabamba resident who played a major role in mobilizing global support for the struggle, visit http://www.americas.org.

36. *ICEM Info 3* (1996) and *ICEM Info 4* (1996); see also *Labor Notes*, October 1994, July 1996, and December 1996.

37. Donald G. McNeil, Jr., "As Devastating Epidemics Increase, Nations Take on Drug Companies," *New York Times*, July 9, 2000, and *Toronto Star*, May 12, 2000.

38. For information on the FAT-UE alliance, visit the UE website at http://www.ranknfile-ue.org/international.html.

39.   Two classic explorations of this dynamic are Robert Michels, *Political Parties: A Sociological Study of the Oligarchical Tendencies of Modern Democracy* (Glencoe: Free Press, 1949), and Sidney Webb and Beatrice Webb, *The History of Trade Unionism*, 2nd ed. (London: Longmans Green, 1902). This process is analyzed by Sartre in terms of the dissolution of a group back into a series. As Alberoni points out, some degree of such re-serialization is probably inevitable, but it can be limited by practices that provide for periodical reconstitution of the group. Those hostile to social movements sometimes maintain that tyranny is their normal or only possible outcome. A classic example is Norman Cohn, *The Pursuit of the Millennium* (London: Seeker and Warburg, 1957). For a discussion of these issues in the context of various left traditions, see Staughton Lynd, "The Webbs, Lenin, Rosa Luxemburg," in *Living Inside Our Hope: A Steadfast Radical's Thoughts on Rebuilding the Movement* (Ithaca: Cornell UP, 1997), pp. 206ff.

38

Globalization and Racialization

MANNING MARABLE

In 1900, the great African-American scholar W. E. B. Du Bois predicted that the "problem of the twentieth century" would be the "problem of the color line," the unequal relationship between the lighter vs. darker races of humankind. Although Du Bois was primarily focused on the racial contradiction of the United States, he was fully aware that the processes of what we call "racialization" today—the construction of racially unequal social hierarchies characterized by dominant and subordinate social relations between groups—was an international and global problem. Du Bois's color line included not just the racially segregated, Jim Crow South and the racial oppression of South Africa; but also included British, French, Belgian, and Portuguese colonial domination in Asia, the Middle East, Africa, Latin America, and the Caribbean among indigenous populations.

Building on Du Bois's insights, we can therefore say that the problem of the twenty-first century is the problem of global apartheid: the racialized division and stratification of resources, wealth, and power that separates Europe, North America, and Japan from the billions of mostly black, brown, indigenous, undocumented immigrant and poor people across the planet. The term apartheid, as most of you know, comes from the former white minority regime of South Africa. It is an Afrikaans word meaning "apartness" or "separation." Apartheid was based on the concept of "herrenvolk," a "master race," who was destined to rule non-Europeans. Under global apartheid today, the racist logic of herrenvolk, the master race, still exists, embedded in the patterns of unequal economic exchange that penalizes African, south Asian, Caribbean, and poor nations by predatory policies of structural adjustment and loan payments to multinational banks.

Inside the United States, the processes of global apartheid are best represented by what I call the New Racial Domain or the NRD. This New Racial Domain is different from other earlier forms of racial domination, such as slavery, Jim Crow segregation, and ghettoization, or strict residential segregation, in several critical respects. These earlier racial formations or domains were grounded or based primarily, if not exclusively, in the political economy of U.S. capitalism. Anti-racist or oppositional movements that blacks, other people of color and

SOURCE: Manning Marable, "Globalization and Racialization," *ZNet* (August 13, 2004). Online: http://www.zmag.org/content/print_article.cfm?itemID=6034&sectionID=1.

white anti-racists built were largely predicated upon the confines or realities of domestic markets and the policies of the U.S. nation-state. Meaningful social reforms such as the Civil Rights Act of 1964 and the Voting Rights Act of 1965 were debated almost entirely within the context of America's expanding, domestic economy, and a background of Keynesian, welfare state public policies.

The political economy of the "New Racial Domain," by contrast, is driven and largely determined by the forces of transnational capitalism, and the public policies of state neoliberalism. From the vantage point of the most oppressed U.S. populations, the New Racial Domain rests on an unholy trinity, or deadly triad, of structural barriers to a decent life. These oppressive structures are mass unemployment, mass incarceration, and mass disfranchisement. Each factor directly feeds and accelerates the others, creating an ever-widening circle of social disadvantage, poverty, and civil death, touching the lives of tens of millions of U.S. people.

The process begins at the point of production. For decades, U.S. corporations have been outsourcing millions of better-paying jobs outside the country. The class warfare against unions has led to a steep decline in the percentage of U.S. workers.

Within whole U.S. urban neighborhoods losing virtually their entire economic manufacturing and industrial employment, and with neoliberal social policies in place cutting job training programs, welfare, and public housing, millions of Americans now exist in conditions that exceed the devastation of the Great Depression of the 1930s. In 2004, in New York's Central Harlem community, 50 percent of all black male adults were currently unemployed. When one considers that this figure does not count those black males who are in the military, or inside prisons, its truly amazing and depressing.

This July, labor researchers at Harvard University found that one-quarter (25 percent) of the nation's entire population of black male adults were jobless for the entire year during 2002. What these nightmarish statistics mean, is that for most low-to middle-income African Americans, joblessness and underemployment (e.g., working part-time, or sporadically) is now the norm; having a real job with benefits is now the exception. *Who belongs to unions, dropping from 30 percent in the 1960s down to barely 13 percent today*. With the onset of global capitalism, the new jobs being generated for the most part lack the health benefits, pensions, and wages that manufacturing and industrial employment once offered.

Neoliberal social policies, adopted and implemented by Democrats and Republicans alike, have compounded the problem. After the 1996 welfare act, the social safety net was largely pulled apart. As the Bush administration took power in 2001, chronic joblessness spread to African-American workers, especially in the manufacturing sector. By early 2004, in cities such as New York, fully one-half of all black male adults were outside of the paid labor force. As of January 2004, the number of families on public assistance had fallen to two million, down from five million families on welfare in 1995. New regulations and restrictions intimidate thousands of poor people from requesting public assistance.

Mass unemployment inevitably feeds mass incarceration. About one-third of all prisoners were unemployed at the time of their arrests, and others averaged less than $20,000 annual incomes in the year prior to their incarceration. When the Attica prison insurrection occurred in upstate New York in 1971, there were only 12,500 prisoners in New York State's correctional facilities, and about 300,000 prisoners nationwide. By 2001, New York State held over 71,000 women and men in its prisons; nationally, 2.1 million were imprisoned. Today about five to six million Americans are arrested annually, and roughly one in five Americans possess a criminal record.

Mandatory-minimum sentencing laws adopted in the 1980s and 1990s in many states stripped judges of their discretionary powers in sentencing, imposing draconian terms on first-time and non-violent offenders. Parole has been made more restrictive as well, and in 1995 Pell grant subsidies supporting educational programs for prisoners were ended. For those fortunate enough to successfully navigate the criminal justice bureaucracy and emerge from incarceration, they discover that both the federal law and state governments explicitly prohibit the employment of convicted ex-felons in hundreds of vocations. The cycle of unemployment frequently starts again.

The greatest victims of these racialized processes of unequal justice, of course, are African-American and Latino young people. In April 2000, utilizing national and state data compiled by the FBI, the Justice Department and six leading foundations issued a comprehensive study that documented vast racial disparities at every level of the juvenile justice process. African Americans under age eighteen constitute 15 percent of their national age group, yet they currently represent 26 percent of all those who are arrested. After entering the criminal-justice system, white and black juveniles with the same records are treated in radically different ways. According to the Justice Department's study, among white youth offenders, 66 percent are referred to juvenile courts, while only 31 percent of the African-American youth are taken there. Blacks make up 44 percent of those detained in juvenile jails, 46 percent of all those tried in adult criminal courts, as well as 58 percent of all juveniles who are warehoused in prisons.

Mass incarceration, of course, breeds mass political disfranchisement. Nearly 5 million Americans cannot vote. In seven states, former prisoners convicted of a felony lose their voting rights for life. In the majority of states, individuals on parole and probation cannot vote. About 15 percent of all African-American males nationally are either permanently or currently disfranchised. In Mississippi, one-third of all black men are unable to vote for the remainder of their lives. In Florida, 818,000 residents cannot vote for life.

Even temporary disfranchisement fosters a disruption of civic engagement and involvement in public affairs. This can lead to "civil death," the destruction of the capacity for collective agency and resistance. This process of depolitization undermines even grassroots, non-electoral-oriented organizing. The deadly triangle of the New Racial Domain constantly and continuously grows unchecked.

Not too far in the distance lies the social consequence of these policies: an unequal, two-tiered, uncivil society, characterized by a governing hierarchy of middle-to upper-class "citizens" who own nearly all private property and financial assets, and a vast subaltern of quasi- or subcitizens encumbered beneath the cruel weight of permanent unemployment, discriminatory courts and sentencing procedures, dehumanized prisons, voting disfranchisement, residential segregation, and the elimination of most public services for the poor. The later group is virtually excluded from any influence in a national public policy. Institutions that once provided space for upward mobility and resistance for working people such as unions have been largely dismantled. Integral to all of this is racism, sometimes openly vicious and unambiguous, but much more frequently presented in race neutral, color-blind language. This is the NRD of globalization.

The anti-globalization struggle must confront this New Racial Domain with something more substantial than tired ruminations about "black and white, unite and fight." The seismic shifts have created new continents of social inequality, transcending nation-states and the traditional boundaries of race and ethnicity. What is necessary is an original and creative approach that breaks with comfortable dogmas of all types, while advancing openly a politics of civic advocacy and democratic empowerment for those most brutally oppressed and exploited. I am *not* suggesting here that the anti-globalization movement play a "vanguard" role for global social change. In the tradition of C.L.R. James, I am convinced that the oppressed, on their own terms, ultimately will create new approaches and organizations to fight for justice that we now can scarcely imagine. Rather, it is our political and moral obligation to provide the critical support necessary for social struggles and resistance that is already being waged on the ground today. Examples of that resistance are in every city and most communities across the country.

The New Racial Domain's reliance on extreme force and the continued expansion of the prison system reshapes how law enforcement is being carried out even in small- to medium-sized towns and cities all over America. The terrible dynamic unleashed against prisoners of social control has expanded into the normal apparatuses and uses of policing itself. There are now, for example, approximately 600,000 police officers and 1.5 million private security guards in the United States. Increasingly, however, black and poor communities are being "policed" by special paramilitary units, often called SWAT (Special Weapons and Tactics) teams. The U.S. has more than 30,000 such heavily armed, military trained police units. SWAT-team mobilizations, or "call outs," increased 400 percent between 1980 and 1995. These trends reveal the makings of what may constitute a "National Security State"—the exercising of state power without democratic controls, checks and balances, a state where policing is employed to carry out the disfranchisement of its own citizens.

The trend toward a National Security State has been pushed actively by the Bush regime, which is aggressively pressuring universities to suppress dissent and to curtail traditional academic freedoms. In early March 2004, the U.S. Treasury Department's Office of Foreign Assets Control stopped 70 American scientists and physicians from traveling to Cuba to attend an international symposium on "coma and death." Some of the scholars received warning letters from the Treasury Department, promising severe criminal or civil penalties if they violated the embargo against Cuba. In late 2003, the Treasury Department issued a warning to U.S. publishers that they would have to obtain "special licenses to edit papers" written by scholars and scientific researchers currently living in Cuba, Libya, Iran, or Sudan. All violators, even including the editors and officers of professional associations sponsoring scholarly journals, potentially may be subjected to fines up to $500,000 and prison sentences up to ten years. After widespread criticism, the Treasury Department was forced to moderate its policy.

In February 2004, U.S. army officials visited the University of Texas at Austin, demanding the names of "Middle Eastern-looking" individuals attending an academic conference on the treatment of women under traditional Islamic law. Subsequently it was learned that two U.S. army attorneys working with the army's Intelligence and Security Commission had actually attended the conference without identifying themselves.

How do we build resistance to the New Racial Domain, in the age of globalized capitalism? It should surprise no one that the resistance is already occurring, on the ground, in thousands of venues. In local neighborhoods, people fighting against police brutality, mandatory-minimum sentencing laws, and for prisoners' rights; in the fight for a living wage, to expand unionization and workers' rights; in the struggles of working women for day care for their children, health care, public transportation, and decent housing. These practical struggles of daily life are really the care of what constitutes day-to-day resistance. Building capacities of hope and resistance on the ground develops our ability to challenge the system in more fundamental, direct ways.

The recently successful "Immigrant Worker Freedom Ride," highlighting the plight of undocumented workers who enter the U.S., represents an excellent model that links the oppressive situation of new immigrants with the historic struggles of the Civil Rights Movement forty-five years ago to overthrow Jim Crow. Many sincere, white anti-globalization activists need to learn more about the historic Black Freedom Movement, and the successful models of resistance— from selective buying campaigns or economic boycotts, to rent strikes, to civil disobedience—which that movement established. You are not inventing models of social justice activism and resistance: others have come before you. The task is to learn from the strengths and weaknesses of those models, incorporating their anti-racist vision into the heart of what we do to resist global capitalism and the nation-security state.

The anti-globalization movement must be, first and foremost, a worldwide, pluralistic anti-racist movement, with its absolutely central goal of destroying global apartheid and the reactionary residue of white supremacy and ethnic chauvinism. But to build such a dynamic movement, the social composition of the anti-globalization forces must change, especially here in the United States. The anti-globalization forces are still overwhelmingly upper, middle-class, college-educated elites, who may politically sympathize with the plight of the poor and oppressed, but who do not share their lives or experiences. In the Third World, the anti-globalization movement has been more successful in achieving a broader, more balanced social class composition, with millions of workers getting actively involved.

There are, however, two broad ideological tendencies within this largely non-European, anti-globalization movement: a liberal, democratic, and populist tendency, and a radical, egalitarian tendency. Both tendencies were present throughout the 2001 Durban Conference Against Racism, and made their presence felt in the deliberations of the non-governmental organization panels and in the final conference report. They reflect two very different political strategies and tactical approaches in the global struggle against the institutional processes of racialization.

The liberal democratic tendency focuses on a discourse of rights, calling for greater civic participation, political enfranchisement, capacity building of community-based institutions, for the purposes of civic empowerment and multicultural diversity. The liberal democratic impulse seeks the reduction of societal conflict through the sponsoring of public conversations, reconciliation and multicultural civic dialogues. It seeks not a complete rejection of neoliberal economic globalization, but its constructive reform and engagement, with the goal of building democratic political cultures of human rights within market-based societies.

The radical egalitarian tendency of global anti-racists speaks a discourse about inequality and power. It seeks the abolition of poverty, the realization of universal housing, health care and educational guarantees across the non-Western world. It is less concerned about abstract rights, and more concerned about concrete results. It seeks not political assimilation in an old world order, but the construction of a new world from the bottom up. It has spoken a political language more so in the tradition of national liberation than of the nation-state.

Both of these tendencies exist in the United States, as well as throughout the world, in varying degrees, [and] now define the ideological spectrum within the global anti-apartheid struggle. Scholars and activists alike must contribute to the construction of a broad front bringing together both the multicultural liberal democratic and radical egalitarian currents representing globalization from below. New innovations in social protest movements will also require the development of new social theory and new ways of thinking about the relationship between structural racism and state power. Global apartheid is the great political and moral challenge of our time. It can be destroyed, but only through a collective, transnational struggle.

39

Hunger on the Border

Interview with Julia Quiñones

DAVID BACON

Today the U.S./Mexico border is the subject of intense political controversy. Most of the fireworks focuses, however, on the idea that more enforcement can keep people from crossing it. Lost in this hysteria is the reality that the border is a huge place, where millions of people live and work. Not only that, but free trade policies hold down living standards and prevent union and community organizing. That, in turn, produces pressure on people to seek a better standard of living elsewhere. To explore the real conditions for border workers, I interviewed Julia Quiñones, coordinator of the Border Committee of Women Workers, the Comite Fronterizo de Obreras, with offices in Piedras Negras, Mexico.

BACON: *In Spanish, the name of the border committee uses the word "obreras," which means women workers. Why?*

QUIÑONES: The Comite Fronterizo de Obreras (CFO) is an organization of rank and file women, led by women and men who work in the maquiladoras. The organization was born out of the particular needs of young women who work in the factories. In the beginning the industry was especially interested in employing women and, even though this situation has changed over time, we continue to maintain a focus on their experiences. We look for a greater level of participation by women inside their unions and at all levels of leadership.

BACON: *What does the Comité do?*

QUIÑONES: The CFO is working in three Mexican states—Tamaulipas, Coahuila, and Chihuahua. Our purpose is to educate and organize workers around their labor rights. We try to engage workers in learning and talking about the impact of free trade and we focus on violence against women. We have a program to build economic self-sufficiency and we've created our own maquiladora, making products and giving employment to women.

SOURCE: David Bacon, "Hunger on the Border: Interview with Julia Quiñones," *Z Magazine* 19 (April 2006), pp. 17–20.

BACON:	*What are the effects of free trade and the North American Free Trade Agreement (NAFTA) in your section of the border?*
QUIÑONES:	Maquiladoras arrived in our region over 40 years ago. With the advent of NAFTA 11 years ago, the working conditions in the maquiladoras got much worse. Even those plants, which over the years had achieved better benefits and wages, began to move south into the interior of Mexico where the salaries were much lower and the conditions worse.
BACON:	*What about the plants that have remained on the border? Have salaries gone up in the years that NAFTA has been in effect?*
QUIÑONES:	The problem of unemployment wasn't resolved at all. Salaries have not gotten better. They're completely insufficient for anybody to live on. Workers continue to live in extreme poverty and many people still arrive in the border region looking for work. The cities are overloaded and don't provide basic services or infrastructure. Look at Ciudad Acuña. It's a disgrace. There are large transnationals, such as Alcoa and Delphi, operating there, yet workers have to build their houses out of cardboard or materials taken from the factories.
BACON:	*What is an average maquiladora factory wage?*
QUIÑONES:	The average salary for a maquiladora worker is $45 a week. This allows workers to buy pasta, beans, rice, potatoes, maybe oil—just the basic things to eat. They can't buy cereals. They buy milk on rare occasions if there are children. No meat.
BACON:	*In a Mexican supermarket on the border, how much does milk cost?*
QUIÑONES:	There is a mistaken idea that just because we live in Mexico all the products we buy are cheaper. In reality the basic food we buy is more expensive on the Mexican side. If you go over to the U.S. side, a gallon of milk will cost about $2.50, or 27 Mexican pesos. On our side of the border, in Piedras Negras, it would be 45 pesos or about $4.50—twice as expensive. It's always the case that in any family two or three people have to work to provide for basic necessities. If there's just one family member working, other members have to supplement this income by selling things like beauty products. Often people cross the border to sell their blood.
BACON:	*What are the conditions in the neighborhoods where workers live?*
QUIÑONES:	It really is a shock, even to workers who come up from the countryside because they are used to living in houses that are bigger, that have patios, that have space. When they arrive, they see there are very few options for workers here. Perhaps the lucky ones can acquire a house through the Mexican housing program, INFONAVIT. But if they do so they're

really in debt to the Mexican government for the rest of their lives. Otherwise, workers are forced to build their own houses out of whatever materials they can find, in places that are completely inappropriate—along the sides of cliffs or in areas prone to flooding, like stream beds.

BACON: *What about basic services, like sewers, running water and electricity? Are the municipal authorities providing those services?*

QUIÑONES: In some of the neighborhoods there are such services. For example, in houses built by INFONAVIT, the government provides electricity. The problem there is that the bills are very high. A monthly electricity bill might get up to 450 pesos, or $45, and a water bill 150 pesos per month, or $15. And the water is not drinkable. In other neighborhoods, where people squat and build their houses the best they can, the government doesn't provide services. People are reduced often to robbing power from electrical lines. When you go into people's houses, you can see the wires running along the ground where kids are walking and playing.

BACON: *Are there unions in the factories?*

QUIÑONES: On the border you have to understand there are many different situations. In Tijuana and Ciudad Juarez, for example, the most common arrangements are known as protection contracts. These are union contracts that the workers don't know anything about, but which protect the company instead. In Tamaulipas or Coahuila most of the maquiladoras have unions, but these are called "charro" unions because they are unresponsive and corrupt and don't support the workers. In Ciudad Acuña, unions are prohibited.

BACON: *The Border Committee was very active helping workers at the Alcoa Fujikura plant in Piedras Negras to improve their conditions and form an independent union. What happened to them?*

QUIÑONES: At Alcoa in Piedras Negras there was a "charro" union there that belonged to the Confederation of Mexican Workers (CTM). It was not responsive to workers so they tried to take control of their own collective bargaining in order to improve their salaries and their benefits. These workers won election to leadership positions on the plant level, but then found that everything they tried to do was undone by higher leaders of the union who made secret agreements with the company.

So they formed an independent union and left the old one. Under Mexican law, they had to get their union registered by the government. They filed the paperwork with the local Conciliation and Arbitration Board, but the agency denied the registration. This case is still not resolved. After appealing within the Mexican legal system, they filed a complaint with the International Labor Organization, accusing the Mexican government of failing to guarantee its citizens the right of freedom of association.

BACON: *What happened to the workers involved in that effort?*

QUIÑONES: Some of the leaders were fired, but others continued organizing. That's really the key to maintaining a movement with an organized rank-and-file base. When the company fires some leaders, other leaders emerge and keep going. Today there are hundreds of workers involved in this movement.

What they went through is a logical evolution and you can see it develop in many factories. First workers begin to make changes in their individual lives and in their individual conflicts. They begin to organize and act together along the same assembly line, and then at a plant-wide level. Ultimately, because they want more say and control, they try to find a union structure that represents them.

BACON: *The story you're telling is very similar to many others. At Sony, in Nuevo Laredo, people were beaten up in front of the factory. It happened at Custom and Auto Trim at the Han Young plant in Tijuana, and at the Duro Bag Company in Rio Bravo. But NAFTA had a labor side agreement that was supposed to guarantee labor rights in Mexico so that this wouldn't take place. What about it?*

QUIÑONES: The labor side agreement supposedly protects the principle of freedom of association. But complaints are filed and after a long process, the only thing that comes is a recommendation, which never translates into actual enforcement. It's not an effective guarantee of anyone's rights.

BACON: *Is there any form of labor protection that can be incorporated into agreements like NAFTA that would guarantee workers rights? Or do you think that workers have to guarantee their labor rights in some other way?*

QUIÑONES: I think both are possible. NAFTA could be renegotiated to include effective and obligatory measures to enforce workers' rights. Holding transnational corporations accountable for complying with the law would be helpful to workers. At the same time, even if you have such protections as part of trade agreements, organizing workers at the grassroots level, forming workers' organizations, is vital. Otherwise, we can't enforce any rights recognized by those agreements.

BACON: *What about support from unions on the other side of the border?*

QUIÑONES: We've been creating alliances with some U.S. labor unions because we're working for the same companies and we need to connect our struggles across the border. At the same time, we want these relationships to respect the autonomy of our organizing style and work. Right now, what's most important to us is developing a greater level .of commitment to Mexican workers among U.S. unions.

BACON: *What about the Mexican labor movement? Is it going to become more effective and responsive to border workers?*

QUIÑONES: I think so. Ultimately we want an independent labor movement in the maquiladoras. Genuine unionism is the best hope for our families and our future. We've been able to build important alliances with other unions and movements within Mexico. We share common objectives with unions like the Authentic Labor Front (FAT), and with the independent union called Alcoa Puebla. This union was formed at the Alcoa factory in Puebla, with the help of the independent union at the Volkswagen plant there. Some groups of miners are part of this network also. We also have an agreement with the National Union of Workers, Mexico's large, new progressive labor federation.

BACON: *What can workers in the U.S. do to be part of this?*

QUIÑONES: The first thing workers can do is organize and fight for their own jobs where they are. This is the first step towards building international solidarity. For the companies, there are no borders anymore or barriers to the movement of capital. We need to take a lesson from their mentality and build the same borderless solidarity and support for one another. If workers in the U.S. understand that Mexican workers face huge economic difficulties when they try to organize themselves, they can contribute economic support. Mexican organizations don't have the same capacity as organizations in the United States.

      Supporting Mexican workers in the United States is important too. The effort of Mexican and other immigrant workers to legalize their status is connected to our rights as workers in Mexico. If workers in the U.S. can't exercise fully their rights, it brings everybody down.

      Ultimately, the economic level of everybody has to come up. Corporations are very good at looking around the world to see where conditions are the worst and move to that place. If we can help each other come up, they won't be able to do that.

40

Students Against Sweatshops

BHUMIKA MUCHHALA

Could you tell us about your background, and how you came to be an anti-sweatshop activist?

I was born in India but grew up in Jakarta from the age of five. My father, an Indian accountant, worked for multinational companies there. I attended an American international school that was completely Eurocentric—they never taught us anything about Indonesian language, culture, or politics. I learned Bahasa from hanging out with the street vendors. I used to sneak out to the street corner and eat *boso*, noodle soup. I came to the U.S. to study at Carnegie Mellon University, in Pittsburgh, and also took a lot of classes at the University of Pittsburgh itself. I had been apolitical as a teenager. But that changed during the ousting of Suharto in 1998—I turned on the TV and watched as my city went up in flames. It shook every fibre in my being to see all these buildings that I knew so well on fire, to see the riots in the streets. I kept hearing references to the International Monetary Fund, and student activists crying out against Suharto's corruption and cronyism, and Chinese-Indonesian dominance of the economy. It was then that I started reading the paper and searching the internet to learn more about the IMF. I learned a lot from the website for Global Exchange, the human-rights and environmental NGO set up in 1988. I decided to double my major, to learn about U.S. foreign policy and U.S. imperialism.

I became involved with Students in Solidarity, a large group of student activists at the University of Pittsbugh, and also got in touch with Robin Alexander at United Electric, an independent union. She had started an organization called PLANTA (Pittsburgh Labor Action Network for the Americas), which introduced me to labour organizing, and all the issues surrounding it. At the same time, Students in Solidarity were doing actions with the janitors at the AT&T buildings in downtown Pittsburgh, as well as with campus workers at the UP. Through an organization called Pugwash, I also got involved in debates on ethical issues in science with students in public policy, science, engineering and environmental departments.

But my main entry into activism came in my senior year, after a summer at Global Exchange. I applied by a fluke, because I really liked the website. They invited me to do an internship, which mostly involved logistical work for the speaking tours of international activists. I translated some information from Bahasa to English for their Gap campaign. I also petitioned outside Gap stores.

SOURCE: Bhumika Muchhala, "Students Against Sweatshops." In *A Movement of Movements: Is Another World Really Possible?* Tom Mertes (ed.). New York: Verso, 2004, Pages 192–201.

Back at college, I launched a Gap campaign in Pittsburgh with Robin Alexander from UE. We did a number of Gap actions, as well as a fair-trade coffee campaign. I began to feel, though, that there were a lot of problems with the whole approach to Gap. Many people wanted to turn it into an all-out boycott, but that means people losing their jobs. Far more important is to find a way to make these corporations accountable, and responsible. We don't want them out of Indonesia, we want them to treat their workers with respect and adhere to codes of conduct. It's more about understanding the power structures—the racism, imperialism and neo-colonialism—than seeing the world in black and white terms. I also remember that my initial reaction to the Gap protests was that my friends in Indonesia would never do anything like this—not simply because they were more interested in hanging out at the mall, but because they don't have the do-gooding impulses of many activists here. I wasn't driven by the same sentiment—I wanted more debate, more facts.

How did you end up in Seattle?

I was going to a lot of Global Exchange workshops, including those on the WTO, that were run in conjunction with JustAct. I got involved with national recruiting and organizing, and decided to go out to Seattle because I knew a lot of my colleagues were going to be there. I didn't do direct action, though, because I'm not a citizen. After that, I compiled an oral history of students who were active around the WTO protests—*Student Voices: One Year After Seattle*—while working at the Institute for Policy Studies. The original plan was to conduct a few interviews as preparation for a workshop we put on with STARC—Students Transforming and Reforming Corporations. That was cancelled, but I was asked to do a report instead. At first, my goal was to interview 15 students, but that turned into 30, then 50—I ended up with 60-plus students. Looking at the report now, what strikes me is how blind it was to the demographics: almost all the students were white. At the same time, that makes it a pretty representative cross-section of the students who were at Seattle, or the April 16 demonstrations in Washington DC, or at the Republican National Convention.

How did you become involved with Students Against Sweatshops?

My involvement began with an eight-month trip to Indonesia. I was approached to do some fact-finding by the Workers' Rights Consortium, an organization set up in 2000 by student activists, labour experts and university administrations to pressure the manufacturers of clothing bearing college logos into adhering to a code of conduct for their workers. It was conceived as an alternative to the Clinton administration's Fair Labor Association, set up in 1996, in which corporate interests predominate. Currently, 116 colleges are affiliated, and pay dues to fund the WRC's activities. Its board of directors consists of five independent labour-rights experts, including Linda Chavez-Thompson of the AFL-CIO and Mark Barenberg, a Columbia law professor; five representatives from universities; and five from USAS. It also has an advisory council, which includes U.S.-based NGOs, academics, George Miller—a Democratic congressman—and international representatives.

I was technically an independent researcher—the WRC provided me with a place to stay in Indonesia, an office and transportation, but no salary. The AFL-CIO's Solidarity Centre was a point of contact for people who would introduce me to the various Indonesian trade unions. I interviewed more than 200 workers, over a period of eight months, in assembly plants for Reebok, Nike, Champion, Gear, Gap, Banana Republic—mostly U.S. multinationals, but also a lot of European labels and knock-offs. Initially, it was a startling experience, but I found it easy to strike up a rapport with the workers, because they were mainly women between seventeen and twenty-six—close to my age—who were curious, fun and energetic. They came from all over Indonesia—from Sumatra, Sulawesi, Kalimantan, Java. Local chiefs often pick young women to be sent to work in industrial centres, and the women consider themselves lucky and honoured. They make friends and learn about Indonesia's patchwork of different cultures. They tend to get 600,000 rupiah a month—around $60—and send up to a third of that home to their villages, which means they're living on little more than a dollar a day. They usually run out of money before the end of the month, but they're very resourceful—friends lend money to each other, women's cooperatives pool their incomes, and everyone makes incredible economies, cutting down an already meagre diet.

What happens to the women after the age of 26?

They just get worn down: they develop arthritis or respiratory problems. In the hat factories, for example, they burn a lot of coal and use thinners and other solvents. Once they can no longer work, they are laid off and return to the village, or else become street vendors, selling vegetables, cigarettes or shampoo from stalls or kiosks; others become homeless. Some dream of returning to school, others get married and settle down to raise a family.

What did the workers think was most exploitative about the process?

They were very aware that they were being exploited—they would work for 15 hours at a stretch and come home with stomach pains and headaches. Some of them told me about having to work till 2 or 6 in the morning, having to clean toilets, about suffering verbal and sexual abuse. They were in no sense passive victims, but they responded to some of the questions with astonishment—they had never really considered air temperature in the factory as something that they were deprived of or denied. They were amused and surprised that somebody cared so much about tiny, inane details of their lives, such as what the bathrooms were like. But when I started explaining to them how the monitoring process works, and how we could negotiate with the retailers who had power over their supervisors, they became more interested.

What did you, and the Workers' Rights Consortium, learn from the experience?

I found we had a common interest, along with Indonesian and U.S. unions and activists, in working out how to channel this power through the current corporate structure—how to effect real changes among the Korean management, U.S.

retailers, contractors and institutional purchasers such as universities, and secure real improvements for the workers. Being able to meet the workers and Indonesian union organizers face-to-face was an important step. The WRC hierarchy are busy flying from one meeting to another, and don't have time to spend weeks or months in one place; activists and union organizers in producing regions, meanwhile, can't easily be reached by phone, and tend not to have email, so it's hard to make people aware of what we're trying to do without making contact in person. I made a few connexions, and after I left, four USAS activists flew in and started operating in some of the same factories, which has helped to keep the momentum going.

You became active in USAS on returning from your stint with the WRC. Can you tell us about the origins and structure of USAS?

There had been various campaigns against sweatshops, starting with groups of immigrant workers in the U.S. garment industry in New York, California and Texas in the late 1980s; there were also campaigns in the early 1990s by the National Labor Committee, and the Coalition to Eliminate Sweatshop Conditions in California. Students began to focus on the issue of sweatshop labour during summer internships at the AFL-CIO in 1996, and at UNITE, the United Needle and Textile Workers Union. Some Duke students, including Tico Almeida, who spent the summer of 1997 researching the question for UNITE, returned to campus and began lobbying the university to make manufacturers of university apparel sign up to a code of conduct. The campaign was successful, and encouraged students who had been thinking along similar lines to start campaigns on their own campuses. USAS was founded in the spring of 1998, as an informal network of campus anti-sweatshop groups.

The first national conference was held in New York in July 1998. Over 200 campus delegates attended the second conference in 1999, when it was decided to set up a permanent office in Washington, DC. Our national conferences—now held in August—last three days, and feature keynote speakers, workshops on labour action and anti–white supremacy training, as well as panel discussions and meetings of the various caucuses.

As far as structure is concerned, the leadership consists of a coordinating committee, democratically elected at the annual conference, plus the student representatives to the WRC. There are seven regional representatives, four from the identity caucuses—women, people of colour, working-class people, LGBTQ—and three members-at-large. In addition to the committee, there are seven regional organizers who report to the committee, and coordinate with organizers on individual campuses. Then, there is the national office with three permanent staff—a field organizer, a programme coordinator, and a person responsible for fundraising and communications. There are also standing committees on individual issues, such as international solidarity, alliance-building and solidarity with farmworkers. As programme coordinator, I primarily liaise with the international solidarity committee—organizing letter-writing and solidarity campaigns with workers in the Gap factory in El Salvador, for instance.

Does the national office determine what your campaigns are going to be, or do groups at particular universities become aware of an issue, which then percolates upwards?

It's primarily campus-based—different groups decide to do different things. Some are involved in mobilizing research or teaching assistants for rallies, some have taken part in local living-wage campaigns for campus workers, notably those employed by Sodexho–Marriott. Other campaigns have come about through contacts initially made by USAS activists. For example, in late 2000 USAS members were part of a delegation that went to the Kukdong factory in Puebla, and which also included people from AFL-CIO, United Electric and the Mexican union Frente Auténtico de Trabajo. The following January around 850 workers at the factory went on strike to protest the sacking of five of their co-workers, who were trying to organize an independent union. That spring, USAS sent out more activists to talk to the workers and find out how best to support their action. Nike and Reebok contract to the Korean-run *maquiladora*, so USAS organized pickets of their stores in the U.S., put pressure on university administrations and commissioned the WRC to investigate. Within two months, the workers had been reinstated, and by September the governor of Puebla made a public promise to give recognition to the Kukdong workers' new union, SITEKIM.

A similar process took place with the campaign at the New Era cap factory in Derby, New York where, in early 2001, two-thirds of the workforce was laid off after affiliating with the Communication Workers of America. USAS sent a delegation there in March 2001, and then began working in tandem with CWA, getting colleges to cut contracts with New Era, putting pressure on major-league baseball teams, for whom New Era are the exclusive supplier. After a long strike, the workers came to a bargaining agreement with the company, and were reinstated—it was a big victory. USAS had a similar success with a cap factory in the Dominican Republic called BJ&B. Apart from that, there have been campaigns against Taco Bell—working with the Coalition of Immokalee Workers in Florida—and Mount Olive pickles, as well as a general campaign against cap producers such as Nike, Adidas, Reebok and Gap over the past few years.

What about the composition of USAS?

The anti-sweatshop coalition is pretty specific to a certain class and culture. There is a considerable level of working-class to upper-class diversity in USAS, but the majority are middle class, suburban—rabble-rousing in actions, but they get good grades. In a 1999 survey of USAS by a researcher called Peter Siu, more than a third of activists stated their family income was over $100,000, and only 8 percent said it was under $40,000. As with the mobilizations at Seattle and elsewhere, it's predominantly a white movement. Though the conditions in sweatshops resonate with Latinos and the Asian diaspora, these people aren't yet as politically active on campuses—perhaps because they don't feel comfortable in organizing culture. Black students are more focused on civil rights, and they often have other priorities that occupy them in their own neighbourhoods—for working-class students of colour, the prison-industrial complex hits home more than the IMF or World Bank. USAS does have good relations with the Prison Moratorium Project, and maintains a presence at their annual conference; but beyond that, it's left up to individual campuses to decide which struggles to adopt. At the moment, USAS is trying to start up dialogue on the subject of race and culture, but so far it's proceeded along the lines of "how can we recruit more people of colour?" Personally, I find that culturally insensitive and tokenistic.

You mentioned working with the CWA. What are USAS's relations with the unions like?

On the New Era campaign we definitely worked hand-in-hand with the CWA —a progressive union compared with some others—but for the most part we work pretty autonomously. The unions have a strong presence at our national conferences, and AFL-CIO and UNITE make important financial contributions—the former gave $50,000 in the academic year 2000-2001. It is quite a contentious issue. The AFL does take up a lot of the centre-ground of our campaign work, and people often ask us if we're being used as their youth wing, pointing to the fact that many USAS students go on to become organizers in the Service Employees International Union or Hotel Employees and Restaurant Employees Union. There's also the question of whether we're being steered in a particular direction—some people in USAS feel the AFL is protectionist, which is not something we would want to be associated with. It's true that part of the struggle at New Era was to protect the workers' jobs, since the company threatened to shift production overseas if they made trouble. But similar conditions applied at the Kukdong factory in Puebla, where we made Nike and Reebok promise not to "cut and run." As I mentioned earlier with regard to the Gap boycott, it's not about shutting down manufacturing in the developing world, it's about making companies treat their workers with respect.

Are there tensions between the unions and USAS with regard to international versus national campaigns?

Yes, because USAS chapters try to address both. A lot of activists look at international campaigns such as Kukdong, and ask, "what about the workers on our own campuses, serving us food, cutting our lawns or cleaning our homes?" Living-wage campaigns have been very prominent on scores of campuses for this reason. Our rank and file are, after all, predominantly white, upper-middle class, liberal college students attending elite institutions; their engagement with labour issues is in many ways the product of privilege, and they make use of their status to focus media attention on those issues.

What other campaigns would you see as models?

One that has brought enormous inspiration is the campaign against the Narmada Dam—Medha Patkar and Arundhati Roy have been very influential. Of U.S.-based campaigns, I would name those focusing on the prison-industrial complex, especially in California. USAS is also a member of the National Student Youth Peace Coalition, which was formed in the wake of September 11, and opposed the assaults on Afghanistan and Iraq.

41

The Rise of Food Democracy

BRIAN HALWEIL

The National Touring Association, one of the largest lobbying groups in Norway, representing walkers, hikers and campers, recently joined forces with the nation's one and only celebrity chef to develop a line of foods made from indigenous ingredients to stock the country's network of camping huts. For instance, someone staying in a mountain cottage in Jotunheimen National Park would dine on cured reindeer heart, sour cream porridge and small potatoes grown only in those mountain valleys. Sekem, Egypt's largest organic food producer, has developed a line of breads, dried fruits, herbs, sauces and other items made entirely from ingredients grown in the country. The brand is recognized by 70 percent of Egyptians, and sales have doubled each of the last five years. In Zimbabwe, six women realized that their husbands, who are peanut farmers, were literally getting paid peanuts for their crop while they bought pricey imported peanut butter. These women decided to invest in a grinder and are now producing a popular line of peanut butter from local nuts that sells for 15 percent less than mainstream brands. In Nebraska, in the United States, a group of local farmers got together and opened a farmers grocery that stocks only foods raised in that state. They found suppliers of bacon and baked beans, sour cream and sauerkraut, and virtually all major grocery items, all from Nebraska.

What ties together these disparate enterprises from around the world? At a time when our food often travels farther than ever before, they are all evidence of "food democracy" erupting from an imperialistic food landscape. At first blush, food democracy may seem a little grandiose—a strange combination of words. But if you doubt the existence of power relations in the realm of food, consider a point made by Frances and Anna Lappé in their book *Hope's Edge* (see UN Chronicle, Issue 3, 2001). The typical supermarket contains no fewer than 30,000 items, about half of them produced by ten multinational food and beverage companies, with 117 men and 21 women forming the boards of directors of those companies. In other words, although the plethora of products you see at a typical supermarket gives the appearance of abundant choice, much of the variety is more a matter of branding than of true agricultural variety and, rather than coming from thousands of farmers producing different local varieties, they have been globally standardized and selected for maximum profit by just a few pow-

SOURCE: "The Rise of Food Democracy" by Brian Halweil, *UN Chronicle Online Edition*, Vol. XLII, #1, January 2005.

erful executives. Food from far-flung places has become the norm in much of the United States and the rest of the world. The value of international trade in food has tripled since 1961, while the tonnage of food shipped between countries has grown fourfold during a time when populations only doubled. For example, apples in Des Moines supermarkets come from China, even though there are apple orchards in Iowa; potatoes in Lima's supermarkets come from the United States, even though Peru boasts more varieties of potato than any other country.

The long-distance food system offers unprecedented and unparalleled choice to paying consumers—any food, any time, anywhere. At the same time, this astounding choice is laden with contradictions. Ecologist and writer Gary Nabhan wonders "what culinary melodies are being drowned out by the noise of that transnational vending machine," which often runs roughshod over local cuisines, varieties and agriculture. The choice offered by the global vending machine is often illusory, defined by infinite flavouring, packaging and marketing reformulations of largely the same raw ingredients (consider the hundreds of available breakfast cereals). The taste of products that are always available but usually out of season often leaves something to be desired.

Long-distance travel requires more packaging, refrigeration and fuel, and generates huge amounts of waste and pollution. Instead of dealing directly with their neighbours, farmers sell into a remote and complex food chain of which they are a tiny part and are paid accordingly. A whole constellation of relationships within the food shed—between neighbours, between farmers and local processors, between farmers and consumers—is lost in the process. Farmers producing for export often find themselves hungry as they sacrifice the output of their land to feed foreign mouths, while poor urbanites in both the First and Third Worlds find themselves living in neighbourhoods unable to attract most supermarkets and other food shops, and thus without healthy food choices. Products enduring long-distance transport and long-term storage depend on preservatives and additives and encounter all sorts of opportunities for contamination on their journey from farm to plate. The supposed efficiencies of the long-distance chain leave many people malnourished and underserved at both ends of the chain.

The changing nature of our food in many ways signals what the changing global economic structure means for the environment, our health and the tenor of our lives. The quality, taste and vitality of foods are profoundly affected by how and where they are produced and how they arrive at our tables. Food touches us so deeply that threats to local food traditions have sometimes provoked strong, even violent, responses. José Bové, the French shepherd who smacked his tractor into a McDonald's restaurant to fight what he called "culinary imperialism," is one of the better-known symbols in a nascent global movement to protect and invigorate local food sheds.

It is a movement to restore rural areas, enrich poor nations, return wholesome foods to cities and reconnect suburbanites with their land by reclaiming lawns, abandoned lots and golf courses to use as local farms, orchards and gardens.

Local food is pushing through the cracks in the long-distance food system: rising fuel and transportation costs; the near extinction of family farms; loss of farmland to spreading suburbs; concerns about the quality and safety of food; and the craving for some closer connection to it. Eating local allows people to reclaim the pleasures of face-to-face interactions around food and the security that comes from knowing what one is eating. It might be the best defense against hazards intentionally or unintentionally introduced in the food supply, including E-coli bacteria, genetically modified foods, pesticide residues and bio-warfare agents. In an era of climate change and water shortages, having farmers nearby might be the best hedge against other unexpected shocks. On a more sensual level, locally grown and in-season food served fresh has a definite taste advantage—one of the reasons this movement has attracted the attention of chefs, food critics and discriminating consumers around the world.

The local alternative also offers huge economic opportunities. A study by the New Economics Foundation in London found that every £10 spent at a local food business is worth £25 for that area, compared with just £14 when the same amount is spent in a supermarket. That is, a pound (or dollar, peso or rupee) spent locally generates nearly twice as much income for the local economy. The farmer buys a drink at the local pub; the pub owner gets a car tune-up at the local mechanic; the mechanic brings a shirt to the local tailor; the tailor buys some bread at the local bakery; the baker buys wheat for bread and fruit for muffins from the local farmer. When these businesses are not locally owned, money leaves the community at every transaction.

This sort of multiplier is perhaps most important in the developing world where the vast majority of people are still employed in agriculture. In West Africa, for example, each $1 of new income for a farmer yields an average increase to other workers in the local economy, ranging from $1.96 in Niger to $2.88 in Burkina Faso. No equivalent local increases occur when people spend money on imported foods. While the idea of complete food self-sufficiency may be impractical for rich and poor nations alike, greater self-sufficiency can buffer them against the whims of international markets. To the extent that food production and distribution are relocated in the community under local ownership, more money will circulate in the local community to generate more jobs and income.

But here's what makes these declarations of food independence, despite their small size, so threatening to the agricultural status quo. They are built around certain distinctions—geographic characteristics—that global trade agreements are trying so hard to eliminate. These agreements, whether the European Union Trade Zone or the North American Free Trade Agreement, depend on erasing borders and geographic distinctions. . . . Multinational food companies that source the cheapest ingredients they can find also depend on erasing these distinctions. . . .

Look around and you can glimpse the change worldwide. Farmers in Hawaii are uprooting their pineapple plantations to sow vegetables in hopes of replacing the imported salads at resorts and hotels. School districts throughout Italy have launched an impressive effort to make sure cafeterias are serving a

Mediterranean diet by contracting with nearby farmers. At the rarefied levels of the World Trade Organization, officials are beginning to make room for nations to feed themselves, realizing that this might be the best hope for poor nations that cannot afford to import their sustenance. Even some of the world's biggest food companies are starting to embrace these values, a reality that raises some unsettling questions and awesome opportunities for local food advocates. Recently, officials at both Sysco—the world's largest food-service provider— and Kaiser Permanente—the largest health care provider in the United States— declared their dependence on small local farmers for certain products they cannot get anywhere else. These changes will unfold in a million different ways, but the general path will look familiar. Farmers will plant a greater diversity of crops. Less will be shipped as bulk commodity and more will be packaged, canned and prepared to be sold nearby. Small food businesses will emerge to do this work, governments will encourage new businesses, and shoppers seeking pleasure and reassurance will eat deliberately and inquire about the origins of their food. Communities worldwide all possess the capacity to regain this control and this makes the simple idea of eating local so powerful. These communities have a choice, and they are choosing instead to eat here.

42

Cyberfeminism, Technology, and International "Development"[1]

RADHIKA GAJJALA

ANNAPURNA MAMIDIPUDI

The simplest way to describe the term "cyberfeminism" might be that it refers to women using Internet technology for something other than shopping via the Internet or browsing the world-wide web.[2] One could also say that cyberfeminism is feminism in relation to "cyberspace." Cyberspace is "informational data space made available by electrical circuits and computer networks" (Vitanza 1999, 5). In other words, cyberspace refers to the "spaces," or opportunities, for social interaction provided by computers, modems, satellites, and telephone lines—what we have come to call "the Internet." Even though there are several approaches to cyberfeminism, cyberfeminists share the belief that women should take control of and appropriate the use of Internet technologies in an attempt to empower themselves. The idea that the Internet can be empowering to individuals and communities who are under-privileged is based on the notion of scientific and technological progress alleviating human suffering, offering the chance of a better material and emotional quality of life. In this article, we make conceptual links between "old" and "new" technologies within contexts of globalisation[3] third-world development, and the empowerment of women. We wish to question the idea of "progress" and "development" as the inevitable result of science and technology, and develop a critique of the top-down approach to technology transfer from the Northern to the Southern Hemisphere. There are two questions of central importance: First, will women in the South be able (allowed) to use new technologies under conditions that are contextually empowering to them, because they are defined by women themselves? Second, within which Internet-based contexts can women from the South truly be heard? How can they define the conditions under which they can interact online,[4] to enable them to form coalitions and collaborate, aiming to transform social, cultural, and political structures?

SOURCE: Radhika Gajjala and Annapurna Mamidipudi "Cyberfeminism, Technology, International 'Development,'" from *Gender & Development* Volume 7, Number 2, p. 8–16. Copyright © 1999 Taylor & Francis Ltd. Reprinted by permission.

THE INTERNET AND "DEVELOPMENT"

Cyberfeminists urge women all over the world to learn how to use computers, to get "connected,"[5] and to use the Internet as a tool for feminist causes and individual empowerment. However, ensuring that women are empowered by new technology requires us to investigate issues which are far more complex than merely providing material access to the latest technologies. The Internet has fascinated many activists and scholars because of its potential to connect people all across the world in a way that has never been possible before. Individuals can publish written material instantaneously, and broadcast information to remote locations. Observers predict that it will cause unprecedented and radical change in the way human beings conduct business and social activities. In much of the North, as well as in some materially privileged sections of societies in the South, the Internet is celebrated as a tool for enhancing world-wide democracy. The Internet and its associated technologies are touted as great equalisers, which will help bridge gaps between social groups: the "haves" and the "have-nots," and men and women.

Since the Second World War, development—in the sense of transferring and "diffusing" northern forms of scientific and technological "progress," knowledge, and modes of production and consumption, from the industrialised north into southern contexts—has been seen by many as the one over-arching solution to poverty and inequality around the world. Much of the current literature, as well as media representations of the so-called under-developed world, reinforces this discourse of "development" and "under-development." As scholars such as Edward Said (1978) have pointed out, this process is also apparent in the context of colonialism, when the production of knowledge about the colonised nations served the colonisers in justifying their project.

What, then, does it mean to say that the Internet and technology are feminist issues for women in developing nations, when the project of development in itself is saddled with colonial baggage? In order to examine whether women in these contexts are indeed going to realise empowerment through the use of technology, we need to understand the complexity of the obstacles they face, by considering the ways in which the conditions of their lives are determined by unequal power relations at local and global levels.

THE FORM OF THIS ARTICLE

In the following, we each describe our engagement with cyberfeminisms, development, and new technology, and discuss some of the problems that we encounter in our efforts. Both of us have interacted quite extensively using the Internet, where our interactions occasionally overlap when we engage in discussions and creative exchanges with others.[6] One of us, Annapurna Mamidipudi, is also involved with an NGO working with traditional handloom weavers in south India. The other, Radhika Gajjala, works within academia, and creates and runs on-line "discussion lists"[7] and websites from her North American geographical location, aiming to create spaces that enable dialogue and collaboration among women with access to the Internet all over the world. This paper was written via the Internet, across a fairly vast geographical distance of approximately 10,000 miles. We have written the article as a dialogue, to make our individual voices and locations apparent. This unconventional form and method seems appropriate for our subject matter: a belief in the possibilities of dialogue and collaboration across geographical boundaries offered by this medium of the future. We do not consider either of us to represent the North or the South, "theory" or "practice," each of us will use her professional and personal experience of technology within both "first-world" and "third-world" contexts. We share caste, class, national, and religious affiliations, but once again, neither of us are representative Indian women.

Annapurna Mamidipudi

As a field-worker in an organisation which focuses on the development and use of environment-friendly dyes for textile production, I am part of a team that has been successfully introducing and transferring the technology of non-chemical natural dyes to clients. The course we offer is comprehensive; it includes training in botany and dye-material cultivation patterns, concepts of eco-friendly technology, actual dyeing techniques and tools, specific methodology for further research, aesthetics, and market research. While the service we provide is similar to that of any professional consultancy, a crucial difference is that we cater solely to traditional handloom weavers; our trainees, sponsors and manufacturers are all artisans, men and women from traditional weaving communities.

The craft of traditional natural dyeing is based on sophisticated knowledge that has been passed down from generation to generation of artisans. The end-product created by these artisans is exquisite hand-loomed cloth, woven of yarn hand-spun from local cotton by women in remote Indian villages, dyed in the vibrant colours of indigo and madder. This has been exported all over the world from pre-colonial times onwards. One might well ask, why should a skill that has been passed down successfully over so many generations suddenly need technical consultants like me for training?

Radhika Gajjala

I am a producer, first, of theory concerned with culture, post-colonialism, and feminism. I am in continuous dialogue with women from non-privileged and non-western locations, examining the experience of activists like Annapurna, and collaborating with men and women from the South. I rely to a large extent on having access to knowledge through Northern technology and power structures, but I am not blind to the fact that these power structures oppress women and men living in poverty in both North and South.

My second role as a producer is in creating electronic "spaces" which are used by people of different identities to express themselves and talk to each other.[8] The Spoon Collective,[9] started in 1994, is "dedicated to promoting discussion of philosophical and political issues" (http://lists.village.virginia.edu/~spoons). The Spoon Collective was started in 1994, and I entered it in 1995, volunteering to co-moderate two "discussion lists." I set up two further discussion lists in 1995 and 1996, which I will mention later in this article.

While members of the Spoon Collective have different individual aims in belonging to the Collective, I believe that all of us are interested in the possibilities of activism through electronic communications. All of us have set up, and continue to moderate, discussion lists that implicitly question the global status quo, in one way or another. One member of the collective said, "One way in which we conceptualise what we do is by talking about thinking [and writing/speaking on-line] as a civic, public activity." As is characteristic of much Internet-based activity (whether activist, personal, or commercial), our goals and our actual output are constantly evaluated. We ceaselessly discuss their impact on society and culture. For example, what determines whether a list "works" or not? The volume of messages exchanged? The quality of information or discussion? But how would "quality" be defined? Do we determine the success of the list by the number of members who subscribe to it? Or by the number of members who participate by sending messages? By the number of websites that have links to our list-archives or the Spoon Collective website? How can we tell from this how many people we really reach?

In order to start up discussion lists, and construct websites, I had to teach myself sufficient programming and computer-related skills to be able to manage the technical side. My background as creative writer and student in the humanities had not trained me for the technical aspects of being an active producer on-line, and my knowledge is mostly self-taught. Later in this article, I will discuss my e-mail lists as part of an effort to try and facilitate collaborations between feminists across vast geographical boundaries. What scope is there for them to discuss and assert their differences on an equal basis, within these electronic social spaces which are themselves based in unequal economic, social and cultural relations? In a sense I suppose my online ventures could be called "cyber-feminist" investigations.

Annapurna

Until the nineteenth century, most of the weaving industry in the area where I work was shaped by the demands of local consumers. Chinnur is a little village in Adilabad, in an interior region of the Deccan plateau in South India. There used to be a large concentration of weavers with a reputation for excellence in this area. Their reputation was based on three things: the skill of the farmers in producing different varieties of cotton; the ability of different groups of people to work together and process the cotton; and, finally, the wealth of knowledge of dyes and techniques that added aesthetic value to utility. Different castes and communities were inter-linked in occupational, as well as social relationships, exchanging services and materials, creating a strong local market economy which was entrenched in the traditions and rituals of daily life. For example, during specific seasons or events, women of leisure from non-weaving communities spun, exchanging spun yarn for sarees (Uzramma 1995).

However, the development of chemical dyes almost 100 years ago in Europe had a calamitous effect on traditional Indian dyeing practices. Processes which were the pride of the textile industry of this country were abandoned and replaced by chemical dyes. Even in remote Chinnur, the spreading wave of modern science changed people's perceptions of traditional technology; they now saw it as outmoded, and this resulted in almost total erasure of knowledge of the traditional processes within these communities.

Europeans had started to document the local dyeing and weaving activities in the eighteenth century; Indians themselves continued this up to the early twentieth century, in a bid to preserve knowledge. But this process meant that knowledge which had been firmly in the domain of practice of the artisans was now converted into textual information, and shifted the ownership of the knowledge to those able to "study," rather than those who "do."

As the outside world mutated into a global village, the organic processes of the traditional artisan weaver turned full cycle, back to popularity when the colour of neeli (indigo) caught the imagination of ecology-conscious consumers in the late 1970s. But even while the self-congratulatory back-patting went on among the nationalists and intellectuals, the weavers had internalised information about "modem" chemical technology. Just as they had begun to find a footing in the market, their practical knowledge was again found wanting. The only available information about vegetable dyes was documented in the language of the colonisers, codified, and placed in libraries or museums, inaccessible to the traditional practitioners from whom the information had been gathered in the first instance. Thus, although it looked as if a demand had been created for their product, in reality this further reinforced the image of weavers' technology as needful of input from outside experts, in the weavers' own minds as well as in those of others.

Today, in most descriptions of the hand-loom industry, the traditional weaver is seen as an object of charity, who can survive only through government handouts or patronage from social elites. Yet their "sunset" industry—as it was referred to by a top official in the Department of Hand-loom and Textiles in charge of formulating strategy for this industry (personal communication, 1999) —has the second largest number of practitioners in India, farming employing the greatest number. For the men and women engaged in weaving in villages across India, the journey from traditional neeli (indigo) to modern naphthol (chemical) dyes has meant a journey from self-sufficiency to dependence, self-respect to subordination; in short, a journey to "primitivity."

Radhika

Most highly-educated women from the third world, whether or not we live in the North, experience a parallel journey to "primitivity" in the sense Annapurna uses above. In part, this happens through acquiring western-style education and professional status, which is not often an autonomous personal choice. No woman of the third world has the luxury of not choosing to be westernised if she aspires to be heard, or even simply to achieve a level of material freedom, comfort, and luxury within global structures of power. Many of us have "made it" within westernised professional systems, and have enjoyed the status of the representative third-world woman within global structures of power. Yet, as a result of our education and professional status, we are not representative, although we are of the third world, and our stories are not those of many truly under-privileged women in third-world locations.

Often, we meet other people's expectations by taking on the role of victims of third-world cultures, or, alternatively, victors who have "survived" our backgrounds. Yet, when we refuse these roles allotted to us, some feminists from Northern backgrounds suggest our experiences don't "count," since we are not "real" third-world women. Even as we demonstrate our potential by attaining the level of education and "westernisation" required to become powerful within global structures we are silenced once again.

Annapurna

Outside the house of one of the weavers in the village of Chinnur is a chalk-written address board in English. It says: "Venkatesh U.S., Weavers' Colony, Chinnur." The initials U.S. after this man's name stand for "Unskilled Labour": a powerful statement on how an expert weaver chooses to categorise himself. This classification in the government records, he hopes, will make him eligible for a low-grade job in a government office.

I first came here as one of a team of field workers from an NGO which offered marketing support to craft groups. Natural dyeing seemed an option which would add value to the cotton cloth, and which would also eventually decrease weavers' dependence on a fickle market and centralised raw material supply systems. We ourselves did not know the technology, but we were optimistic about the chance of reviving it, provided there was active participation on the part of producer groups.

Transferring the technology of natural dyeing to the field presented many challenges. The sources of information available were texts—some of them 300 years old—noting original processes of artisan practice. Some scholars had researched fragments of the old processes, and some practitioners recalled parts of them. We needed historians to access information from libraries where the documentation was kept; we then needed dyeing experts to interpret the recipes, botanists to participate in the process of identifying materials, engineers to create appropriate technology to ensure fastness and brightness in colours, and chemical technologists to interpret the techniques and demystify processes that had been inter-linked with ways of life that were sometimes centuries old. Making scientists of the weavers, we had to help them reinterpret information to suit their changed environment and resources. We did not want to impose on them—in the name of traditional technology—processes that would place demands on them which would be more oppressive than toxic chemical dyes. The innate capability and skill of the weavers made this seemingly impossible task feasible, and success came five years later, when we produced a range of colours and dyeing techniques that withstood the most stringent of quality tests. A group of dyer-weavers now acted as resource people in workshops held by us to train other groups.

Our clients today are confident weavers who come back to us time and again, to participate in the effort to empower more and more artisan groups by sharing information on a technology that has emerged from their efforts on the field.

In Chinnur lives Venkatavva, whose husband is one of a group of six weavers who decided that they would take the risk of inviting an outside agency to help them become self-sufficient. When we first visited Chinnur eight years ago, Venkatavva was unable to offer us any hospitality. Her three-year-old daughter's staple drink was weak coffee, drunk without milk. There was no food to be offered to visitors who turned up once the morning meal was past. Today, she entertains buyers from Europe, while listening to her husband tell the story of his successes. Her eyes are bright with laughter when she remembers less successful experiments which resulted in pale and fugitive colours, and irate customers. She points proudly to the shirt that her husband Odellu wears today, which he himself has woven. The journey from chemical technology to the indigo vat, from dearth to bounty, from apathy to laughter—this is her journey. In this context, which technology is traditional and which modern? Who is to decide which one is the road to empowerment and self-respect?

Radhika

My journey to "modernity" began with an increasing awareness of my own ignorance, and of the contradictions and injustices which exist within the Northern educational system. I refuse to be either a "victim" or a "victor," and continue to hope that through dialogue, women, men and children from different backgrounds throughout the world can work together to overcome injustice.

In late 1995, when I started my first Internet discussion list, access to the Internet was limited mainly to men and women from the North. (This is the case even now, although there are more men and women from the South who use the Internet). I started the third-world women discussion list partly as a result of my frustration with what I saw as a lack of political commitment and exchange within some women-centred lists. I was frustrated with the way in which topics were discussed. Even in those instances where women and men from the South had access, they came from a particular class background. I was also frustrated with the way people represented themselves. In my opinion, some were too eager to be "ideal native informants" for Northern audiences. Southern participants used the Internet as an opportunity to perform to a Northern audience and receive favours for sufficiently western, or appropriately exotic, performances. Even discussion lists and websites that claimed to be critical and feminist sometimes fell into this trap (possibly, my own lists and websites do so, too).

It was important to me at the time I started the third-world-women list, and continues to be now, that a conscious effort should be made to be critical and self-reflexive: My second list, Sa-cyborgs, was started with a similar goal in mind, but the focus of this list is an interactive exchange of creative writing in relation to gender, race, class and geographical location. Both lists were formed in the recognition that acts of representation are political.

One of the main purposes of both my Internet discussion lists is to facilitate connections between third-world activists and scholars located within, and outside, U.S. academic institutions. I hope that this dialogue will result in collaborative work by and for women living in under-privileged and oppressive conditions, in North and South. My lists are humble efforts which form a small section of the larger efforts being made by women all over the world. Whether they have been successful in any sense is not for me to say. There are many feminists and activists using the Internet in far more effective ways, and examples of these can be seen all over the world-wide web (see http://www.igc.apc.org /vsister/res/index.html for some examples).

Annapurna

Women who tussle with the question of how to define their class and Northern or Southern identity on the Internet are a privileged few. Questions relevant to women to whom Internet technology is being touted as the route to empowerment, might ask: "but who has the Internet empowered? How has this happened? How relevant is this process for women like Venkatavva?"

Venkatavva has seen the advent of roads, cars, telephones, and television in the short 30 years of her life, and understands the advantages, the disadvantages, and the illusion of access that these give her. In a land of faulty cables and unpredictable electricity supply, her children only drink milk on the days that the bus doesn't run, because on those days the milk in the village can't be taken to the city to be sold, and isn't worth any money. Modem technology holds no bogies for her; she has choices that many women in the north don't have access to. On days the electricity fails she watches the traditional story-telling enacted in the village square instead of the distant Santa Barbara on television. The quality and quantity of the choices available to her are based as much on the failure of technology, as on its success. So would modem technology be working towards more quality and quantity in choice, or less?

As an activist working in developing technology for her I can only say this: let her have access to the Internet—why should this be barred when other aspects of modern life are imposed, from Western consumer goods, to twentieth-century diseases such as HIV/AIDS. But let it not be assumed that the Internet will empower her. Otherwise this too will do what other imposed technology has done: the exact opposite of what it purports to do.

The Internet will be a more colourful, exotic place for us with women like Venkatavva flashing their gold nose pins, but what good will it do them? As it is at present, the Internet reflects the perceptions of Northern society that Southern women are brown, backward, and ignorant. A alternative, kinder, depiction of them which is also widespread is that they are victims of their cultural heritage. Is being exposed to such images of themselves going to help Southern women by encouraging them to fight in dignity and self-respect, or will it further erode their confidence in their fast-changing environment?

What, then, is the process by which a woman like Venkatavva could be empowered by the Internet?

Radhika

Venkatavva should be free to decide how the Internet and other related technologies might be used to benefit and empower her and her community. The tools and access should be provided unconditionally, not as a way of selling a so-called superior life-style modelled on the "civilised" and urban centres of the world. Women like Venkatavva are perfectly capable of making the decisions needed to empower themselves according to their everyday needs. Policies designed to be empowering should aid and enable, not impose and preach while fostering further inequalities and inadequacies.

I would like to paraphrase (not without reservations similar to those voiced by Annapurna in her rejoinder) a contribution made to the Gender and Law thematic group[10] at the World Bank. For Spivak, the speaker, the key question that emerges in the context of her work with women in Bangladesh is "How do we approach the bottom?" That is, "How can we learn from below?" The idea is to enter into a society and learn its traditions from inside, seeing what traditions can be worked with to slowly improve the situation, and to ensure that new developments are initiated from the inside so that the changes are accepted. Spivak sees a need to do "invisible mending" of the native fabric, by weaving in the different positive threads which exist in the fabric (moderatorgl@worldbank.org, 20 April 1999).

Annapurna

How do we resolve the contradictory sentiments of seeing the Internet as a panacea to the problems of the south; of thinking that on the contrary, it may even be bad for us; and of asserting that this doesn't mean we don't want it? We need to study processes of empowerment and work out how it is to be done in the context of the Internet. While case studies abound for the failure of this process, development workers in particular would not regard it as fair (or politically correct) to down-play the potential of the Internet to empower many women like Venkatavva in South and North. We cannot say, "I won't give you the Internet, for your own good."

Radhika

My experience of observing the development of the Internet, and using this mode of communication, is that while there are hierarchies of power embedded in the very construction and design of Internet culture, there is still potential for using it in ways which might subvert these and foster dialogue and action on various unexpected fronts, in unpredictable ways. However, it remains true that the NGOs who speak with and for women living in poverty throughout the world, as well as the women themselves, have to negotiate and engage in dialogue with the powerful in the North from positions of lesser power. This situation of unequal economic and social power relations between the North and the South presents challenges for people such as myself who are trying to design electronic spaces of dialogue and activism.

Therefore I reiterate the questions central to our discussion in this article, and ask readers to think deeply and honestly about the issues they raise, beyond those we have addressed here. Will women all over the world be able (allowed) to use technologies under conditions that are defined by them, and therefore potentially empowering to them? Within which Internet-based contexts will women of lesser material and cultural privilege within "global" power relations be able to develop collaborative work, and coalitions, to transform social, cultural, and political structures?

These questions cannot be addressed only in relation to women of the third world. Women from the first world need answers to these questions too. The Internet has its "headquarters" in the first world, but this does not mean that it is contextually empowering to all women in that context. Whether located in the Northern hemisphere or the South, whether rich or poor, global structures of power (through their "invisible" control of the market, Internet service providers, software design, language and so on) clearly determine women's use of the Internet. If cyberfeminists want to ensure that the Internet is empowering, it is not enough to "get connected" and set up websites and maintain e-mail–discussion lists. The latter tasks, while necessary, are only a miniscule part of the battle.

ENDNOTES

1. The writers thank Dr. Melissa Spirek, Dr. A Venkatesh, and the editor of *Gender and Development*, Caroline Sweetman, for commenting on several drafts of this article. Radhika Gajjala also wishes to thank all the Spoon Collective members as well as the members of the various lists that she (co-)moderates. They contribute significantly to our understanding of on-line existence. Several "real-life" bodies also commented on this article, including family members of both writers.

2. The Internet is a world-wide network of computers which communicate via an agreed set of Internet protocol. The world-wide web is a subset of the Internet: which uses a combination of text, graphics, audio and video material to provide information on many subjects.

3. I use this term to denote "the rapidly developing process of complex inter-connections between societies, cultures, institutions and individuals worldwide. It is a process which involves . . . shrinking distances through a dramatic reduction in the time taken—either physically or representationally—to cross them, so making the world smaller and in a certain sense bringing human beings 'closer' to one another. But it is also a process which 'stretches' social relations, removing the relationships which govern our everyday lives from local contexts to global ones" (Tomlinson 1997).

4. The term "on-line" refers to activities carried out via the Internet or e-mail.

5.   Getting "connected" means acquiring the necessary technology (computer, Internet browsing software, telephone modem, connection to an Internet Service Provider) to access the Internet.

6.   Even as we collaborate on projects such as this article, we are exchanging non-traditional creative writing, in relation to our personal/professional/political conflicts and dilemmas, on sa-cyborgs. For information on sa-cyborgs and third-world women, see http://lists.village.virginia.edu.

7.   Electronic networks whose participants discuss a particular topic or topics.

8.   See http://ernie.bgsu.edu/~radhik

9.   The Spoon Collective is operated through the Institute for Advanced Technology in the Humanities at the University of Virginia.

10. Quoted from a post to the gender-law discussion list, gender-law@jazz.worldbank.org, received on 29 April 1999.

REFERENCES

Said, E. (1978). *Orientalism*, New York: Pantheon Books.

Tomlinson, J. (1997). Cultural globalisation and cultural imperialism, in Mohammadi, A., (ed). *International Communication and Globalisation*, London: Sage.

Uzramma (1995). *'Cotton handlooms—industry of the future,'* paper presented at a seminar on Indian textiles in 1995.

Vitanza, V, (1999). *Cyberreader*, Boston: Allyn & Bacon.

REFLECTION QUESTIONS FOR CHAPTER 10

1.   Can "bottom-up" social movements actually be successful? Have there been instances in U.S. history where the powerful have been thwarted by the powerless? Any examples from world history? What appear to be the conditions present when these movements have been successful? Are any of those conditions present now in the United States? In other parts of the world?

2.   David Brooks, writing in the *New York Times* (November 11, 2004), stated that globalization, for the most part, is working to reduce poverty. He quotes a World Bank report that economic growth is producing a "spectacular" decline in poverty in East and South Asia. Other areas are improving (except for sub-Saharan Africa) but not as rapidly. His explanation: globalization, with lower trade barriers, ensuring property rights, and free economic activity, is causing international trade to surge. In his words: "free trade reduces world suffering." If this is true, then why all the protest by the critics of globalization? Social movements for worldwide economic justice and environmental safeguards should be drying up. Is Brooks right, or is he

missing something(s)? How would Brecher, Costello, and Smith respond to Brooks?

3. Marable introduces the concept *global apartheid*. What does this mean? What does his analysis mean for anti-globalization struggles?

4. What parallels with Marable do you find in David Bacon's interview with Julia Quiñones?

5. How does Muchhala's account of his own experiences in the Students Against Sweatshops movement reflect some of the problems of transnational organizing and resistance?

6. How does Halweil account for the success of local food markets? Do you agree that food democracy can promote justice in the new global age?

7. What is *cyberfeminism?* After reading the dialogue between Gajjala and Mamidipudi, what do you think of the Internet as a tool for women's collaboration across geographical boundaries?

Websites

www.g8alternatives.org.uk
G8 Alternatives is a Scottish broad-based coalition that coordinates protests against the various G8 (leaders of the leading governments in transnational economic activities) summits.

www.developmentgap.org
The Development Gap is an organization promoting a just and sustainable alternative to free trade.

www.clc-ctc.ca
The Canadian Labor Congress represents most Canadian labor unions. It publishes a newsletter on the effects of NAFTA.

www.globalsolutions.org
The goal for Citizens for Global Solutions is to build peace, justice, and freedom in a democratically governed world.

www.cleanclothes.org
The Clean Clothes Campaign is an effort based in the Netherlands to promote fair labor practices in the apparel industry.

www.canadians.org
The Council of Canadians is devoted to advancing alternatives to corporate-style free trade and other issues facing Canada.

www.epinet.org
The Economic Policy Institute publishes reports on international and domestic economic issues.

www.equalexchange.com
Equal Exchange is a worker-owned cooperative dedicated to fair trade with small-scale farmers in the developing world.

www.50years.org
The 50 Years is Enough Network is a coalition of U.S. citizens groups linked to groups in 50 countries whose goal is to reform the World Bank and International Monetary Fund.

www.foe.org
Friends of the Earth focuses on the environmental impact of globalization.

www.globalexchange.org
Global Exchange is an organization that works across borders to help build democracies, battle racism and inequality, and evolve a sustainable future.

www.foodfirst.org
The Institute for Food and Development Policy (FoodFirst) is an education-for-action organization working to reduce hunger and poverty throughout the world.

www.ifg.org
The International Forum on Globalization sponsors education and research on the global economy.

www.imf.org
The International Monetary Fund provides information about the processes of globalization, focusing on the positive consequences.

www.laborrights.org
The International Labor Rights Fund is an advocacy group focused on strengthening enforcement of international labor rights.

www.essential.org
Multinational Monitor is a magazine that focuses on the negative consequences of transnational corporations.

www.tni.org
The Transnational Institute is an international network of activists working on solutions to global problems.

www.forumsocialmundial.org.br
The World Social Forum is an open meeting place where groups and movements of civil society opposed to corporate global domination and imperial military domination come together to share their experiences, debate ideas, formulate proposals, and network for effective action.

www.worldwatch.org
The Worldwatch Institute promotes environmentally sustainable development by providing information on global environmental threats.